P9-CBQ-100

ULTIMATE SPEED SECRETS

THE COMPLETE GUIDE TO HIGH-PERFORMANCE AND RACE DRIVING

"I've used the *Speed Secrets* books since I was a kid just learning to drive racing go-karts, and I continue to use the "secrets" to this day. There are so many great tips, techniques, and secrets in this book that if you apply the things you read you can become faster and more complete as a driver!"

—Colin Braun, NASCAR and Grand-Am race winner

ROSS BENTLEY

ACKNOWLEDGMENTS

Thank you. It's people like you who read my books that drives me to share what I've learned about driving and racing.

Of course, without all the drivers I've raced against, taught, coached, and studied, I never could have written this book. I also want to acknowledge all of the mechanics, engineers, team owners, team members, marketing and PR people, race series and track personnel, sponsors, instructors, coaches, car club members, friends, fans, and anyone else I've come across through the years I've been in motorsports.

Thanks to MBI Publishing for helping me share my *Speed Secrets* with the world.

As always, thanks to my family for letting me do what I love. I'm a lucky guy to have you.

—Ross Bentley

Quarto is the authority on a wide range of topics.

Quarto educates, entertains and enriches the lives of our readers—enthusiasts and lovers of hands-on living.

www.quartoknows.com

First published in 2011 by Motorbooks, an imprint of Quarto Publishing Group USA Inc., 400 First Avenue North, Suite 400, Minneapolis, MN 55401 USA. Telephone: (612) 344-8100 Fax: (612) 344-8692

quartoknows.com
Visit our blogs at quartoknows.com

ISBN: 978-0-7603-4050-9

Digital edition published in 2011
eISBN: 978-1-61058-273-5

Editors: Chris Endres and Zack Miller
Design Manager: Kou Lor
Layout by John Sticha
Designed by Cindy Laun and John Sticha

On the back cover: Shutterstock

Printed in China

CONTENTS

PREFACE

It's a rewarding experience when someone tells me how much they've learned from my *Speed Secrets* books. But when the idea of combining *Speed Secrets 1* through *6* into just one book first came up, my thought was, "Why would I want to do that?" The more I thought about, though, the more it appealed to me. After all, when I wrote my first *Speed Secrets* book, I didn't plan to write any of the others, and I've learned a lot since then. Writing *Ultimate Speed Secrets* allows me to do what I would have done in the beginning—almost 15 years ago—if I'd known then what I know now.

Of course, I couldn't have written this book 15 years ago because I didn't have the knowledge or experience to write it. Now I do. Does that mean that everything you'll ever need is in this one book? I can only wish. I'm still learning.

In writing *Ultimate Speed Secrets* I've tried to do more than just make a "Best of . . ." album. I've attempted to combine my books in a way that would be most efficient for drivers and readers. My hope is that *Ultimate Speed Secrets* will be more than all my books simply put together. By combining them, rewriting sections, and putting them in the order that I would have wanted from the beginning (if I'd known I was going to write them all), I can do more. I can make this one book more concise, clarify some things that I've received feedback on, and add a little more of what I've learned through the years. I've also written new material, clarified some old material, and bridged old and new (and even old and old) in a way that I think will help you.

Who is this book written for? Primarily drivers of all levels, aspirations, and goals. Amateurs and professionals. Young and old. Track-day drivers, club racers, Indy car and NASCAR drivers. Road racers and oval racers. Even rally drivers, autocrossers, drifters, drag racers, and motorcycle racers. While there are differences between these various motorsport disciplines, they have more in common than you might think.

I said the book was written for drivers, primarily, but there are other people who can learn from the information in this book: engineers, parents of drivers, team managers, instructors, mechanics, sponsors, and fans—anyone who wants to know more about how to drive a car at speed and how to race. Some of the book applies to driving better on the street (but I must emphasize right now that I do not recommend breaking any laws to test out what I say in this book).

INTRODUCTION

Driving a race car is not something you can do "by the book." You have to learn mostly through hands-on experience, but you can learn many of the basics by reading and studying a book. In fact, this may allow you to learn more quickly once you're behind the wheel. If you understand the theory (and can picture it clearly in your head before you start to drive), you will be more sensitive and able to relate to the experience. That means you will learn to drive at the limit much sooner. You may save years of trial and error learning by simply reading and understanding this book.

For the beginner, I hope this book serves as a reference for a long time. Some of the information may not make sense until you've gone past the basics and begun working on fine-tuning your techniques. But I hope it will help you start on the right foot and that you can refer to it again later.

For the experienced driver, there may be a lot of information you already know. You may be using some of that information, and you may not even understand why you are using some of it. I suggest you read it again and really think it through. It's surprising how a fresh approach can sometimes make it all click for you, resulting in a dramatic increase in speed.

So this book is written for both the novice racer and the experienced driver, particularly if you've reached either a plateau or a point where you can't seem to go any faster. My hope for this book is it will do more than just teach you the basics of how to drive a racetrack quickly. I want to give you the tools or background to continue to analyze how to go faster at all times. And not only how to drive fast, but more important, how to be a winner in any class or level of racing and have a successful and enjoyable career.

Parts of *Ultimate Speed Secrets* are written with one focus: driving a car at the limit, getting the maximum speed out of it, no matter what the surface or layout of road or track in front of you. Another part of the book is written in the pursuit of winning races. No matter how fast you are, there comes a time in racing where the wheel-to-wheel competition, or racecraft, takes precedence. It's one thing to drive fast and another entirely to out-race the competition. And because some people either want to make a living racing professionally, or at least want someone else to offset their expenses, I touch on what it takes to be a professional racer, from career moves to sponsorship.

There's often been an argument about how much of driving is mental and how much is physical. I've even instigated the argument more than once. It's a silly discussion, because your body doesn't do anything without your brain telling it to, so I can argue that driving is 100 percent mental. At the same time, it's physically challenging. And that's why I write about both the mental and physical skills in *Ultimate Speed Secrets*.

And that's where writing a book like this gets challenging. Driving, racing, mental skills, physical skills, techniques, learning. It doesn't happen sequentially. It doesn't happen one after another. To learn a relatively advanced topic, you need to know the basics, and then you need other basics that have nothing to do with that particular topic before we can come back to the topic again. For that reason, I will be juggling multiple topics and often come back to repeat and enhance what I've written about earlier in the book.

I've been fortunate to learn from many incredibly smart, talented, dedicated, knowledgeable, experienced, and interesting people. I can't claim to have developed every piece of information in this book. Far from it, in fact! But over the past 40 years or so, I've been able to study driving from many perspectives: from behind the wheel, racing against some of the best in the world; from an engineering viewpoint; from the sides of thousands of tracks; from the passenger seat; and from video, data acquisition, and television. I've studied coaching techniques from every angle, how humans learn, sports psychology, kinesiology, performance, vehicle dynamics and engineering, and just about every other angle that I think might give me or you an advantage in driving and racing. It helps to be a learning addict, because I can't stand a day that I don't learn something. I hope to share my learning with you in this book.

Over the years I've had hundreds, maybe thousands of drivers tell me that they take their *Speed Secrets* books to the track with them and use them as reminders. That's the reason I place the key messages, the Speed Secret Tips, throughout the text. That way, you can flip open to most pages and pick up something that's meaningful and helpful to you. The one thing I know for sure is that if you don't use what I write about in this book, it won't help you at all. It's one thing to read a book, and it's an entirely other matter to actually put it into practice. This is not a book to read once and place on a shelf. My hope is that you'll refer to it regularly, keeping it nearby, at home and at the track.

While what I write is accurate to the best of my knowledge, there will be some topics, such as vehicle dynamics and chassis setup issues that might be explained in a slightly different way if you were to talk with an engineer. I'm not an engineer and may not use the same technical language they might. What I will do is explain it in language that you can understand and use when driving a car. My goal is to make everything in this book useable for you.

During some of my driving career (which I hope never ends!), I drove cars that were less competitive than others, usually due to a lack of financial support. I look at that now as one of the best things that ever happened to me. It made me work harder to figure out ways of getting an edge on my competitors without spending money. Without that drive, I suspect I would not have learned enough to have written this book.

I've also had the pleasure of coaching thousands of drivers, many of whom have gone on to great levels in professional racing. I've learned just as much from coaching an older, amateur track-day driver as I have from a NASCAR or Indy-car driver. And because I've taught and coached motorcycle racers, rally drivers, firefighters, police and military drivers, teens, and just about every other type of driver you can mention, in about a dozen countries around the world, the varied knowledge and experience I've gained has contributed to information I'll be sharing with you in this book.

Every time I think I've got a handle on what driving is really all about, I realize I don't. It's much like what my wife said to me one day about parenting, "The more I learn, the less I know for sure." I suppose that is why I love doing what I do.

I hear it all the time: "All I need is a little more seat time and then I'll be really fast. I just need a bit more time to develop the feel and skills to drive consistently at the limit." Sure, there is some truth in this statement. But to simply "sit around" (in the driver's seat) and wait for the seat time to give you the feel and the skills is time wasted. Call me impatient, but I don't like to wait for things to happen. I like to make things happen, and that includes developing skills. I would rather use strategies to develop these skills in a big hurry.

Perhaps the one thing that has made me more successful as a driver coach is my approach. There are a lot of instructors who teach drivers where the apex of a corner is, will advise where the line is or where to brake and get on the throttle, or even how to set up the car's handling. But adding the mental game of driving, and most important, practice strategies, is what has worked for me. And it's what I've included in this book.

Many drivers, once they've been taught the basics, believe that all that separates them from a world championship is more seat time, more practice. Oh, and the best car. But a message you'll hear from me a lot throughout this book is that practicing the same thing over and over doesn't guarantee success. As Albert Einstein once said, "A sure sign of insanity is doing the same thing over and over again and expecting something to change." But, that's what drivers do all the time. They drive around and around, getting seat time, and expecting to improve. The fact that they sometimes do improve is more luck than anything else. It's also why many drivers do not improve as much as they could, and in some cases actually get worse. In fact, practicing the wrong thing will only make you better at doing the wrong thing. And driving around a racetrack like you're racing is not an effective practice strategy.

If football and basketball teams practiced like race drivers, they would show up and play a game each time. But they don't, do they? Instead, their coaches break the game down into deliberate practice exercises, called drills, and only every now and then do they play a scrimmage game. This book is meant to help you do that same thing with your driving. It's meant to break the act of driving down into deliberate practice strategies, and then when you put these drills together in a race or lapping session, you'll perform better.

I use examples from real racing life as often as possible in this book, picking on the styles and techniques of some of the world's great drivers—and some not so great—to demonstrate what works and what doesn't when it comes to driving race cars fast and winning races. But because time doesn't stop, some drivers I refer to will either be less successful or no longer driving by the time you read this. That's okay. There's still something to be learned from them, even if they are now very old.

Even if you are a beginner, reading and using the information in this book may help you develop your basic skills without acquiring any bad habits. That will give you an edge on your competition, who often spend more time dealing with their bad habits than working on improving.

As an example, a funny thing has happened with a number of road racers I have coached through the years: They have become very, very good oval racers. Why? Without trying to sound as though I'm blowing my own horn, the reason was me. To be truthful, it could have been any good coach. Many drivers I coach have some experience on road courses and little to none on ovals. So I spend a great deal of time correcting bad road-racing habits. The first time they ever drive an oval, I'm there to help them learn the basics and develop the right habits. They literally have no bad habits, and so learn quickly to be great oval racers. That's what this book may do for you; it may help you improve without developing bad habits along the way.

My main goal is to help you learn more in a shorter period of time. Left on your own, you will gain experience and improve your abilities. My hope is that this book will speed up that process, so that you can learn in one season what would have taken four or five on your own.

I'm frustrated by drivers who would rather spend thousands of dollars on making their cars faster than they would on themselves, on their own driving. When I see drivers who will spend $2,000 on a set of tires to shave off half a second from their best lap time, rather than spend half that much to gain a second in lap time through developing their driving, I can't help but shake my head. Tires have a limited life, but the improvement a driver makes will last a lifetime.

I suppose I'm preaching to the converted here, since you probably wouldn't be reading this book if you didn't think there was something more to be gained from improving your driving. For that, I congratulate you, and I strongly encourage you to never lose that mindset, that there is always more to learn, always more to improve in your driving, and always more speed and fun to be had.

If I could wish one thing for you, it would be have fast fun!

1 BEHIND THE WHEEL

Being comfortable in the car is critical. If you're not comfortable, it will take more physical energy to drive and affect you mentally. A painful body will reduce your concentration level.

If you want to drive a race car well, whether to win an IndyCar, Formula One, or NASCAR race, or just have fun competing in the middle of the pack in an amateur race, you must be seated properly in the car. If you are uncomfortable, it will be overly tiring and difficult to concentrate. Many races have been lost simply because a driver lost concentration due to discomfort from a poorly fitting seat.

Top drivers in IndyCars, Formula One, sports cars, and NASCAR will spend dozens of hours working to make their seat fit just right, and then fine-tune it all year long. When I first started racing, I was told a seat that fits well could be worth half a second per lap. I can't tell you the number of times I've proven that to be true over the years. I recall two races in my career where I lost positions simply due to a seat that caused me so much pain I could not drive effectively. The first was a Trans-Am race in Portland, Oregon, where the seat bracket broke, allowing the seat to flex and move. I had to use so much effort and energy just trying to keep my body stable that I couldn't concentrate on what I was doing. The second time was in an IndyCar race at Long Beach in 1993. We hadn't yet been able to build a seat that gave my lower back and hips enough support; by 30 laps into the race I had pinched a nerve in my hip, causing my right leg to go entirely numb.

The race car seat, and your position in it, is more important than most racers ever think, especially when first starting their racing career. Many drivers are so wrapped up in getting prepared for their first few races, and in making the car fast, they forget to pay attention to making the seat fit properly.

You receive much of the feedback from the car through the seat. When you are sitting properly in a well-built seat you will be more sensitive to the various vibrations and g-forces you need to interpret what the car is doing. Think about it. Your body has only three contact points with the car: the seat, the steering wheel, and the pedals.

You should use a seating position that puts as much of your body in contact with the car as possible. You want to sit *in* the seat, not *on* it, with as much lateral support as possible, the limiting factor being the ability to move your arms freely.

You should sit as upright as possible, with your shoulders back (not hunched forward) and your chin up. Of course, the lower you sit in the car the better. This is the most efficient way of driving a race car. It's where you are the strongest and most sensitive to the car. It's also the safest.

This seating position should allow you to turn the steering wheel 180 degrees without any interference or moving of your hands on or from the wheel. To do this, you should be able to place your hand at the top of the steering wheel (at the 12 o'clock position) and still have a bend at the elbow without pulling your shoulder off the seat back. Check this with the seat belts/safety harness done up tight. Many drivers sit too far away from the steering wheel with their arms totally straight. This doesn't allow you the leverage to turn the steering wheel properly. It's also very tiring to drive in this position.

While seated, check to see if you can reach the shifter comfortably. You may have to modify or adjust the shifter to suit.

You should also be able to fully depress the pedals and still have a slight bend in the legs. This is not only the least tiring, but allows for ideal modulation of the pedals as you will be able to depress them by pivoting your foot at the ankle, not moving your entire leg in midair.

Whenever possible, I highly recommend that you have a custom-fitted seat built for you. The best way is to have someone who specializes in custom seat building make one. With a little thought and preparation, however, you can mold a seat yourself using expandable foam. This is a simple operation, which can greatly improve your driving performance. Use a two-part foam (available at fiberglass shops), which forms up like a solid Styrofoam-type material. It is poured into a plastic bag between your body and the seat shell or monocoque tub. Before pouring, be sure to cover everything—and I mean everything!—with plastic garbage bags, as the foam is practically impossible to remove after it's set onto something. Upon removing the plastic bags, you can trim off the excess and cover it with tape or material (preferably fire retardant), or it can then be used as a mold to make a carbon-fiber or fiberglass seat.

Any time you build, modify, or adjust your seat or seating position in the race car shop, you have to realize you will know for sure how it feels only on the track. Every time I've had a perfect fitting car in the shop, it's needed modifications after being driven on the track. Consider this before spending a lot of time and money on covering the seat. Wait until it's been track-proven.

Speaking of covering seats, don't bother with a lot of soft padding. It will only crush and distort with the g-forces of your body against it, and result in a loose-fitting seat. Besides, you need to feel the vibrations and forces from the car. Thick padding will only reduce your sensitivity. If you do use padding, use only a thin layer of high-density foam rubber.

Part of the seat's job is to provide support so your feet can do their part, working the pedals accurately and deliberately. The dance your feet do on the pedal is what allows you to dance with the car at the limit, on the edge, at speed.

You want to use the balls of your feet on the pedals. They are strongest part of the foot, as well as being the most sensitive. When you are not using the clutch, the left foot should be on the dead pedal (the rest pad area to the left of the clutch pedal), not hovering above the clutch pedal. This will help support your body under the heavy braking and cornering forces you will experience. However, there are some single-seater race cars that are so narrow in the pedal area that it is almost impossible to have a dead pedal. Do everything you can to make even a very small one. But if you can't, it's even more important to have a well-built seat. Ensure there is good support in front of your buttocks to stop your body from sliding forward under heavy braking.

Before getting into your car and heading out onto the track, make sure that both the pedals and the bottom of your shoes are dry and clean. Many drivers have crashed because their feet slipped off the brake pedal approaching a corner. Have a crew member wipe your shoes with a clean rag before getting into the car.

Ensure that every part of the roll cage is covered with the type of padding specifically designed to protect you in the event of a crash, no matter how unlikely you think you would hit it during an accident. Unfortunately, you'll be surprised by how far your body can stretch—and what it can hit—with a heavy impact. *Shutterstock*

The first time I watched a Formula One Grand Prix was in Montreal, and it rained a lot that weekend. One thing I remember most was watching the drivers being taken to their cars on a cart and then being lifted straight from the cart to the cockpit of the car so that their feet wouldn't get wet. I also saw other drivers wearing plastic bags over their shoes.

The safety harnesses in a race car are not only there in case of a crash, but also to help support your body. Only use the best seat belts in your car, and then take good care of them. Keep them clean and inspect them often for wear and damage. Adjust them so they hold your body firmly and comfortably. And remember, they will stretch and loosen throughout the course of a race, particularly the shoulder harnesses, so make sure you can reach down and tighten them while driving. Also make sure you have some form of head restraint behind your helmet.

Ensure any part of the roll cage or cockpit that you could come in contact with during a crash is covered with a high-density foam rubber. Many drivers have been seriously injured just by impacting the roll cage. You might be amazed at how much a driver moves in the cockpit during a crash, even when tightly belted in. Some drivers' heads have actually made impact with the steering wheel.

And finally, do everything possible to help keep the cockpit cool. Have air ducts installed to direct air at you. The cockpit of a race car can get extremely hot, which will negatively affect your stamina, and therefore, your performance.

2 THE CONTROLS

A race driver has a number of controls to help achieve the desired goal of driving at the limit: the steering wheel, shifter, gauges, clutch pedal, brake pedal, throttle, and mirrors. Everything you do with these controls should be done smoothly, gently, and with finesse.

I often see racers, particularly at the back of the pack in amateur races, *trying* to go fast, with their arms flailing around, banging off shifts, jerking the steering into a turn with feet stabbing at the pedals, the car usually in massive slides through the turns. It may feel fast, and even look fast, but I'll guarantee it's not. In reality, the car will be unbalanced, and therefore, losing traction and actually going slower. If the driver would only slow down, the car would actually go faster. It reminds me of the saying, "Never confuse movement for action."

Steer, shift, and use the pedals smoothly, and with finesse—not with blinding speed and brute force.

SPEED SECRET

The less you do with the controls, the less chance for error.

GAUGES

A typical race car has four main gauges to which you need to pay attention if you want to drive reliably at the limit. They are tachometer, oil pressure, oil temperature, and water temperature, the four most important ones. The tachometer helps you go fast; the others help you ensure the car keeps running. You may also have to deal with other gauges such as fuel pressure, ammeter, turbo-boost pressure, exhaust temperature, and so on.

It's important that the gauges are mounted so that you can see them easily and read them at a glance. Normally, you should only have to take a quick glance at the gauges, checking more for a change of position of a needle, rather than the absolute number it is pointing at.

The amount of information presented to you in some race cars can be overwhelming. Make sure you practice being able to read it so you can take everything in with just a glance. *Shutterstock*

Often, it is best to mount the tachometer and other gauges so the range that you must see is in good view; the redline or ideal needle position should be at 12 o'clock. This way, with a quick glance, you know when to shift or if the temperatures or pressures are okay. Also, make sure the gauges don't reflect the sunlight into your eyes or have so much glare that you can't read them.

I like to use the tach at the exit of most corners to judge how well I did in that particular corner. It's my "report card." I pick a spot on the track and check how many revs the engine is at. If I'm pulling 50 more revs than the previous lap, I know that what I did differently worked on that lap. Also, I try to glance at my gauges at least once a lap on the straightaway. Otherwise, I depend on the warning lights to advise me of any problems.

Warning or "idiot" lights can prove extremely valuable. These are usually set to only come on if one of the critical engine functions reaches an unacceptable level, such as if the oil pressure drops below 40 psi or the water temperature reaches 240 degrees. Using these lights allows a driver to only have to check the gauges when it's convenient, such as on the straightaway. They warn the driver only if there is a major problem.

A simple dashboard layout is best, with as few gauges as possible. More and more race cars are now using computerized dashes that are linked to data acquisition systems. These are useful as they can tell you your lap time, your minimum or maximum speed at various points on the track, and other information that you can use to help determine where you may be able to improve. However, don't let yourself get so caught up in reading all the information that it takes away from your driving.

BRAKE PEDAL

When braking, think of "squeezing" the brake pedal down and easing off it. The smoother you are with the brakes, the better balanced the car will be, enabling you to drive at the limit. Three-time World Driving Champion Jackie Stewart claims the reason he won so many Grand Prix was because he eased off the brakes more smoothly than any of his competitors. Hard to imagine how that could affect the outcome of a race that much, isn't it? But it allowed him to enter corners a fraction of a mile faster because the car was better balanced. Obviously, this squeezing on and easing off the brake pedal must be done quickly—and it can be done very quickly with practice—but always emphasize smoothness.

This is one technique you can safely and easily practice every day on the street. Every time your foot goes onto the brake pedal, think of the word "squeeze," then think of the word "ease" when releasing the brakes. Practice it so that quickly squeezing and easing becomes second nature or habit.

Because how you use the brakes is so critical, you'll find that I come back to talk about braking technique many times throughout this book. Did I mention that how you use the brakes is critical to being a successful racer?

THROTTLE

Always use the throttle (gas pedal) gently. As with the brakes, progressively squeeze on more throttle as you accelerate and quickly ease off as you slow down. Anytime you pounce on the gas pedal or abruptly lift off, it unsettles the car, which reduces traction. The smoother you are with the throttle, the better balanced the car will be, and ultimately the more traction and speed you will have.

SPEED SECRET

The throttle is not an on-off switch.

If you find yourself having to back off the throttle after you begin accelerating in a corner, you must have applied the gas too soon or too hard in the beginning. Ease on the throttle. It takes time and practice to develop a feel for how quickly and how much throttle you can squeeze on.

When you are moving your foot from the throttle to the brake pedal, or vice versa, it must be done as quickly as possible. Your right foot should always be either on the throttle (even if it's a light, steady throttle) or the brakes. Don't waste time doing nothing, with your foot in between the two. You should never be coasting.

STEERING WHEEL

Use a firm, but relaxed, grip on the steering wheel, with your hands in the 9 and 3 o'clock positions. Lightly hook your thumbs over the spokes of the wheel if that's comfortable. By always holding the wheel in the same position, you'll always know how much you've turned it and where straight ahead is. You will see how

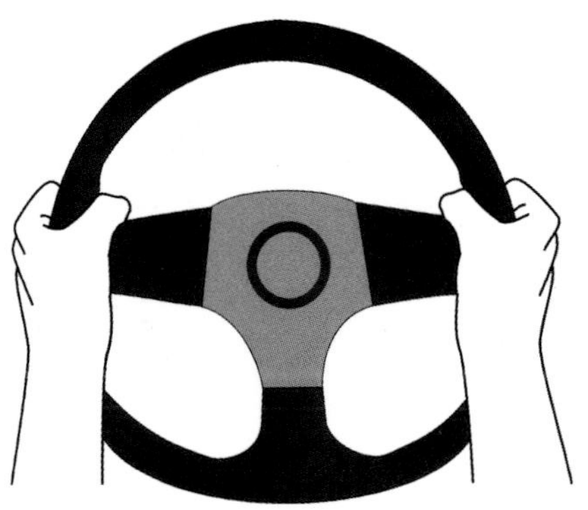

The proper hand placement on the steering wheel, at the 9 and 3 o'clock position.

important this is when the car begins to spin and you don't which way is straight ahead.

With the 9 and 3 grip you should be able to steer through almost every corner without moving your hands from this position. This will result in smoother, more controlled steering. Perhaps, in some large, production-based racing sedans, you may not be able to turn sharply enough for some tight hairpin corners with this grip. In that case, reposition your hands slightly before the corner (e.g., to the 8 and 2 position for a right-hand corner), to allow you to make one steering action without sliding the hands around the wheel.

When turning the steering wheel, allow both hands to do an equal amount of work. While one hand pulls down on the wheel, the other pushes up smoothly. Keep both hands on the wheel at all times (except when shifting, obviously, but even then, get your hand back on the wheel between gear changes). Make small steering corrections with the wrist, not the arms. Every movement with the wheel must be made smoothly and progressively, never jerking the steering into a turn. Feed in the required steering input to generate a gentle, smooth arc through the corner.

Think about it. Every time the front tires are at an angle to the road they are scrubbing off speed. Pretty obvious, right? But what does this really mean? How can you get around a corner without turning the steering wheel? Look and think farther ahead, planning your path or line through a corner, so that you will be able to turn the steering as little as possible, straightening the corner out as much as possible. If you feel or hear the front tires scrubbing or squealing through a turn, try to unwind your steering input (straighten it out).

SPEED SECRET

The less you turn the steering wheel, the faster you will go.

Once you've turned into a corner, try to unwind the steering as soon as possible. Of course, this means you have to use up *all* the road available. You can even practice this on the street (within limits of the law), steering smoothly into and out of corners, keeping the front wheels pointed as straight as you can.

MIRRORS

Mirrors play a critical role in the race driver's job, and you must be comfortable using them. In racing, it is just as important to know what's behind and beside you as it is to know what's in front. You should use your mirrors enough to always know who's around you, and exactly where they are. A competitor should never take you by surprise by being somewhere you didn't expect (like to the inside of you on the approach to a corner). Take time to adjust all your mirrors properly, and make sure they don't vibrate so much that you can't see out of them.

The lower you sit in your car, the better, because doing so lowers the overall center of gravity of the car. The limiting factors are your visibility—to the front, to the sides, and in the mirrors—and your comfort. *Shutterstock*

Don't constantly look in the mirrors while driving, however. Some drivers have caused more problems doing that than they would have if they never looked in the mirrors. I've seen drivers veer off the track while looking in the mirrors.

I take a quick glance in the mirrors each time I come onto a straightaway of any decent length. If I adjust them properly (aimed a little to the sides so that I can see to either side), I don't actually have to turn my head to look in the mirrors to see other cars. I automatically notice them with my peripheral vision, which minimizes the chances of being surprised by a faster overtaking car.

The mirrors on some modern formula cars have gotten smaller over the years. Fortunately, I think they've gotten as small as they will ever get. If you are using a small mirror, make sure it is convex to help increase your vision to the rear and sides.

SPEED SECRET

Check your mirrors as often as it takes to always know where everyone else is around you.

3 SHIFTING

Proper shifting technique is an often overlooked racing skill. Many drivers feel they have to bang off their shifts as fast as possible to go quickly. Wrong! In fact, the amount of time you can save is minimal, especially compared to the time you can lose if you miss one single shift. A shift should be made gently and with finesse.

SPEED SECRET

A shift should be made gently and with finesse.

UPSHIFTS

Simply *place* the shifter into gear as smoothly as you can. A shift should never be felt. You may be surprised at just how slowly and relaxed the world's top drivers shift.

DOWNSHIFTS

Downshifting is one of the most misunderstood and misused techniques in driving. And it is a must for extracting the full potential of your car. It's not always easy—it requires timing, skill, and practice—but when mastered, it will help you drive at the limit.

What is the real reason for downshifting? Many drivers think it's to use the engine to help slow the car down. Wrong again! The engine is meant to increase your speed, not decrease it. In fact, by using the engine to slow the car you can actually hinder accurate brake modulation and balance. Race drivers, and good street drivers, downshift during the approach to a corner simply to be in the proper gear at the optimum engine rpm range and to allow maximum acceleration out of the corner.

Again, the reason for downshifting is *not* to slow the car. I can't emphasize this enough. That's what brakes are for. Too many drivers try to use the engine compression braking effect to slow the car. All they really achieve is upsetting the balance of the car, the braking effectiveness (if the brakes are right at the limit before locking up, and then you add engine braking to the rear wheels, you will probably lock up the rear brakes) and more wear and tear on the engine. So brake first, then downshift.

SPEED SECRET

Brake first, then downshift.

HEEL-TOE

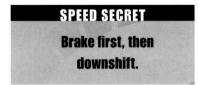

To complicate things a bit, in racing you must downshift to a lower gear while maintaining maximum braking. This must be done smoothly, without upsetting the balance of the car. But if you simply dropped a gear and let out the clutch while braking heavily, the car would nose-dive, upsetting the balance, and try to lock the driving wheels because of the extra engine compression braking effect.

The smoothest downshift occurs when the engine revs are increased by briefly applying, or stabbing, the gas pedal with your right foot. This is called "blipping" the throttle. This matches the engine rpm with the driving wheels' rpm.

The tricky part is continuing maximum braking while blipping the throttle at the same time. This requires a technique called "heel-and-toe" downshifting. To get a basic feel for this technique, practice it while the engine is turned off (see Illustration 3-1). Then you can begin to practice it on the road or racetrack.

It's important to apply consistent brake pressure all the way through this maneuver. You are simply pivoting the right foot to blip the throttle while braking at the same time.

This blipping of the throttle is one of the most important aspects. You want to match the speed of the engine with the

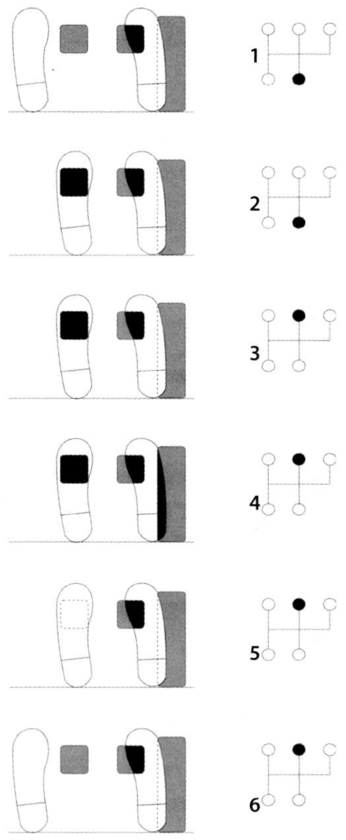

ILLUSTRATION 3-1 Here is a step-by-step explanation of how to heel-and-toe downshift:
1. Begin braking, using the ball of your right foot on the brake pedal while keeping a small portion of the right side of your foot covering the gas pedal but not pushing it yet.
2. Depress the clutch pedal with your left foot, while maintaining braking.
3. Move the shift lever into the next-lower gear (from fourth to third in the illustration), while maintaining braking.
4. While continuing braking and with the clutch pedal still depressed, pivot or roll your right foot at the ankle, quickly pushing or "blipping" the throttle (revving the engine).
5. Quickly ease out the clutch, while maintaining braking.
6. Place your left foot back on the dead pedal, while continuing braking, now in the lower gear.

speed of the gear you are selecting, and you can't watch the tachometer. Your eyes must be looking ahead. So, correct blipping of the throttle and matching of revs depends on practice and input from the ears and the forces on the body. If you don't blip enough, the driving wheels will lock up when the clutch is reengaged. That'll cause big problems. If you blip too much, the car will attempt to accelerate when you are supposed to be slowing down.

The best way is to rev up the engine slightly higher than required, select the required gear, and quickly engage the clutch as the revs drop. It's going to take practice, constant practice. It may seem like a lot to do all at once, but once you get the hang of it, it will become second nature.

To heel-and-toe properly your pedals must be set up correctly. When the brake pedal is fully depressed, it should still be slightly higher and directly beside the gas pedal. In a purpose-built race car, take the time to adjust the pedals to fit. If racing a production-based car, you may have to bend or add an extension to the throttle to suit you. Do *not* modify the brake pedal by bending or adding to the pedal. This will weaken it.

There isn't a successful race driver in the world who doesn't heel-and-toe on every downshift. And, again, it can be practiced every day on the street. In fact, that's the only way to drive all the time.

TIMING

Now that we've talked about how to shift, what about when to shift? First, downshifting. Remember: "Brake first, then downshift." If you don't follow this rule, you will end up badly over-revving the engine. Think about it. If you are at maximum rpm in fourth gear and you downshift to third without slowing the car, you'll over-rev the engine. And remember again, downshifting is not a means of slowing the car, unless you have no brakes.

Make sure you always complete your downshifts before you turn into a corner. One of the most common errors I've seen drivers make is trying to finish the downshift while turning into a corner. As the driver lets out the clutch (usually, without a smooth heel-and-toe downshift), the driving wheels begin to lock up momentarily, and the car starts to spin. Time your downshift so that you have completed it, with your left foot off the clutch and over onto the dead pedal area, before you ever start to turn the steering wheel into the corner.

When upshifting, for absolute maximum acceleration, you need to know the engine's torque and horsepower characteristics. Talk to your engine builder, or study the engine dyno torque and horsepower graphs to determine at what rpm you should be shifting. It makes a huge difference. With many engines you're better off to shift before reaching the redline. You want to shift at an rpm that allows the engine to stay in the peak torque range.

Let's look at an example using the Torque and Horsepower versus Engine rpm graph in Illustration 3-2. Assuming a 2,000-rpm split between gears (an upshift from one gear to another, dropping the engine speed by 2,000 rpm), if you shifted from first to second gear at 7,000 rpm, you would then be accelerating from 5,000 rpm back up to 7,000. As the graph shows, from 5,000 rpm the torque curve is on

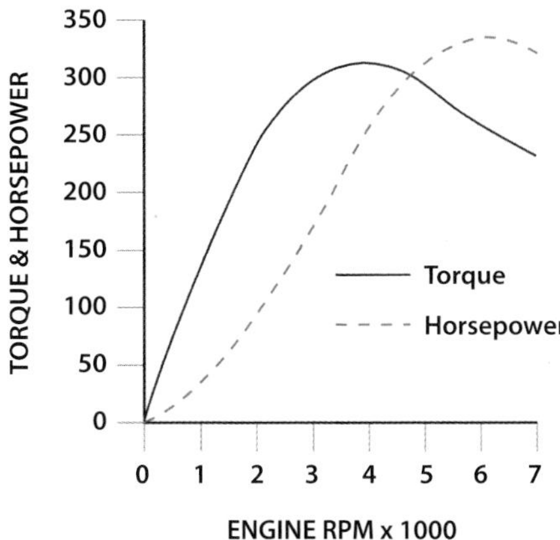

ILLUSTRATION 3-2 Torque and horsepower versus engine rpm graph.

a decline. However, if you shifted at 6,000 rpm, the engine would be accelerating through the maximum torque range to maximum horsepower. In fact, an engine will operate most effectively—resulting in the maximum acceleration—when the rpm is maintained between the torque peak and horsepower peak.

Notice I talk more about the engine torque than horsepower. As they say, "Horsepower sells cars; torque wins races." Torque is what makes the car accelerate; horsepower maintains that.

When you are proficient at smooth, well-timed downshifts, try skipping gears when downshifting. Instead of running through all the gears (for example, from fifth to fourth, fourth to third and third to second) shift directly to the required gear (from fifth to second). Obviously, this takes the right timing, using the brakes to slow the car, then downshifting just before turning into the corner. You must slow down the car with the brakes even more before dropping the two gears.

This goes back to what I was getting at earlier: The less you do behind the wheel, the faster you will go. Every time you shift, there is a chance you may make a small error that will upset the balance of the car. So shift as little as possible. In fact, the less downshifting you do while approaching a corner, the less likely it is you will make a mistake. It will be easier to modulate the brakes smoothly.

Now, with some cars, it seems the gearbox doesn't like it when you skip gears. Often, it is difficult to get a perfect match of the revs, therefore making it hard to get a good, clean downshift without "crunching" it into gear. Obviously, with this type of car, you're better off not skipping gears.

DOUBLE-CLUTCHING

What about double-clutching? I believe double-clutching is unnecessary in any modern production car (anything built in the last 20 to 30 years or so), but *may* be useful in some real race cars with racing gearboxes.

What is double-clutching? Basically, you depress and release the clutch twice for each shift. The routine goes like this for a downshift: You are traveling along in fourth gear and begin to slow down for a corner. You then depress the clutch pedal, move the shifter into neutral, release the clutch, rev the engine (blipping the throttle using the heel-and-toe method), depress the clutch again, move the shifter into third gear, and release the clutch. Your downshift is now complete.

The reason for double-clutching is to help evenly match the rpm of the gear you are selecting with that of the engine to allow a smoother meshing of the gears. In a non-synchromesh transmission, such as a racing gearbox, it may make gear changing easier. And that's why I say it may be *unnecessary* to double-clutch in production-based race cars with their synchromesh transmissions. But, if the synchros in your car's transmission are beginning to wear out, double-clutching can extend their life a little longer and make it easier to get it into gear.

You may be able to go racing for many years and never have to double-clutch. But a complete race driver knows how and is proficient at it. In endurance races, a driver may want to double-clutch to save wear and tear on the gearbox. At other times it is more a matter of driver preference.

CLUTCHLESS SHIFTING

Another option with a pure racing gearbox is not to use the clutch at all when shifting. This takes practice, as it is more critical that the engine and gearbox revs are matched perfectly when downshifting. The advantage to not using the clutch is that it may save a fraction of a second on each shift. The disadvantage is that it usually causes a little extra strain on the gearbox, perhaps wearing it out a little sooner or risking a mechanical failure in the race. Also, there may be more chance for you to make an error this way. Again, I think it's important for a driver to know how to drive without the clutch. You never know when you're going to have a clutch problem and be forced not to use it.

More and more race cars are being built with sequential shifters. This is much like a motorcycle shifter, in that the shift lever is always in the same position. You simply click it back to shift up and forward to shift down. With this type of shifter it is impossible to skip gears on a downshift. You have to go through all the gears. Also, it may work better if you do not use the clutch. On an upshift, you just ease up on the throttle (as you would with a normal gearbox) and click the shifter back into the next gear. On a downshift, it works the same way only you heel-and-toe blip the throttle as you click it down a gear.

Throughout my career, with most cars, I have usually used the clutch when shifting. I've found it puts less wear and tear on the gearbox. But when I started driving cars with sequential gearboxes, I found they shifted much quicker and easier without using the clutch. It took a few laps to get used to not using it—and to not being able to skip gears on downshifts—but once comfortable with it, I realized it was the only way to go with the sequential shifter. With a regular gearbox, though, I still prefer to use the clutch.

4 CHASSIS AND SUSPENSION BASICS

Understanding chassis and suspension adjustments and what they mean to you as a driver is a critical part of your job. There are many good books that deal with race car dynamics in great detail. At the back of this book I've listed the ones I think are mandatory reading for any driver. If you don't understand something, go back to these books or ask someone to explain it. If you want to win, you must know this information.

I don't intend to go into great detail, but the following is a brief overview of some of the key areas of chassis and suspension adjustments that you have to know to reach any level of success. I hope this piques your interest to go out and learn more.

CAMBER

Camber angle is the inclination of the wheels looking from the front or rear of the car. A wheel inclined inward at the top is said to have "negative camber"; a wheel inclined outward at the top has "positive camber." The angle is measured in degrees.

It is important to keep the entire tread width of a tire, which is generally very wide and flat, in complete contact with the track surface as much as possible. When a tire leans over, part of the tread is no longer in contact with the track, drastically reducing traction. Therefore, the suspension must be designed and adjusted to keep the tire flat on the track surface during suspension movement.

Understand that as a car is driven through a corner, it leans toward the outside of the turn. This causes the outside tire to lean outward, creating more positive camber, while the inside wheel tends toward more negative camber. Therefore, to keep the outside tire (as it's the one that is generating most of the cornering

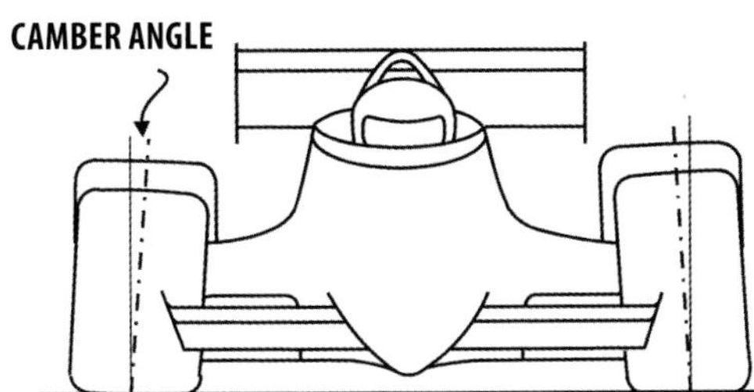

ILLUSTRATION 4-1 Camber is the angle of inclination of the wheels when viewed from the front or rear. This shows negative camber.

force) as flat on the road surface as possible, generally the suspension is adjusted to measure negative camber when at rest or driving down a straightaway.

Your goal in adjusting the camber angle is to maximize cornering grip by having the tire close to 0-degree camber during hard cornering. This can take a fair bit of adjusting and testing to come to the best static setting that will result in the optimum dynamic camber angle.

CASTER

Caster angle provides the self-centering effect of the steering (the tendency for the car to steer straight ahead without holding the steering wheel). It is the inclination angle of the kingpin, or upright, looking from the side. Positive caster is when the top of the kingpin or upright is inclined to the rear. Negative caster is never used.

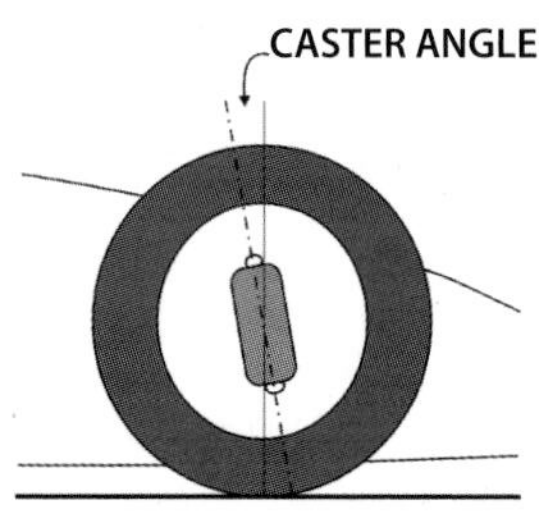

ILLUSTRATION 4-2 Caster is the angle of the inclination of the suspension upright.

The more positive caster, the more the steering will self-center, which, generally, is a desirable effect. However, the more positive caster, the more effort it takes to turn the steering against this caster. There has to be a compromise between easy self-centering and heavy steering.

Caster also affects the camber when the steering is turned. The more positive caster, the more negative camber on the outside tire during cornering. This must be kept in mind when adjusting for the optimum camber setting. Perhaps, instead of dialing in more static camber, you may be better off adjusting in more caster. Remember, this will result in more negative camber on the outside tire during cornering. This can be an important factor. Learn and understand caster.

TOE

Toe can be either "toe-in" or "toe-out." It is the angle of either the two front or two rear tires looking at them from above. Toe-in is when the front of the tires are closer together than the rear; toe-out is the opposite. The front of the tires are farther apart than the rear. Toe can always be adjusted at the front and can be adjusted at the rear on cars with independent rear suspension.

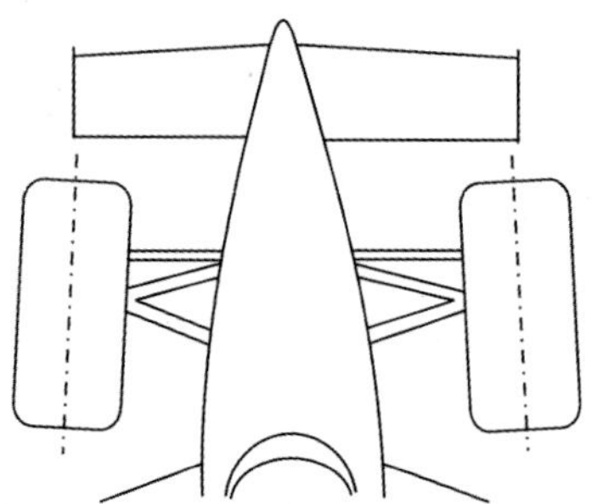

ILLUSTRATION 4-3 Toe is the angle of the wheels looking from above; in this case, toe-in.

Toe plays an important role in the car's straight-line stability, as well as its transient handling characteristics (how quickly the car responds to the initial turn into the corner). Generally, front-wheel toe-in results in an initial understeer;

toe-out results in an initial oversteer (more about understeer and oversteer in the next chapter).

Rear wheel toe-out must be avoided. It causes instability and unpredictable oversteer.

ACKERMAN STEERING

The inside wheel of a vehicle driving through a corner travels on a tighter radius than the outside wheel. Therefore, the inside front wheel must be turned sharper to avoid it scrubbing. The geometry of the front suspension is designed to achieve this. This is called Ackerman steering.

Some race cars have been designed or modified to have anti–Ackerman steering. This means the inside tire is actually turned less than the outside tire. The reasoning is that the inside tire has so little of the cornering load that some tire scrub will not hurt. Other cars have increased Ackerman geometry to the point where the inside wheel is turned more than would be necessary to track the inside radius. Both of these variations are designed to help the car's initial "turn-in" characteristics.

BUMP STEER

Bump steer should be avoided. This is when the front or rear wheels begin to either toe-in or toe-out during the vertical suspension movements caused by a bump or from body roll (sometimes called "roll steer"). Although it has been used to help "Band-Aid" a handling problem, generally bump steer makes a vehicle very unstable, particularly on the rear wheels.

ANTI-DIVE

When you apply the brakes, the front end of the car has a tendency to dive. The suspension geometry is designed in such a way as to reduce this tendency. Generally, this is something designed into the car and requires—or even allows—little or no adjustment.

ANTI-SQUAT

When a car accelerates, the rear tends to "squat" down. As with anti-dive, the suspension geometry is designed to limit this. And again, little adjustment is required or available.

RIDE HEIGHT

The ride height is the distance between the road surface and the lowest point on the car. Often, this is different at the front than the rear. This difference is called "rake"—usually with the front lower than the rear. Adjustment of the ride height, particularly the rake, is used to tune the handling.

The ride height is usually determined by running the car as low as possible without the chassis bottoming out (or, at least, just barely touching) on the road surface, or the suspension running out of travel. Usually, the lower the car is run, the better the aerodynamics. Additionally, the lower center of gravity is advantageous.

SPRING RATE

Choosing the optimum spring rate is one of the most important setup factors you'll have to deal with. The spring rate is the amount of force needed to deflect a spring at a given amount and is usually measured in pounds per inch of deflection. The diameter of the spring wire, the overall diameter of the spring, and the length or number of coils determines this rating.

It's your goal in developing the car to find the optimum spring rate for the front and rear suspension. Generally, it's a compromise between having a soft enough spring to allow the suspension to handle the undulations in the track surface, while being stiff enough to keep the car from bottoming out when hitting a bump. There are many more factors involved, such as your driving style or preference, the amount of aerodynamic downforce you are running, the weight of the car, the shape and condition of the track surface, and so on. Perhaps most important though is the balance front to rear. Generally, it's best to use the softest spring possible on the rear, to help the rear tires achieve maximum traction under acceleration, and then balance the handling with the optimum front springs.

WHEEL RATE

The wheel rate is the amount of force needed to move the wheel a given distance. It is also measured in pounds per inch of deflection. It is determined by the geometry of the suspension and spring-mounting location and the spring rate. Understand that even though you have the same spring rate on the front and rear suspension (or two different cars), the wheel rate may differ due to the amount of leverage a suspension system applies to the spring.

ANTI-ROLL BAR

An anti-roll bar (sometimes, wrongly referred to as a sway bar) is used to resist the vehicle's tendency to lean during cornering. The anti-roll bar, usually a steel tube or solid bar, is used to alter the front or rear roll resistance. This affects the car's handling characteristics. Many cars have adjustment controls in the cockpit, so you can make changes as the track conditions, fuel load, and tire wear change throughout a race.

Adjusting the anti-roll bars is probably the easiest and quickest change you can make to the suspension setup. Therefore, it's important to try the car at full stiff and full soft settings to see what effect it has. When beginning to dial in the setup of the car, I like to do a "bar sweep." This is where I will adjust the front bar from full soft to full hard, then do the same with the rear bar while noting the change in handling. That gives my engineer and I a good indication as to which direction we will have to go to develop a good balance in the car.

As a general rule, to improve the grip on the front of the car (to lessen understeer), you should soften the front bar or stiffen the rear bar. To improve grip on the rear (lessen oversteer), you should soften the rear bar or stiffen the front bar. However, that's not always the case (as I've discovered a few times), so be prepared to try the opposite.

ROLL STIFFNESS

Roll stiffness is the total amount of resistance to the car leaning or rolling provided by the springs and anti-roll bars. This is measured in pounds per inch of spring travel at the wheel. This is a function of the spring rate and the anti-roll bar stiffness.

The distribution of the vehicle's roll stiffness between the front and rear suspension is called the roll stiffness distribution, and is expressed as a percentage front to rear. Generally, it's the roll stiffness distribution that we use to fine-tune the handling balance of the car, using the springs and anti-roll bars. Adjusting the front roll stiffness (with springs or anti-roll bars) in relation to the rear, and vice versa, is the most common method of altering the handling balance of the car.

SHOCK RATE

The purpose of a shock absorber is to slow down and control the oscillations of the spring as the suspension absorbs undulations in the roadway. Actually, a shock absorber is a damper; it damps the movement of the springs.

Shocks work in both directions: Compression is called bump; extension is called rebound. A shock absorber, therefore, is rated by the rate of deflection at a given shaft speed, both in the bump and rebound direction. If the car's springs are force sensitive, the shocks are velocity sensitive.

You can also use the shock absorbers to alter the transient handling characteristics (how responsive the car is to your inputs). If the springs and anti-roll bars determine the amount of body roll and the distribution front to rear, then how quickly that body roll occurs is determined by the shock absorber rates.

So the shock absorbers are another important suspension-tuning component. And, as with the spring rate, finding the optimum shock setting is a delicate compromise. It takes some experience before you have the sensitivity as a driver to be able to find that perfect setting.

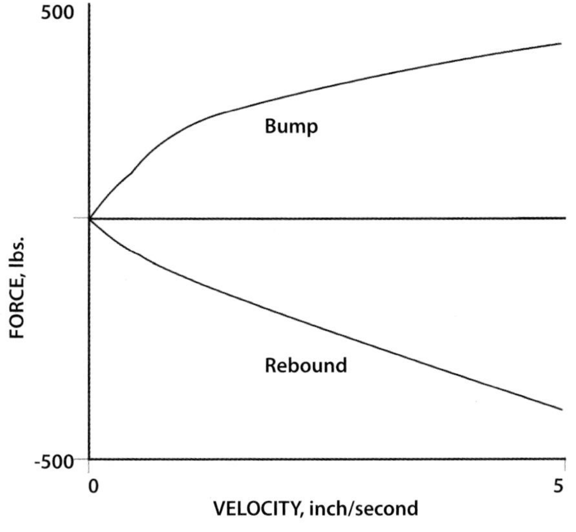

ILLUSTRATION 4-4
A shock-absorber dyno produces a graph that relates the force it takes to stroke the shock, in both bump and rebound, versus the velocity at which it moves. Learn to read and understand shock dyno graphs and especially how their data relates to what you feel when driving.

CORNER WEIGHT

If you place the four tires of a vehicle on four separate scales, they will give you the corner weights of the vehicle. From there, you can determine the front-to-rear and left-to-right weight distribution, as well as total vehicle weight.

Ideally, for a road course, the left-to-right corner weights should be identical. In practically any midengine car the rear corner weights will be higher than the front. For oval tracks, often the setup will be biased to one side or corner.

Adjusting corner weights is one of the most important suspension-tuning tools, one that is often overlooked by many inexperienced racers.

TIRES

One of the most effective ways of checking and optimizing chassis adjustments is by "reading" the tires. Evaluating tire temperatures will indicate if the tire pressures are correct, if the alignment settings are correct, how the overall handling balance of the car is, and to some extent how close to the limit you're driving.

All tires are designed to operate within an optimum tread temperature range. In this optimum range, the tire generates its maximum traction (as shown in Illustration 4-5). Above or below that optimum range, the tires will not grip the track surface well. Also, if they are operated above the optimum range for too long the tread may begin to blister, chunk, or wear quickly. An average temperature range for a high-performance street radial is in the 180 to 200 degrees Fahrenheit area; for a racing tire, it's 200 to 230 degrees Fahrenheit.

To determine tire temperatures use a tire pyrometer, an instrument with a needle that is inserted just under the surface of the tire's tread, generally at three points across the tire—the inside, the middle, and the outside of the tread.

Tire temperatures taken after the car has come into the pits are an average of the corners and straightaways. If it's after a long straightaway or a slow cool-off lap, the temperatures may be misleading, as part of the tread may have cooled more than others. So, it's important to take temperatures as close to a corner as possible. They must also be taken as soon as the car has come to a stop as they will begin to cool after about a minute.

The optimum camber angle is indicated when the temperature near the outside of the tread is even with the temperature near the inside of the tread. If the temperature near the inside of the tread surface is significantly higher than the outside, there is too much negative camber. The inside is heating up too much. If the outside temperature is hotter than the inside, there is too much positive camber.

If the temperature in the middle of the tread is equal to the average of the inside and outside of the tread, then the tire pressure is correct. If it's too hot in the middle of the tread, then the tire pressure is probably too high. If it's too cool in the middle, then the pressures are too low. Ideally, the tire temperatures should be even all across the tread.

If the temperatures on the front tires are even with the rear tires, then the overall balance of the car is good. If they are hotter than the rears, then the fronts are sliding more than the rear, and a spring, shock, or anti-roll-bar adjustment may be necessary. The reverse is true, as well.

ILLUSTRATION 4-5

Tire temperature versus traction graph. In this graph you can see that the tire gains traction as its heat builds, until it reaches a point where it then begins to lose traction.

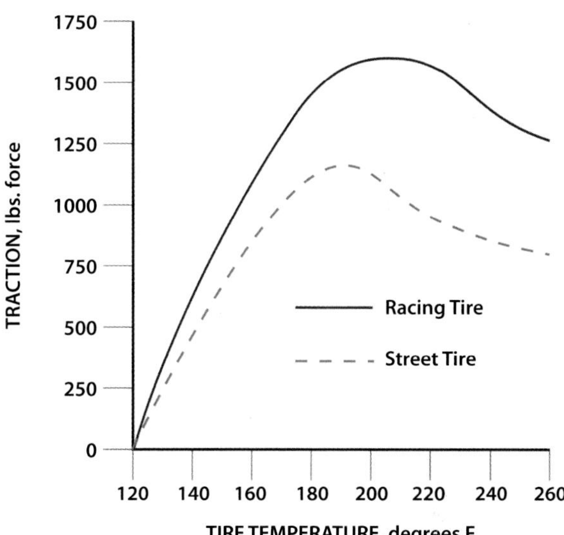

If all four tires are not running in the optimum temperature range, it means one of two things: Either the tire compound is not correct for the application or it has something to do with your driving. If the temperature is too low, you're not driving the car hard enough. You're not working the tires. If the temperature is too hot, you're driving too hard. You're sliding the car too much. There is more about this in the next chapter.

Get used to reading a tire. If you can look at the tread surface in relation to how the car felt and the tire temperatures, and then determine what to do to make improvements, it may make the difference between you and your competitors.

Generally, the surface should be a dull black all across the tread. There should not be any shiny areas. If there is, it probably means that part of the tire is being overloaded. Also, if you are driving the car hard enough (using the tires), the tread surface will show a slightly wavy grained texture. It should be this same texture all across the tread.

A couple of notes on how to treat new tires: When starting with a new set of tires, it is best to break them in. First, "scrub" them in by weaving back and forth (if safely possible) to clean the mold release agent off the surface. Second, don't destroy them on the first lap by putting the car in huge slides through the corners and getting massive wheelspin under acceleration. Instead, gradually build up the heat in them by progressively increasing your speed. Their overall grip will last longer this way.

5 RACE CAR DYNAMICS

The more you understand about the car, the more successful you will be. All the driving talent in the world will not guarantee a win. Take the time to learn and fully understand everything you can about how the car works, how it is set up, and what each change should and does do. Even if you don't work on the car yourself, being able to tell your mechanic what the car is doing is the only way of getting the maximum performance from the car. As with many other aspects of racing, read, listen, and learn as much as you can. At the end of this book, I've listed some additional books I strongly suggest you read.

Before making vast changes to the car's setup, be sure that you first know the track well, are comfortable with it, and are driving well. I've seen drivers (myself included) get so caught up in the idea of making the car work better, they forget about their own driving. Also, when making changes to the setup, only make one adjustment at a time. If you make more than one, how do you know which one made the difference?

I bought my first Formula Ford from a driver who had been racing for a number of years, and who I knew was knowledgeable about the setup and mechanics of the car. I knew the car was pretty good. So I decided I wouldn't try to out-trick myself. I promised myself I wouldn't make any drastic changes in the car for at least the first season. I was just going to concentrate on learning to get 100 percent out of the car as a driver and only fine-tune the suspension. The second year I raced it, I made some serious modifications to the car. By that time, I felt like I knew enough to do that.

TIRE TRACTION

You've looked at the tires from the perspective of how they relate to chassis adjustments in the last chapter. Now, let's get back to how to drive them. In fact, to get the most from your tires, you really do have to understand them. You can be somewhat successful in racing without knowing many of the suspension basics I talked about previously, but you must understand how tires work.

Every force that affects your car, and your performance, is transmitted through the four tires. Absolutely everything. So, you better know how they work and be sensitive to them.

SPEED SECRET

You will never win a race without understanding how tires work.

There are only three factors that determine the amount of traction you have available from the tires. The first is the coefficient of friction between the tire and the track surface, which is determined by the road surface itself and the rubber compound of the tire. The second is the size of the surface of the tire that contacts the track surface. Obviously, the more rubber in contact with the road surface, the more traction available. And the third is the vertical load on the tires. This load comes from the weight of the vehicle and the aerodynamic downforce on the tires.

Tires do not reach their limit of traction and then all of a sudden break away into the land of skidding and sliding. Sometimes it may feel like that, but they always give you some warning signs. As they reach their limit of adhesion or traction limit, they gradually relax their grip on the road.

In fact, primarily due to the elasticity of the rubber, tires have to slip a certain amount to achieve maximum traction. The term used to describe this tire slippage in cornering (lateral acceleration) is called "slip angle" and is measured in degrees. As your cornering forces and speed increase, the tire ends up pointing in a slightly different direction than the wheel is actually pointing. The angle between the direction the tire is pointing and the path the wheel is following is the slip angle.

When accelerating or braking, the amount of tire slippage is measured in percentages.

The tire's traction limit, and therefore its cornering limit, is achieved within an optimum slip angle range, as shown by the "Slip Angle vs. Traction" graph on page 35. That range may vary slightly for different tires (radial tires slip less than bias-ply tires), but the basic characteristics remain the same. Up until that optimum slip angle range is reached, the tire is not generating its maximum traction capabilities. If the cornering speed or steering angle is increased, slip angle will increase along with tire traction until it reaches a point where tire traction then begins to decrease again.

How quickly the tire reaches its optimum range and then tapers off determines the "progressivity" of the tire. A tire that is too progressive (one that takes too long to reach its limit, and then tapers off very slowly) is not responsive enough. It feels sloppy. A tire that is not progressive enough will not give the driver enough warning when it has reached its traction limit and is going beyond it. It doesn't have enough feel. This tire is difficult to drive at the limit since you never know

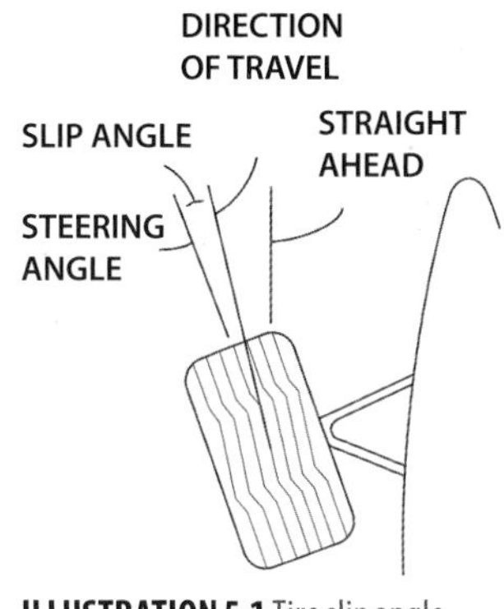

ILLUSTRATION 5-1 Tire slip angle.

precisely when you're going beyond it. Typically, a street tire is more progressive than a racing tire. A racing tire is less forgiving than a street tire.

On a dry track, maximum traction—and therefore maximum acceleration, braking, and cornering (maximum slip angle)—occurs when there is approximately 3 to 10 percent slippage (as shown in the "Percent Slip vs. Traction" graph in Illustration 5-3), depending on the type of tire. This means a tire develops the most grip when there is actually a certain amount of slippage.

Fortunately, as I said earlier, when tires reach their traction limit and then go beyond, they don't lose grip completely and immediately. They actually lose grip progressively. And even when they are beyond the limit, completely sliding, they still have some traction. Think about it. Even when you have locked up the brakes and you are skidding, you still slow down, not as fast as when the tires are still rotating, slipping 3 to 10 percent, but you do slow down. The same thing applies during cornering. When the car starts to slide, the tires are still trying to grip the road. And, as they grip the road, they are scrubbing off speed down to the point where the tires can achieve maximum traction once again.

This is a reassuring fact to remember. It's possible to go slightly beyond the limit without losing complete control and crashing. We'll talk more about driving at and beyond the limit later.

ACCELERATION

When accelerating, think of squeezing the gas pedal. Don't pounce on it. Again, the throttle is not an on-off switch. It should be used progressively, squeezing it down and easing off it. This must be done quickly, but smoothly.

As I said before, there is a limit to your tire's traction, which should be approximately 3 to 10 percent slippage on dry pavement and somewhat less on wet pavement. Should the tires exceed this percent slippage, leading to wheelspin, it will result in less than maximum acceleration. At that point simply ease off the throttle slightly, "feathering" it until you have controlled traction and maximum acceleration again.

BRAKING

The braking system on most race cars is more powerful than any other system in the car. In other words, the car is capable of stopping much quicker than it can accelerate. Take full advantage of this.

As with acceleration, maximum braking occurs with approximately 3 to 10 percent slippage. This means the wheels are actually turning slightly slower—3 to 10 percent slower —than they should be for any given car speed. Exceeding this limit leads to lock-up, 100 percent slippage, and loss of steering control. Braking at the limit, or threshold of traction, is called "threshold braking." It's the fastest, most controlled way to slow, or stop, a vehicle. This is what I mean by maximum braking.

If you brake too hard and lock up the front wheels, you will lose all steering control. If this happens, you will have to ease your foot off the brake pedal slightly to regain control, back to threshold braking. If you do this, you will most likely "flat-spot" the tires. This happens when the tires have skidded along the roadway and worn a patch of tire to the point where the tire is no longer perfectly round. You'll know exactly when you've done this. You'll feel a thumping or vibration in the car as the flat spot rotates.

SLIP ANGLE

Let's take a closer look at slip angles. If you notice in the "Slip Angle vs. Traction" graph in Illustration 5-2, the peak traction limit, or lateral acceleration, is when the tires are in the 6 to 10 degrees of slip-angle range. Let's look at four hypothetical drivers to see where on the graph it's best to drive.

Our first driver is probably inexperienced and definitely a little conservative. He consistently drives through the corners with the tires in the 2- to 5-degree slip-angle range. As you can see from the graph, the tires are not at their maximum traction limit. Driver 1 is not driving at the limit, and therefore will be slow.

Driver 2 has a bit more experience and is known to be a little on the wild side. He consistently overdrives the car. But what does that mean? Well, he always drives through the corners with a slip angle above 10 degrees. In other words, he is sliding the car too much. It may look great, with the car in a big slide all the way through the corner, but the graph shows that in this range, the traction limit of the tires has begun to decrease from maximum. Plus, all this sliding about will increase the temperature of the tires to the point where they are overheated, further reducing the traction capabilities of the tires.

Our final two drivers are consistently cornering in the 6- to 10-degree slip angle range. Both are very fast. Both are cornering at about the same speed. Both are driving the car with the tires at the limit. So, what's the difference? Driver 3 is cornering in the upper end of the 6- to 10-degree range—about 9 or 10—while Driver 4 is around 6 or 7 degrees. Again, the cornering speed is the same, but Driver 3 is sliding a little more than Driver 4, causing more heat buildup in the tires.

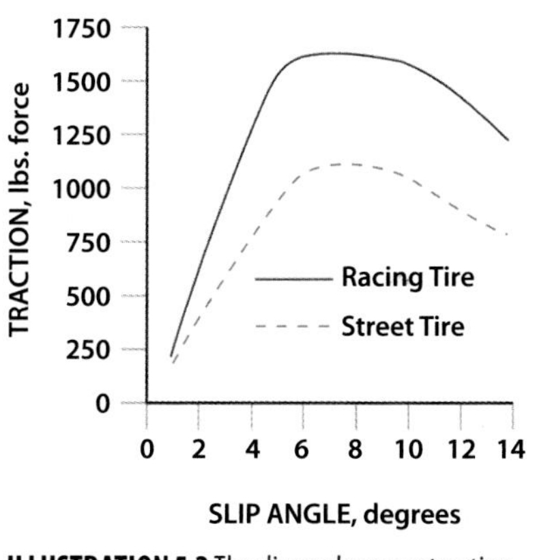

ILLUSTRATION 5-2 The slip angle versus traction graph shows that a tire gains traction as it "slips," up to a certain point, at which it begins to lose traction.

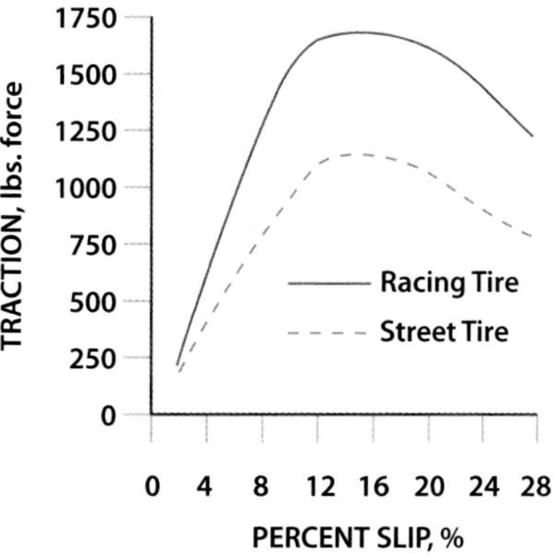

ILLUSTRATION 5-3 Percent slip versus traction graph.

Both drivers will run at the front of the pack early in the race, but eventually Driver 3's tires will overheat and he will fade. He's the one complaining at the end of the race about his "tires going off." Meanwhile, our winner—Driver 4—has gone on, consistently driving with the tires in the 6- or 7-degree slip-angle range and is praising the tire manufacturer for making a "great tire" and his crew for a "great handling car."

The goal, as this example demonstrates, is to consistently drive at the lowest possible slip angle that maintains maximum traction.

Understand that the difference in speed between cornering with a slip angle of 2 degrees and 12 degrees may be 1 or 2 miles per hour, or even less. So you can imagine how much skill and practice it takes to be able to control the car well enough to stay between 6 and 7 degrees of slip angle!

Now, I'm going to contradict myself. Sometimes you have to drive in the upper end of the ideal slip-angle range. If the tires are too hard a compound for your car (perhaps they were designed for another type of car), or the track temperature is low, you may have a difficult time getting the tires to their optimum temperature range. In this case, you may want to slide the car a little more, drive in the upper end of the optimum slip-angle range to generate more heat in the tires to achieve maximum traction. Consistent winners have learned to "feel" this and interpret their tire temperature readings, then adapt their driving style to suit.

SPEED SECRET

Drive at the lowest possible slip angle that maintains maximum traction.

TIRE CONTACT PATCH

I want you to really understand this, as this is the basis for much of what we'll be talking about for a while and what will allow you to drive at the limit. There are only four small tire contact patches (the actual patch of tire or footprint that is in contact with the road at any one particular time) that are actually holding you and your car on the road. The larger the contact patch, the more grip or traction that tire has. Increasing the tires' width obviously puts more tire footprint on the road. The result is more traction. Unfortunately, tire size on race cars is usually limited by the rules.

ILLUSTRATION 5-4 The tire contact patch, or "footprint," is the part of the tire that makes contact with the track surface as it rotates.

VERTICAL LOAD

A factor not limited by the rulebook, but one that has a great effect on the tire

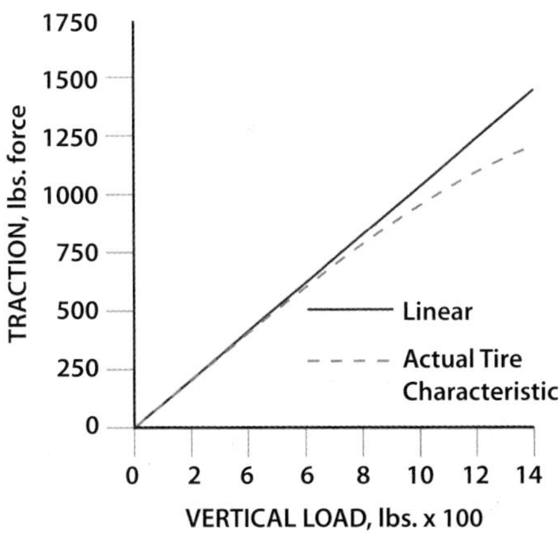

ILLUSTRATION 5-5 Vertical load versus traction graph.

contact patch and the traction it offers, is vertical load, or pressure applied downward on the tire. By increasing this load on a tire, you increase the pressure applied on the contact patch. Thus (up to a certain point where the tire becomes overloaded), you increase the traction limit of the tire.

Now, before you get any ideas of adding a 2,000-pound lead weight to your car, believing that all that extra load will put more pressure on the tires and give them more traction, think about this. Yes, the extra load increases the traction capabilities of the tire, but the work required by the tires to grip the road while carrying that extra load also increases. In fact, it increases even faster. It's not a linear relationship, as noted in Illustration 5-5.

So traction increases with an increase in vertical load, but the work required of the tire increases faster. The result is an overall decrease in lateral acceleration, and therefore, cornering capabilities.

However, there is a way of getting something for nothing here. Aerodynamics. Aerodynamic downforce increases the vertical load on the tires without increasing the work required of the tires. That is why an increase in aerodynamic downforce will always improve the cornering capabilities of a car.

WEIGHT TRANSFER

One of the keys to driving a car at the limit is controlling the balance of the car. In this case, "balance" describes when the car's weight is equally distributed over all four tires (see Illustration 5-6). When the car is balanced, you are maximizing the tires' traction. The more traction the car has, the more in control the car is and the faster you can drive around the track.

I'm sure you already know that as a car accelerates, the rear end tends to squat down. That's because a percentage of the car's weight has now transferred to the rear (see Illustration 5-7). When braking, the car nose-dives. The weight has transferred forward (see Illustration 5-8). In a corner, the weight transfers laterally to the outside, causing the car to lean, or "body roll" (see Illustration 5-9). The total weight of the car has not changed, just the distribution of it has changed.

So, as a car accelerates and weight is transferred to the rear (the back-end squatting down) the pressure, or load, on the rear tires' contact patch increases, resulting in the rear tire traction increasing. During braking, the exact opposite happens. The car nose-dives (weight is transferred to the front) and front tire

ILLUSTRATION 5-6 The car, when balanced, has equal traction capacity on each tire.

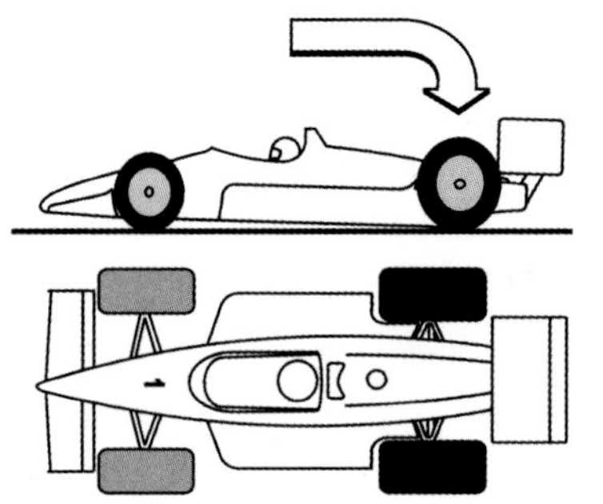

ILLUSTRATION 5-7 Under acceleration, weight transfers to the rear, increasing rear tire traction.

ILLUSTRATION 5-8 Under braking, weight transfers to the front, increasing front tire traction.

traction is increased. While going around a corner, weight transfers to the outside tires, increasing their traction.

However—and this is very important to understand—when the weight transfers onto a pair of tires, increasing their traction, weight is being taken off the other two, *decreasing traction*. Unfortunately, the overall effect to the car is a decrease in *total vehicle traction*.

You can, and must, control this weight transfer to your advantage. Again, as the weight transfers onto a pair of wheels, pushing them into maximum contact with the road, we achieve better traction with those tires. Conversely, the tires that become unweighted lose traction.

TRACTION UNIT NUMBER

Let me explain it this way. If you were to quantify the amount of traction each tire has, and give it a corresponding number, that would be what I call the tire's "traction unit number."

Let's take a look at an example see illustration 5-10. With a car sitting at rest, or traveling at a constant speed, each tire has, let's say, 10 units of traction for a total of 40 traction units gripping the car to the road. Now, when you corner, weight is transferred to the outside tires, increasing the vertical load on them, and therefore their traction, giving them 15 units of traction. But at the same time, weight is transferred away from the inside tires, reducing their vertical load and traction, resulting in only three units of traction each. The total traction for the car is now 15 + 15 + 3 + 3 = 36, which is less than you had before you caused the weight transfer by turning.

As we have already seen in Illustration 5-5, vertical load versus traction is not a linear relationship. As load is increased on a tire, traction increases but not at the same rate as the weight increase. As load is decreased from the opposite tire, traction is reduced at a faster rate. The more the weight transfers, the less the total vehicle traction will be.

BALANCE

Obviously, it is impossible to drive a car without causing some weight transfer. Every single time you brake, corner, or accelerate, weight transfer takes place. However, the less weight transfer that occurs, the more overall traction the car has.

So your goal then is to drive in such a way as to keep the weight of the

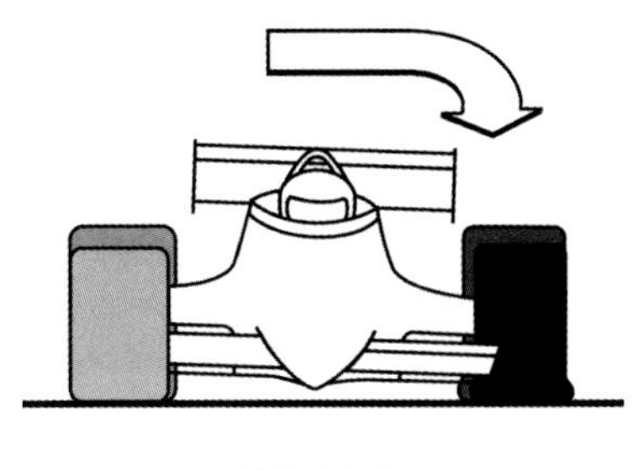

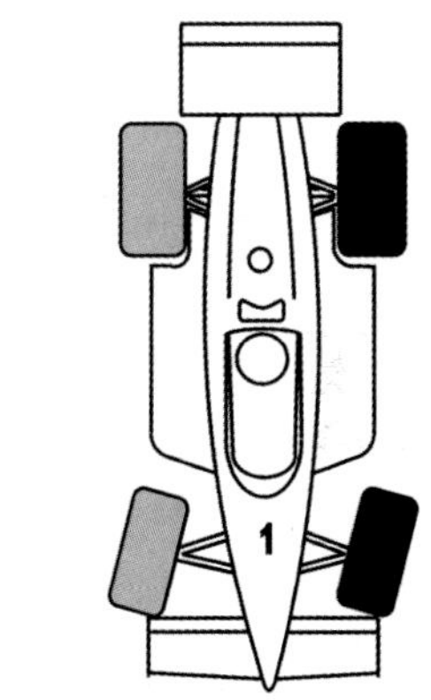

ILLUSTRATION 5-9 Weight transfers laterally, to the outside of the turn when cornering, increasing the traction of the outside tires and decreasing the inside tires' traction.

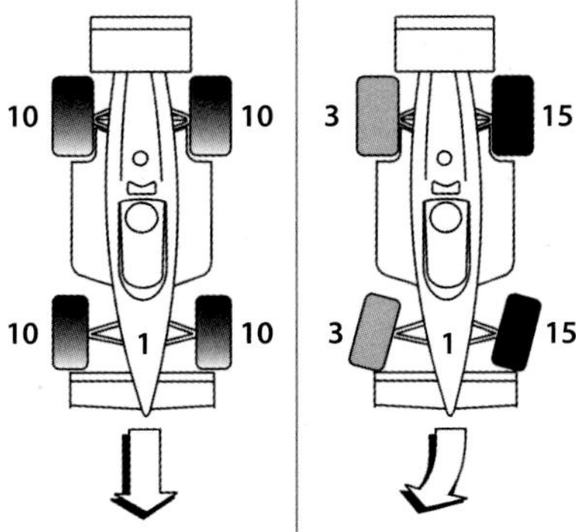

ILLUSTRATION 5-10 The Traction Unit Number example demonstrates that as weight transfer occurs, the car's overall traction limit is reduced. In other words, the better balanced you keep the car, the more traction it will have, and the faster you can drive through the turns.

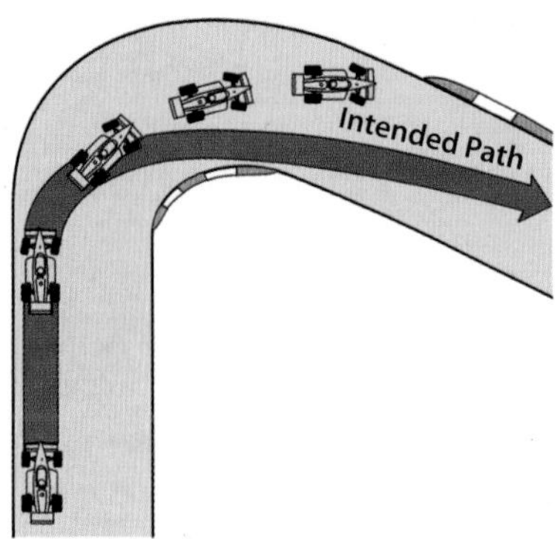

ILLUSTRATION 5-11 An understeering car does not steer, or turn, as much as you want along the intended path. This is also called "pushing" or "tight."

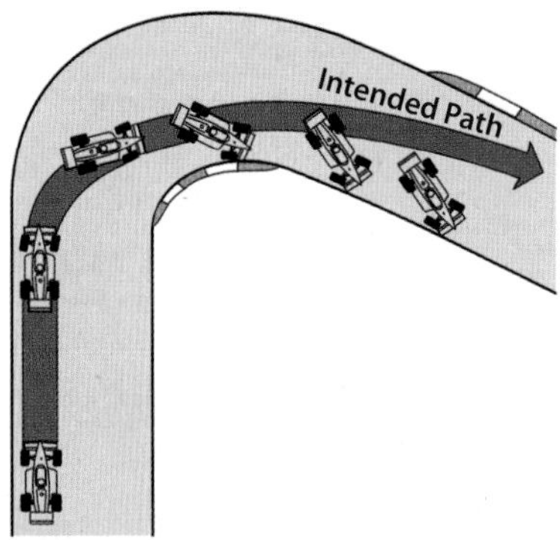

ILLUSTRATION 5-12 An oversteering car steers, or turns, more than you want along the intended path. This is also called "loose."

car as equally distributed over all four tires as possible. In other words, balance the car. How? By driving smoothly. Turn the steering wheel as slowly and as little as possible. If you jerk the steering wheel into a turn, the car leans, or transfers weight a lot. If you gently turn into a corner, the car does not lean as much. Squeeze on and ease off the brakes and gas pedal. Never make a sudden or jerky movement with the controls.

Now you see why it is important to drive smoothly and how it can affect the balance and overall traction of your car. Again, the greater the weight transfer, the less traction the tires have. You play the major role in controlling weight transfer and maximizing traction.

Weight transfer and balance also has an effect on your car's handling characteristics, contributing to either "understeer," "oversteer," or "neutral steer."

UNDERSTEER

Understeer is the term used to describe the handling characteristic when the front tires have less traction than the rears, and regardless of your steering corrections, the car continues "plowing" or "pushing" straight ahead to the outside of the turn. Think of it as the car not steering as much as you want, so it is "understeering." Understeer, in effect, increases the radius of a turn.

Accelerating too hard or not smoothly enough through a corner transfers excessive weight to the rear, decreasing traction at the front and causing understeer.

Most drivers' first reaction to understeer is to turn the steering wheel even more. Don't! This increases the problem because the tires were never designed to attack the road at an extreme angle. The tires were meant to face the road with their full profile, not with the sidewall. So the tires' traction limit has now been further decreased.

To control understeer, decrease the steering input slightly and ease off the throttle gently to transfer weight back to the front. This increases the traction limit of the front tires and reduces speed. Once you have regained front tire traction and controlled the understeer, you can begin squeezing back on the throttle. Obviously, this easing off and getting back on the throttle will destroy your speed on the following straightaway and upset the balance of the car. So make sure you accelerate smoothly the first time.

OVERSTEER

Oversteer is the handling characteristic in which the rear tires have less traction than the fronts, the back end begins to slide, and the nose of the car is pointed at the inside of the turn. The car has turned more than you wanted it to, so it has "oversteered." This is also called "being loose," "fishtailing," or "hanging the tail out." Its effect is to decrease the radius of a turn.

Turning into a corner with the brakes applied, or lifting off the throttle in a corner ("trailing throttle oversteer") causes the weight to transfer forward, making the rear end lighter, thus reducing rear wheel traction. The result: oversteer.

Also, if you accelerate too hard in a rear-wheel-drive car, it will produce "power oversteer." What you have done is used up all of the rear tires' traction for acceleration and not left any for cornering. To control excessive power oversteer, simply ease off the throttle slightly.

To control excessive oversteer, just look and steer where you want to go. This forces you to turn into the slide, or to "opposite lock," thereby increasing the radius of the turn. At the same time, gently and smoothly ease on slightly more throttle to transfer weight to the rear, and thus, increase traction. Whatever you do, avoid any rapid deceleration. This will most likely produce a spin as you decrease the rear-wheel traction even more.

NEUTRAL STEER

Neutral steer is the term used to describe when both the front and rear tires lose traction at the same speed or cornering limit and all four tires are at the same slip angle. Sometimes described as "being in a four-wheel-drift," this is ideally what a driver is striving for when adjusting the handling of the car and trying to balance it.

I love the feeling when I'm controlling the balance of the car with the throttle, driving through a fast, sweeping turn at the limit. If the car begins to oversteer a little, I squeeze on more throttle to transfer a little weight to the rear; if it starts to understeer, I ease off slightly, giving the front a little more grip. When it's done just right, all four tires are slipping the same amount (the car perfectly balanced, neither oversteering nor understeering) in a perfect neutral steer attitude through the turn.

In terms of how the car is set up, however, most drivers prefer a little understeer in fast corners, as it's a more predictable, safer characteristic, and oversteer in slow corners to assist in pivoting the car around the tight turn.

TAKING A SET

"Taking a set in a turn" describes when the car has finished all of its weight transfer. It is the point in a turn where all the weight transfer that you are going to cause has occurred. The car is most stable when it has taken a set and can be more easily driven to its limit then.

How quickly the car takes a set in the turn is largely a matter of how the shock absorbers are adjusted and how you drive. As you turn into a corner, the quicker the weight transfers, the quicker the car takes a set. The sooner the car takes a set, the sooner you can drive the car at its limit, and the faster you will be.

Why? Remember the Traction Unit Number example. As weight transfers, the tire traction available is reduced. Once all the weight transfer that is going to occur has occurred, and the car has taken a set, you can then work with the traction available and drive at the limit. If you don't make the weight transfer happen quickly enough, you spend most of the corner waiting for the car to take a set. Therefore, you wait a long time before you really know what traction limit you're working with. If you don't drive smoothly—causing a little weight transfer, then a lot, then less, then more again, all through the same corner—the car will never take a set. It's difficult to drive at the limit when that limit is constantly changing.

Before you get any ideas about making the weight transfer occur too quickly, however, think about the Traction Unit Number example again. If you quickly transfer weight by jerking the steering into a corner, the effect will be more overall weight transferred and therefore less overall traction.

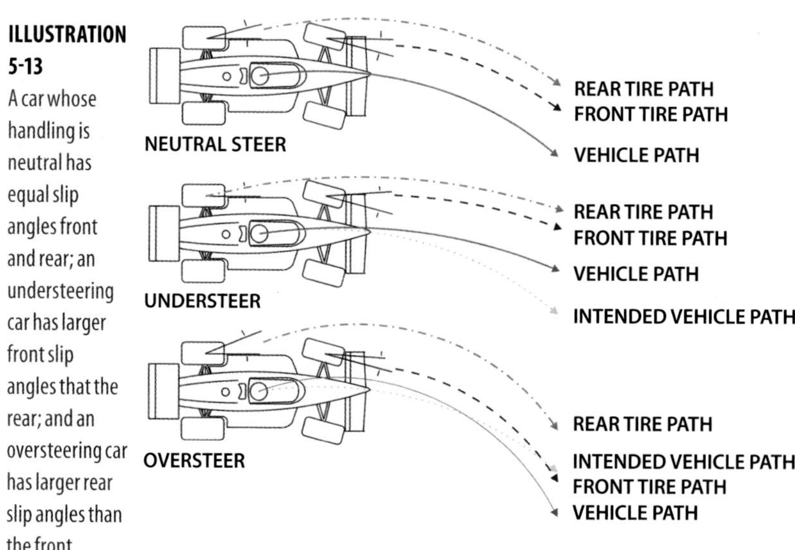

ILLUSTRATION 5-13
A car whose handling is neutral has equal slip angles front and rear; an understeering car has larger front slip angles that the rear; and an oversteering car has larger rear slip angles than the front.

So your goal is to make the car take a set in the turn (get to its maximum weight transfer and stay there) as quickly as possible without causing any more weight transfer than necessary. That means use smooth, precise, and deliberate actions with the controls.

DYNAMIC BALANCE

Getting back to balancing the car, there is also what I call "dynamic balancing." Few cars have a perfect 50/50 weight distribution to begin with. Most purpose-built race cars are midengine with a weight distribution around 40 percent front and 60 percent rear, as this is close to ideal for a race car. Production-based front-wheel-drive cars are usually closer to 65 percent front and 35 percent rear. Only production-looking tube-frame race cars (Grand-Am GT, NASCAR, etc.) are close to 50/50 weight distribution.

Realizing this, a driver must compensate by controlling the weight transfer to balance the car into a neutral handling state (no understeer or oversteer). To do this, the driver may have to effect the weight transfer so that statically there would be more weight on the front or rear, but dynamically the car is perfectly balanced.

Look at it this way: Let's assume your race car's static or at-rest weight distribution is 40 percent front, 60 percent rear, and it is set up to oversteer at the limit (either on purpose or because you haven't been able to find the right setup). While driving through a 100-mile-per-hour corner, you know you could go quicker if the car oversteered less—if it was neutral handling. To make the car oversteer less, you will have to cause some weight to transfer rearward by squeezing on the throttle. This would change the weight distribution to approximately 35 percent front and 65 percent rear. At speed through a corner, dynamically, this is balanced.

BRAKE BIAS

Keeping this weight transfer in mind, an important factor in braking is how the brake bias is set or adjusted. Braking forces are not equally shared by all four wheels. Due to the forward weight transfer under braking, and therefore more front tire traction, most of the braking is handled by the front brakes. So the brake forces will be biased toward the front. This is why all vehicles have larger brakes on the front wheels than on the rear.

Actually, you want to adjust the brake bias so that the front wheels will lock up just slightly before the rears. This is a more stable condition, as it gives you more warning of a skid. You will feel it in the steering immediately if the front tires begin to skid. Plus, if the rear tires lock up first, the car will tend to skid sideways.

Different conditions will require a different ratio, or bias, of front-versus-rear braking forces. In the rain, because there is less forward weight transfer to the front (because traction limits are lower, heavy braking is not possible without locking up), you will have to adjust the brake bias more to the rear. Some cars also change dramatically as the fuel load lessens during a race. This is where a driver-actuated brake-bias adjuster is beneficial.

Practically all purpose-built race cars have an adjusting mechanism for changing the bias. Learn how to "read" and then adjust your race car's brake bias. With production-based race cars, you will pretty much have to live with the bias that the factory built into the car.

AERODYNAMICS

Aerodynamics really only come into full effect at relatively high speeds. Only very sensitive, experienced drivers will feel the effects of aerodynamics at anything under 60 miles per hour. Beyond that, aerodynamics play a big role in the handling of a race car, and therefore, you must learn as much as possible about how to adjust and feel the effects.

In the simplest terms, a race driver is only concerned with two aspects of aerodynamics: drag and lift (both negative and positive). Drag is the wind resistance or friction against the body of the vehicle that effectively slows down the car. Lift is the effect the air has on the weighting of the vehicle: Positive lift is the lifting up of the body (which is what airplanes like), and negative lift is the downforce on the body (which is what cars like, or should I say, what drivers like). This keeps the car in contact with the road.

Aerodynamics can influence the balance of a car and cause it to either understeer or oversteer. This is referred to as the car's "aerodynamic balance." Sometimes, a car that understeers at relatively low speeds will begin to oversteer at higher speeds. The low-speed understeer is a result of suspension design. But as the speed increases, bodywork design (including wings if present) begin to affect the situation. A vehicle with more downforce on the front-end than on the rear (possibly due to spoilers, wing adjustments, etc.) will have more traction on the front tires as the speed increases, resulting in high-speed oversteer. It is important for you to understand the difference between suspension-induced and aerodynamic-induced handling characteristics.

The balancing of the suspension and aerodynamics is what setting up the car is all about. Many hours are spent developing the handling in the slow corners with suspension adjustments, then changing aerodynamics for the ultimate balance between downforce (front or rear) and drag. Unfortunately, the increase in downforce (resulting in higher cornering speeds) means more drag, resulting in less straight-line speed. The lift (downforce) to drag ratio is quite a compromise.

Another important factor for a driver is how a car leading another will affect the following car's speed and handling. When a leading car blocks the air, reducing the wind resistance for the second car, it is called "drafting" or "slipstreaming." This allows the second car to travel faster, to perhaps pass the leading car or even back off the throttle slightly to conserve fuel.

Another, often forgotten, factor comes into play here as well. Particularly with winged or ground-effect cars, the car relies on a certain airflow for downforce. When that airflow is blocked by a leading car, the second car's cornering ability will be reduced. That is why you will see a car catch up to another quite quickly and then struggle to get past. When it's running by itself, it is quicker; but when its airflow is reduced, it is no faster than the leading car. As a driver, you must recognize this and not overdrive while following closely behind another car. Perhaps, the best strategy is to "take a run at the leading car." In other words, hang back a little until you get enough momentum from the draft to quickly pull out and pass on the straightaway.

The first time I drove an Indy car on an oval track, I couldn't believe the effect other cars around me had on the handling of my car. If there was a car in front of me, it took away a lot of the air flowing over my car, causing it to understeer. If there was a car close to my tail, it seemed to make the airflow over the rear wing less effective, causing my car to oversteer. I learned quickly to make note of the other cars' positions and to predict what they would do to my car. By the way, this doesn't just happen with Indy cars. Any car that relies on aerodynamic downforce for grip will be affected to some extent.

Any car that depends on ground effects for downforce has another little trick for you to consider. The faster the car, the more downforce, and therefore traction, it has. This can make for an uncomfortable situation when you begin to drive a ground-effects car. If you reach a point when it feels like the car is at its limit, you may have to drive faster to get more downforce. Once you go faster, the car has more grip, and it feels like you're nowhere near the limit, and you're probably not!

SMOOTHNESS

As I mentioned earlier (a number of times!), balancing the car is one of the most important and probably most difficult aspects of driving. But, again, it is the key. Whenever one or two tires become unweighted due to weight transfer from braking, cornering, or acceleration, they lose traction. So, obviously, you want to cause as little weight transfer as possible. How? By driving *smoothly*! The less abruptly you apply the brakes, turn the steering wheel, or use the gas pedal, the smoother you will drive, and the more overall traction the car will have. In other words, don't abuse the traction the tires give you.

We've seen how important controlling the weight transfer in the car is and how to do this with your controls, but you also have to accomplish this with extreme smoothness. If you jerk the steering into a turn, it immediately transfers excessive weight to the outside of the car, reducing traction. Now you have to wait until the car's weight settles down and is balanced again—taken a set—before being able to corner at the limit and accelerate out of the corner. This wastes time.

SPEED SECRET

Smooth is fast.

The car should be driven absolutely as smoothly as possible all the time. Practice this in your everyday driving. Don't pounce on the gas pedal; squeeze it on and ease off gently. Don't slam on the brakes; squeeze them smoothly and progressively to the threshold braking limit. Don't yank or jerk the steering wheel; smoothly and gently feed in the required steering input that your eyes looking well down the road tell you. Don't bang the shifter into gear; simply place it in gear—with finesse.

Keep in mind that each tire has a specific, limited amount of traction. If you exceed that traction limit, the car will begin to skid or slide. The smoother you drive, the easier it will be to stay within those traction limits. A tire achieves a higher traction limit if it is gradually built up to that limit. In other words, if you enter a corner and quickly jerk the steering wheel into the turn, or jab at the brake pedal when trying to panic stop, you haven't given the tires a chance to gradually build up their traction forces. They will not be able to hold on, and a skid or slide will result.

Think of the tire's traction limit as the force it takes to snap a piece of string. If you gradually and gently pull two ends of the string, it requires a lot of force to break it. However, if you quickly jerk the string apart, it snaps with much less force, just like a tire's traction limit.

So, everything you do behind the wheel must be done smoothly. When turning into a corner, turn the steering wheel as gently and *slowly* as possible. This will make the turn smooth! When braking, *squeeze* the brake pedal, don't jab at it. Believe me, if you squeeze the brakes, you will stop faster and with more control than if you quickly jabbed at the pedal. So, always think *squeeze* when applying the brakes or the throttle. Progressively squeezing the gas pedal down will give more controlled acceleration, even when trying to accelerate in a hurry.

It is better to be smooth than to be fast. Speed will come with practice, practicing smooth driving. Trying to drive fast before learning to be smooth is a mistake. You will never be as fast as if you learn to be smooth first and let your speed pick up naturally.

Once again, the slower and smoother *you* move behind the controls, the more in control you will be, and the faster your car will be.

TRAIL BRAKING

"Trail braking" is a term used to describe the technique of continuing braking while turning into a corner. In other words, you brake and turn at the same time. There is a specific reason for doing this, as explained by *understanding* the Traction Circle.

TRACTION CIRCLE

The traction circle is a simple graphic way of showing the performance of any driver in any car. Basically, it is an X-Y axis graph produced by a computer data acquisition system of the g-forces during braking, cornering, and acceleration that the car experiences while being driven around a track. See Illustration 5-14.

First of all, 1 g-force is equal to the force of 1 times the weight of the vehicle; i.e., if a 2,000-pound car is cornering at 1.0 g, there is a centrifugal force of 2,000 pounds pushing outward on the car.

Consider that a tire has relatively equal traction limits in each direction—braking, cornering, or acceleration—say 1.1 g, for example. In other words, the car and tire combination is capable of braking at 1.1 g, cornering at 1.1 g, and

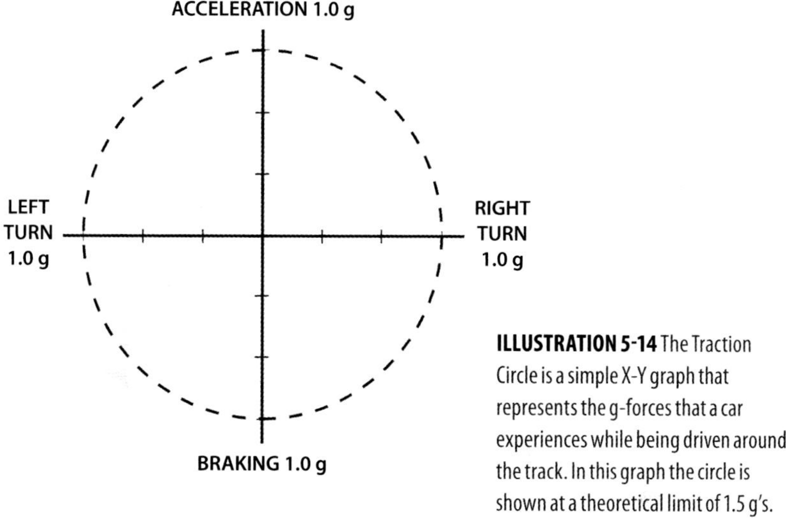

ILLUSTRATION 5-14 The Traction Circle is a simple X-Y graph that represents the g-forces that a car experiences while being driven around the track. In this graph the circle is shown at a theoretical limit of 1.5 g's.

accelerating at 1.1g before the tires begin to break away and start to slide. If you exceed the tires' traction limit, they will begin to slide, slowing you down and, if not controlled, resulting in a spin. On the other hand, if you do not use *all* the tires' traction available, you will be slow.

These g-forces can be measured and graphed as you drive through the corner. If you use the proper driving technique, the graphed line will somewhat follow a circle—the traction circle—telling you that you are using the tire's full potential.

In the transition from one directional force to another, say from braking to cornering, there are two ways to get from one limit of traction to the other. You may, upon reaching the end of the braking zone (where you braked at 1.1 g), suddenly lift off the brakes, and then turn the steering wheel into the corner (building up to 1.1 g of cornering force). The second option is to gradually ease off the brakes while progressively applying more and more steering angle and overlapping some of the braking and cornering. This is called *trail braking.*

In the first scenario, the car goes through a short period (perhaps only a fraction of a second) where the tires are doing little work. They are not being used to their full potential. This wastes time, no matter how short, because the car cannot instantly change from straight-line braking to a curved path. The second scenario, which keeps the tire and car on the outside edge of the traction circle graph, is a much faster way of driving a race car. It is also the smoothest way of "building" traction forces, which as we know, generates higher cornering speeds.

So what you must do (what the traction circle is telling you to do) is to continue the braking into the corner-entry phase (trail braking) so that, while the tires are in the process of building up cornering force, they are still contributing braking force. Or brake at 100 percent of the traction limit (1.1 g) along the straightaway up to the corner, and begin to ease off the brakes as you turn in, trading off some of the braking force for cornering force (90 percent braking, 10 percent cornering; then 75 percent braking, 25 percent cornering; then 50 percent, 50 percent; and so on),

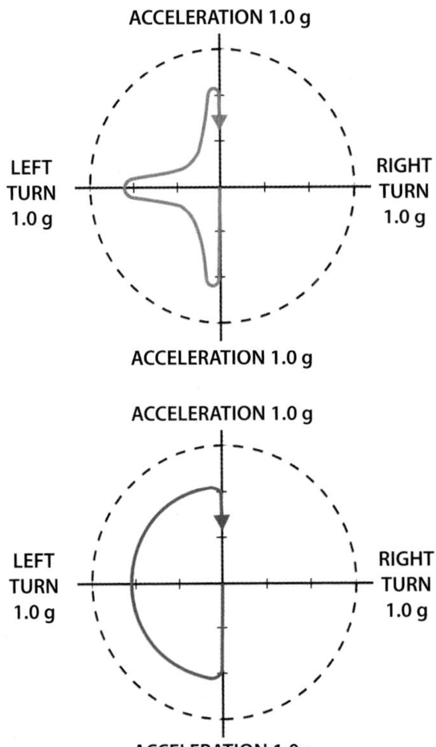

ILLUSTRATION 5-15 These two Traction Circle graphs represent two ways of driving the same corner. In the graph on top, a lot of the tires' potential is not being used, and time is being wasted. The graph on the bottom shows the correct way to take the corner; the tires' full traction potential is being used.

until you are cornering at the limit (using 100 percent of the traction for cornering, at 1.1 g). Then, you start to straighten the line through the corner, "unwinding the car" out of the turn early, so that the tires have traction capacity for the acceleration phase (90 percent cornering, 10 percent acceleration; 75 percent cornering, 25 percent acceleration; 50 percent cornering, 50 percent acceleration; and so on).

The real key to the traction circle is the smooth progressive overlap of braking, cornering, and acceleration. If you follow the old advice "Do all of the braking in a straight line, go through the corner at maximum cornering force, then accelerate in a straight line," you are going to waste a lot of the car's potential and a lot of lap time. You must "drive the limit" by balancing and overlapping the braking, cornering, and acceleration forces to keep the tires at their traction limit at the edge of the traction circle. This will lead to the fastest possible lap and to another type of circle, the winner's circle.

As I said before, tires do have a limit to their traction. If you are using 100 percent of that traction for cornering, you can't use even 1 percent for acceleration. The traction circle demonstrates how a tire's traction limit can be used and shared. It shows that if you are using all of the tire's traction for braking, you can't expect to use any for cornering without easing off the brakes. If you are using all the traction for cornering, you can't use any for acceleration until you begin to "unwind" or "release" the steering (straighten the wheel). If you are using all the traction for acceleration, you can't still be cornering near the limit.

Think of the throttle and brake pedal as being connected to the steering wheel. More steering angle means less brake or throttle pedal pressure. More pedal pressure means less steering angle. Too much steering angle combined with too much pedal pressure puts the tires beyond their traction limit.

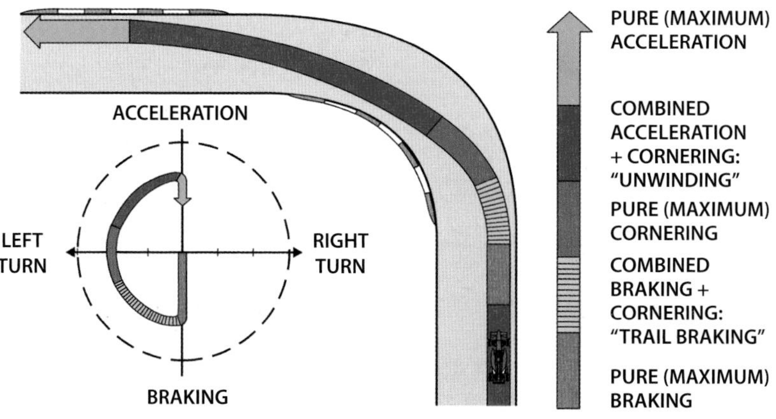

ILLUSTRATION 5-16 This illustration shows the relationship between what the driver is doing in the corner and the Traction Circle graph.

Too much steering angle for the amount of braking or acceleration (or vice versa) will cause the car to exceed the traction limit, usually at one end of the car before the other (understeer or oversteer). This can sometimes "trick" you into believing there is a handling problem with the car, when it most likely is your technique, asking either the front or rear tires to do more than they are capable of doing.

When I attended my first racing school, I was taught to do all my braking in a straight line on the approach to a corner, then turn into the corner. Over the next couple of years I gradually learned by trial and error to trail brake. But when I started to race a Trans-Am car a few years later, I had to improve my trail braking. It was the only way to go fast in one of those cars. So over the next couple of weeks, at night in my street car, I would practice trail braking well into the corners of a deserted industrial park. I didn't have to go fast. I just practiced the technique of trailing off the brakes while turning into the corner, then squeezing back on the throttle while unwinding the steering out of the corner. It really was an effective way of improving my technique.

The traction circle really demonstrates the key to driving fast is balancing the pedal application with the steering angle. Learn how to overlap the braking, cornering, and acceleration, and you will drive the limit.

6 DRIVING THE LIMIT

As we've seen in the last chapter, to be a winner, you have to use the tires' traction limit. Once you have built up the tires' braking, cornering, or acceleration forces, keep them there. Drive the limit.

I know it sounds easy to say, but that's what it takes. Entering the corner, brake at the traction limit. That's threshold braking. As you reach the point where you begin turning into the corner, ease off the brakes as you turn the steering wheel. The more you turn the wheel, the more you ease off the brakes (trail braking), until you are completely off the brakes. At this point your vehicle should be at the tire's maximum cornering traction limit. As you start to unwind the steering coming out of the corner, you start to increase the acceleration until you are at full throttle onto the straight. (See Illustration 6-1.)

What you want to do is brake at the traction limit, then trade off braking for cornering as you enter the corner. Then corner at the traction limit, and then trade off cornering for acceleration as you unwind out of the corner. Then use full acceleration traction onto the straight.

This overlapping of forces *must* be done with extreme smoothness, resulting in one flowing drive through the corner *at the limit.*

SPEED SECRET

Overlap your braking, cornering, and acceleration forces.

If it's not done smoothly, the car won't be balanced, and the limit will be reduced, possibly at one end of the car sooner than the other, causing oversteer or understeer. If done smoothly, though, you can control the oversteer or understeer at a higher limit or speed to your advantage to help control the direction or "line" of the car. You do this by what I call "steering the car with your feet": controlling the balance of the car.

I still remember the first time I experienced steering the car with the throttle. It was at my first racing school course, driving a Formula Ford. As I drove through a fast sweeping turn, I eased off the throttle. The car began to oversteer, making it turn a little more into the inside of the corner. I then applied more throttle, and it understeered, causing it to point more toward the outside edge of the track. The whole time I kept the steering in the same position. I was thrilled. I could change the direction of the car with my right foot as much as I could with the steering wheel. Of course, what I had learned was the effect weight transfer had on the car when driving at the limit and how to use that to my advantage. It's still one of the most enjoyable parts of driving for me.

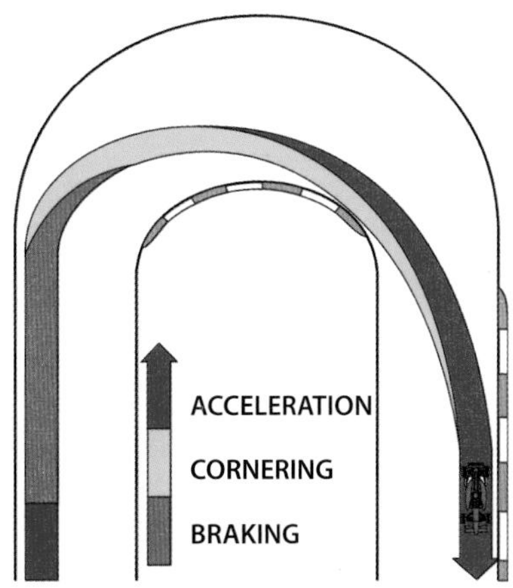

ILLUSTRATION 6-1 This illustration shows the overlap of braking, cornering, and acceleration.

Before I go any further I want to define exactly what I mean by "driving at the limit." When I say "at the limit" I mean having all four tires of the car at a point along the slip angle-versus-traction graph (see Illustration 5-2) where they are producing their maximum amount of traction. It is when the car is being driven at a speed dead in the middle between two extremes:

- At one extreme, the car is not being driven fast enough, and not all of the tires' traction is being used up. The car is being driven below the limit.
- At the other extreme, the car is being driven beyond the limit. The tires, and therefore the car, are sliding too much.

When I talk about "the limit" I'm not talking about some theoretical thing, mind you. No, I'm talking about the very real, physical limit or threshold of the tires gripping the track.

Although the limit is a very real, physical thing, it can change. That is, the way you drive the car will determine to some extent at what speed your tires and car reach the limit. That is why one driver can drive the car at the very limit, only to have another driver hop into the same car and go even faster. Was the first driver not driving the limit? He may very well have been. The point is, though, his driving technique may have produced a slightly lower limit than the second driver. How does that happen? Mostly by driving in such a way that the car is not as well balanced as it could be.

Of course, defining what driving at the limit is, and even doing it, is much easier than telling someone how to do it. Which is what I'm trying to do in this book.

How do you really know when you are driving the car at a speed where all four tires are right at their limit of traction? Start by asking yourself some questions. "Is the car sliding?" If not, you can drive faster. "Is the car sliding too much?" If so, you are scrubbing off speed and possibly overheating the tires. An excessive slide or drift may feel good, and look great, but it is usually not the quickest way around the track.

So, if no sliding is not enough and too much is really too much, just how much is enough? And how do you get there? Well, I could say that it just takes experience—seat time—honing in from too little to too much to just right. And that's the start, but you didn't buy this book to hear me say you just need more seat time, so let me try to explain.

Most drivers, when they first begin racing, do not slide the car through the turns enough. It's like the car is on rails. Then, with experience, they begin to slide the car more and more, eventually learning to slide it too much. They are driving slightly beyond the limit. Finally, they learn to fine-tune the amount of sliding, homing in on the ideal slip angle.

As I said, the key is to keep the tires right at the peak of the slip angle-versus-tire traction curve. Without great traction-sensing skills and awareness, you will never be able to tell when you are there and when you are on either side of the peak. And remember, we are not just talking about one part of a corner, for example, standing on the throttle exiting a turn with the car in an oversteer drift. No, we are talking about having all four tires at the ideal slip angle from the nanosecond that you turn into a corner to the exit point—every corner of every lap.

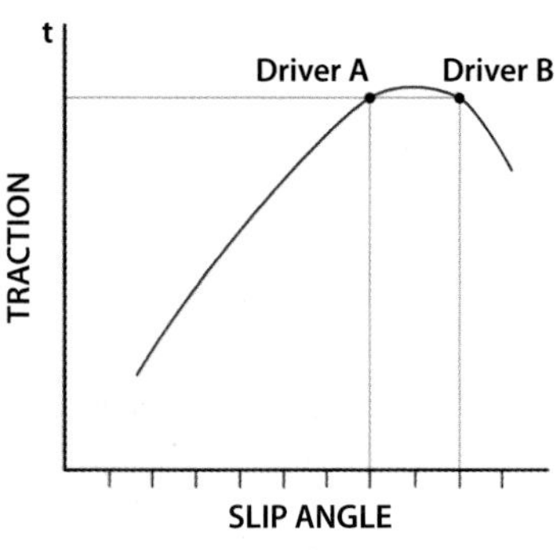

ILLUSTRATION 6-2 Here are two drivers, Driver A and Driver B, on the same Slip Angle versus Traction graph. Both drivers are generating the same level of traction (t), but Driver A is driving the car with 7 degrees of slip angle, while Driver B is using 9 degrees of slip angle. Both will turn the same lap time, as their cornering speeds will be the same, but Driver B has a bigger challenge: Driver B's point on the graph is less forgiving. If Driver A makes a small error, he or she will either be too slow or will actually generate more traction. If Driver B makes a small error, he or she will either generate more traction or exceed the tires' traction limit and spin.

People would talk about Michael Schumacher—or Jackie Stewart, Richard Petty, Jacky Ickx, Dario Franchitti, or any of the greats—and how he just seemed to be able to drive faster than everyone else. And yet, they don't talk about what he did to do that. Primarily, it was his ability to balance the car better than everyone else—therefore, his limit was higher—and his ability to sense the peak of the tire traction curve and keep the car there.

When drivers ask me how I know when I'm driving at the limit, I immediately know one thing: They are trying to do too much at one time. The reason they don't know when they are at the limit is they are trying to drive the entire track or corner at the limit, all at once. A driver's mind is not capable of taking that much information in, and focusing on it, at one time. If the driver focused on driving at the limit in one phase, say the exit, only at first, then the entry, and so on, he or she would be more successful at driving the limit.

Some people's response to the question is, "If you have to ask, you'll never be a real race driver." That's silly. These people either have never driven near the limit themselves or were one of the lucky ones who stumbled onto the strategy I suggested, without knowing that is what they are doing.

Of course, the other response is that you have to drive over the limit, to the point of spinning or crashing, and then dial it back a bit from there. In my way of thinking, that is not the ideal way to reach your goal. It is dangerous, expensive, and not the quickest way of learning since you will spend so much time getting your car back on track or repairing it.

The key is to have a strategy and specific objectives. Break the task down into manageable bites, and then focus on only two or three of them at a time.

BITES AT THE LIMIT

Having spent most of my life studying the driving styles and techniques of thousands of drivers who I have raced against, coached, or just (intently) watched, I have come to the following conclusion. The art of driving a race car really fast—driving at the very limit—does not come from just one thing (which, I'm sure, surprises no one). No, when you break the craft of driving a race car down to the basics, there are four separate but related things that a driver must do to drive as fast as possible:

- Identify and then drive the race car along the ideal path or line around the racetrack; this is called the Line.
- Drive the race car at the limit at the exit of every corner, through the exit phase.
- Drive the race car at the limit when entering every corner. through the entry phase.
- Drive the race car at the limit in the middle of every corner, through the midcorner phase.

Simple and obvious enough. And by the way, these four stages are exactly what every driver, from novice to world champion, naturally works on when trying to drive at the limit and in the order most drivers naturally approach it.

When someone begins to race cars, the first thing they learn is how to determine and then drive the ideal line. As the driver gains a little more experience, he or she begins to work on the exit phase of the corner, getting on the throttle earlier and earlier to maximize the ensuing straightaway speed. In most cases, at the club and minor professional levels, the driver who drives the best line and gets on the throttle at the exit of the corner first wins most often.

Drivers at the upper levels of professional racing have all but perfected the line and the exit stages of driving, and now the difference between the winners and losers is all in the entry phase. Closely watch the speed that the winner in an Indy-car race carries into the turns. It is definitely quicker than the drivers who do not win. Going back a few years, Juan Montoya, who was driving an Indy car, was visibly quicker at the entry of every corner than the drivers finishing farther back in the field. Yes, I know that some of it has to do with the car and its setup, but the driver is the final determining factor, and it's all in the corner-entry speed.

Finally, what separates the truly great drivers from everyone else is the speed carried through the middle of the turns.

The ultimate objective of the first step, perfecting the ideal line, is obvious: Just drive the line around the track that minimizes the amount of time spent lapping the track. There are practically an infinite number of possible lines a driver could drive through a particular corner, however. Combine that with the number of lines used to connect each turn on the track, and you begin to see the enormity and challenge of just this part of driving a race car.

If the only goal of selecting the ideal line were to maximize your speed through each corner, the job would only be difficult. But each turn on the track cannot be considered in isolation. They are all connected and often affected by each other. The line chosen will determine, to some extent, your success with the other three priorities, the exit, entry, and midcorner speed.

Learning the line is the first step every race driver goes through, and it is often the difference between winning and losing in the early stages of a racing career. Let's just say it is highly doubtful that a driver is ever going to win if he or she hasn't figured out how to identify and drive the ideal line on a racetrack. Of course, this is the first area a good driver coach can help you with.

In terms of the exit phase of the corners, the ultimate goal is to begin your acceleration as early and as hard as the car can possibly take. This must still be done as smoothly as possible; otherwise, it will delay the acceleration. And again, the line you have chosen will play a big factor in your exit phase.

Practically every race driver that wins races at the amateur or professional level has pretty much perfected the line and exit phases of race driving.

The one place that is obvious that the real stars and champions outshine the rest is in the entry phase of the corners. They are able to carry more speed into the turns, without it negatively affecting the line or exit phase. Of course, any driver can carry lots of speed into a corner. The key is to be able to do it without it hurting your corner exit speed.

The key to the entry phase is carrying enough speed but not too much. That is why this phase often separates the winners from the also-rans.

The last step in becoming a real superstar is the midcorner phase. This is what separates World Champions from all the rest. Yes, I'm sure the entire F1 grid can drive as good a line as the champion does and probably accelerate out of the turns as well as he does. And yes, a few others can carry as much speed into the turns as a champion. But that is where the similarities start to diminish. I wish it was easier to compare, but if you watch carefully, you can see for yourself that the champion is able to consistently carry more speed through the middle of the corners than anyone else.

So the driver who drives the ideal line, gets on the throttle the earliest and is the hardest exiting the turns, carries the most speed into these turns, and manages to maintain the momentum or speed through the middle of the corners is going to be the quickest. Simple enough (I wish!).

As I said, each and every one of these stages are related and interrelated. You can never think that once you have mastered one stage that you will never have to go back and work on it again. It is a constant game of getting one just right, only to have to modify it once you get another stage right, and then another, and so on.

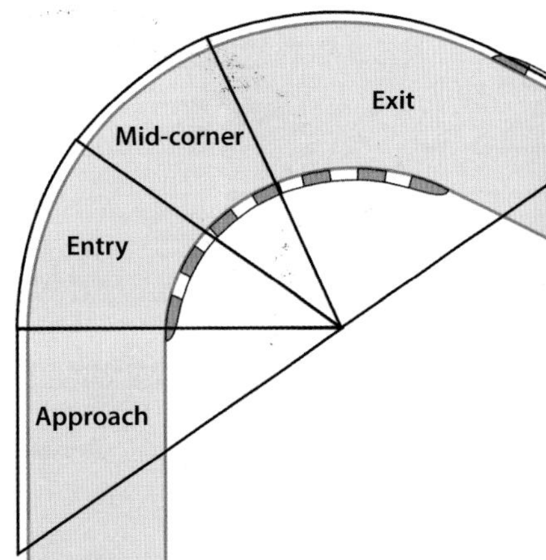

ILLUSTRATION 6-4 By breaking your job of driving at the limit into manageable bites, the odds of consistently doing it are better. You should start by breaking each corner into phases: the approach, entry, midcorner, and exit.

No matter who the driver is, it is not just a matter of learning each step and then never having to go back to relearn or improve one of the steps. In fact, it is a continual cycle. The driver learns the ideal line, then works on getting on the power earlier and earlier until the exit phase is under control. Then the driver works on maximizing his corner-entry speed; and finally, on perfecting the midcorner speed. At that point, the driver usually has to go back and alter the line slightly, which results in having to work on the exit phase again, then the entry and midcorner again. Then the cycle starts all over again.

In reality, it is not something that you will go through once, perfecting them at one shot, and then moving on to the next. No, it is almost a continuous loop, sometimes not even going back to the beginning, but hopping around from one stage to another. In the beginning you will work on getting the line down just right, then accelerating early out of the corners, carrying more speed into the turns, and getting the midcorner right. Then it's possibly back to the line again as all this focus on gaining speed in the other phases means altering the line slightly. Or you may get the corner-entry phase just right and then have to go back and work on getting back on the throttle early in the exit phase; or once the midcorner speed is up, the entry phase needs to be worked on. In fact, it is an endless pursuit, the pursuit of the perfect corner, then the perfect lap, and ultimately the perfect race. Can it be done? Perhaps not. But the pursuit of it is the real challenge, and thrill.

The truly great drivers are doing this all the time, on every lap on the track, whether they are consciously aware of doing it or not. For the greats, this whole process occurs at a subconscious level where they are not actually thinking through what they are doing, they are just doing it.

How do you do that? How do you perfect each step, each piece of the puzzle? I hope to answer those questions throughout the rest of the book, as I attempt to explain what it takes to maximize your performance and drive the car at the limit in each of these four stages. Of course, I will only do that at a

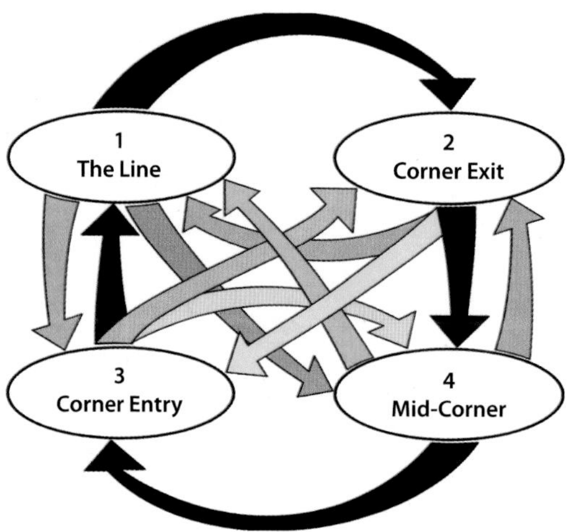

ILLUSTRATION 6-5 There are four stages every driver goes through in learning to drive consistently at the very limit. But as the illustration shows, once you've gone through each stage once, you should continually go back and fine-tune each one again and again. The learning never ends.

conscious level. I will help you become aware of what is required to drive at the limit. It will then be up to you to take that knowledge, that understanding of the process, and turn that into an ability to do it on the racetrack, at speed, at a subconscious level.

There may be times where you are looking for the last few tenths of a second in lap time, but you're not sure where it is going to come from. At this point, someone may suggest that it is not going to come from one place, but rather from a tiny little bit from a bunch of places, perhaps from each corner on the track. And they are right.

The key to finding the last few tenths most often comes from ensuring the car is being driven to the very limit, the ragged edge, in every segment of every turn on the track. Many drivers drive the car at the limit through one or two segments of a turn, for most of the corners. You need to drive at the limit through all three segments of every corner to be really fast.

SPEED SECRET

Drive the car at the limit for every segment, for every turn, for every lap.

Before we go any further, I think it is important that you understand what I mean by some common terms:

Turn-in

The term "turn-in" is used to describe what the car is doing during that fraction of a second that you initially turn the steering wheel at the beginning of a corner. Ultimately, you want the car's turn-in to be "crisp," meaning the car changes direction immediately when you turn the steering wheel. At the same time, the turn-in can be too crisp. The opposite of a crisp turn-in is a lazy turn-in. A lazy turn-in means there is some amount of delay from the time you turn the steering wheel until the time the car changes direction.

Of course, how *you* turn in is going to vary, as well, depending on the type of corner you're faced with. More about this later.

Corner Entry

The entry of a corner is just after the initial turn-in until the midcorner section. Think of it as the section of the corner between the turn-in and the point in the corner where the car is in a steady state. In the corner entry phase, you are continuing the motion of winding in more steering input.

The entry phase can also be thought of as beginning just after turn-in and continuing until your right foot begins to apply some throttle.

Midcorner

The midcorner is usually when you have wound in all of the steering input required to get the car aimed toward the apex, and you are not yet unwinding the steering. The car is on a consistent radius, not decreasing or increasing. Some corners do not have a midcorner phase, as the second you have dialed in enough steering to aim the car toward the apex, you immediately begin to unwind it toward the exit.

You can also use the throttle application to define the midcorner: from the second your foot touches the throttle, to the point it begins to really "hammer down" (smoothly) on it. Therefore, the midcorner may be non-existent if the second you touch the throttle you squeeze it all the way down. It may be very short if you have a brief period of time where you are using a maintenance throttle, not increasing or decreasing the throttle. Or it could be relatively lengthy in a long, fast sweeper.

Corner Exit

The exit of the corner is the section where you are unwinding the steering, increasing the radius of the line the car is following. Typically, it is from the apex to the exit or track-out point of the corner. Again, the exit phase is also defined as when you begin squeezing the throttle down to wide open.

Trail Braking

Braking can actually be broken down into "approach braking" and "trail braking." Approach braking is just as it sounds, the braking you do on the approach to the corner. The second you begin to turn the steering wheel into the corner, approach braking ends. Trail braking begins as soon as approach braking ends, at the turn-in point. It is the physical act of easing, or trailing your foot off the brake pedal. Where you finish trail braking, and how much you trail brake is entirely dependent on the specific corner, the type of car you are driving and your driving style.

I know there are some people who say they never trail brake. Some racing schools actually teach this. They say a driver should never trail brake. They are dead wrong. Every successful driver trail brakes to some extent, in some corners.

Off-Throttle

In theory, you should never be coasting in a race car. You should either be braking or applying the throttle. In reality, it is sometimes necessary. There are some cars that reward a short period of coasting prior to beginning to accelerate out of the corner (but this is an exception to the rule). Any moment where you are neither braking nor applying the throttle, you are coasting or off-throttle.

Maintenance Throttle

This is where you are not accelerating or decelerating. You are simply maintaining your speed. Think of it in terms of driving down a highway at a constant 55 miles per hour. Not all corners and cars require maintenance throttle. You may directly and immediately go from off the throttle while braking to squeezing down the throttle, accelerating out of the corner. Other cars and corners require a short period of maintenance throttle.

Acceleration

Accelerating is when you are progressively increasing the speed of the car by either squeezing down on the throttle or holding it to the floor.

DRIVING ON THE EDGE

To many, the model of the perfect race driver is one where the driver never makes a mistake and is perfectly smooth at all times. While this is not a bad model to have, it's not completely accurate. Yes, I know, I have stressed it myself, over and over. The reason for this is that many drivers need to be smoothed out. They need to realize how much of a "finesse" sport racing is. However, once they learn this, they sometimes get the image of this perfect driver being so smooth and tidy that they are not quick. They take the idea too far.

Winners (read "fast drivers") make mistakes. They are not always perfectly smooth. They often drive over the limits of the tires and car. They even crash on occasion. There is nothing wrong with you doing that, especially if you win. In fact, it is doubtful you will win without driving like that a majority of the time.

I'm not saying the goal is to crash (obviously), but sometimes it is a side effect of driving the car at the limit. The only way you can consistently drive the car at the limit is to overdrive at times. Actually, you will end up overdriving it as much as you underdrive it—over, under, over, under, over, under, over—and the average is the limit.

If you are consistently smooth, tidy, and never making mistakes, your average is most likely slightly below the limit—under, under, over, under, under, under, over, under, over, under, under, over . . .

In doing so—in overdriving the car at times—you will get even better at overdriving the car and getting away with it. You know that sometimes when you overdrive, you don't get away with it. Again, that's okay. It's part of being fast and part of being a winner. With experience, though, the results of overdriving will be fewer and fewer offs, spins, and crashes. The results may be running a little wide, a bit of a lock-up, or a half-spin-and-go, but that's okay too. In other words, you will get even better at controlling an "overdriving off-line experience" (a "moment"), to the point where it will seem to others that you don't make mistakes. Sure, you're still making mistakes, but they are so small that they are hard to notice.

This concept of overdrive, underdrive, over, under . . . averaging out to being at the limit is critical. In the beginning the difference between over and under is quite large. With experience, the difference becomes quite small. Some drivers see the overdriving part as being "wrong" or unnecessary. If that's you, you need to recalibrate your impression of the model. In other words, you need to spend just a little more time overdriving the car than you currently do. You need to raise the average.

The truly fast race driver is one who is over the limit at times. He's a bit wild at times. He hangs it out there. He's aggressive. He makes mistakes every now and then, but is confident enough in himself to know that, on average, he's driving at the limit and that's why he's fast, and that's why he wins. That's okay. In fact, it's better than okay. It's what makes him a winner. It's especially okay when the driver is young and wins. If a young driver is smooth, tidy, and never makes a mistake, do you know what people think of him? They think he's slow.

Why? Because we all know that with a little more experience and maturity, he will smooth out, get more consistent, and make fewer mistakes. He will rarely get faster, just more consistent. We know that. If the driver is one who can win and who is blindingly fast, we know that will make him the perfect driver. Start out a little wild, fast, and mistake-prone, add some experience and maturity, and you've got the perfect driver. Start out with a driver who is smooth, tidy, and mistake-free, add in some experience and maturity, and you've got a driver who is conservative, a driver who is a little bit slow, and one who will not win unless the race is given to him. An also-ran.

Imagine that you're about to enter the toughest race season of your life. With your talent and skills, you have what it takes to win. You have the ability to wring every last ounce out of the car this season. But even though you have the ability to do it, you don't always do it. Sure, you can average driving the car at the limit—being over, under, over, under, over—you just need to do this more often. The only reason it doesn't happen all the time is because your mental image is not of that. It's close, very close in fact, but not dead on there.

Develop a mental image of the perfect race driver—you—as being able to drive over the limit at times, bring it back, hang it out there, dance with the car at the ragged edge.

Having said all that, it's time to talk about drivers who aren't interested in driving at that ragged edge, who race for fun, who want to drive just under the limit, having the time of their lives. That's okay. It really is. In fact, that's what I'm talking about throughout this book: driving at your limit. But if your goal is to drive at eight-tenths or nine-tenths or seven-tenths, that's great. Just make sure you drive at that limit consistently. If your goal is to drive at eight-tenths, don't drive sometimes at six-tenths, sometimes at nine-tenths, sometimes at seven-tenths. Drive at the limit, the limit you've set for yourself.

7 FANCY FOOTWORK

We've looked at the basics of how to use the brakes, throttle, and steering wheel; now's the time to talk about the advanced stuff, the stuff that makes the car dance on the edge, the stuff that makes the difference between fast and very fast.

BRAKING

Proper braking actually starts with how you take your foot off the throttle. Many drivers, due to the habits they've developed from driving on the street, make a huge mistake on the track: They get to the end of a straightway, gradually lift their foot off the throttle, wait a second or two, and then begin to brake. At the end of the straightaway, the transition from throttle to braking should be immediate. Smooth, but immediate, with no gap in between, no coasting in between.

Squeeze the brakes on, quickly, until you are at maximum braking. This is threshold braking. If you exceed the limit for threshold braking and begin to lock up, ease up slightly on the pedal; think of curling your toes back and feel for the tires to begin rotating at the limit of traction again. In other words, you may have to modulate the pedal pressure slightly, using the feedback from the tire noise, the forces on your body and the balance of the car, to achieve maximum braking.

When approaching a corner, squeeze the brakes on smoothly and firmly. And then, as you reach the corner, release it gently as your foot goes to the throttle, so that you don't actually feel the point at which they are fully released. Remember when I mentioned what made Jackie Stewart so successful: It was how he eased off the brakes.

Most drivers, due to street-driving habits—but also because they've been advised to be smooth with their braking—brake too gently at first, gradually building up pedal pressure. Now, I'm not saying you shouldn't be smooth with your braking, but at the same time I am saying apply hard initial pressure to the brake pedal. Many drivers "ramp up" their braking pressure over the duration of most of the braking zone, starting reasonably soft, gradually building up pressure, and having the most pressure on the brakes near the end of the brake zone. This is backward. It should be a hard but smooth initial application of the brakes, maintain the pressure throughout the brake zone, and then gradually release the pressure toward the end.

Not all cars will be the same. Cars with little or no aerodynamic downforce will not be able to handle as much initial pressure; cars with lots of downforce can take lots of initial brake pressure. With aero cars, the faster they're moving, the more traction the tires have, therefore allowing you to brake hard at the beginning. As the car slows, it loses downforce and traction, and you have to bleed off brake pressure. But to some extent, all cars are like this: The faster the car is moving, the harder you can brake; the slower the car is moving, the lighter you have to brake.

ANTI-LOCK BRAKING SYSTEMS

Anti-lock braking systems (ABS) are perhaps the most important safety device to ever be developed for street vehicles. However, as of this writing, ABS has not found much use on purpose-built race cars (Indy, Formula One, prototype sports car, and so on). Why? Well, mainly because of the rules. All of these series prohibit the use of ABS, mainly as a cost-controlling measure. About the only use it saw in purpose-built race cars was in Formula One, where a couple of teams used it in 1992 and 1993. It was banned from the 1994 season on.

However, when ABS is standard equipment on a production car, it is sometimes allowed to be used on production-based race cars such as the showroom stock class. Here, ABS can be both an advantage and disadvantage. It is a wonderful safety device, stopping a driver from ever being able to lock up the brakes. This is particularly useful in endurance racing where it's more important to be consistent and never flat-spot a tire.

At the same time, ABS can be difficult to get used to and maybe even a disadvantage. Often, a driver wants to "pitch" a car into a turn by going slightly beyond the threshold of traction on the rear wheels while turning into the corner. With ABS, however, this is not possible.

A big part of the magic of driving at the limit is managing the brakes to keep the tires at the threshold just before they lock up, as the inside front tire in this photo has done. A little lock-up like this—especially since it involves the unloaded inside tire—is okay every now and then. But more than one tire locking up means you have to ease up on the pedal just slightly. *Shutterstock*

It's important, if you're going to race a car with ABS to get comfortable with the feel of it. Get used to the feeling of the brake pedal pulsing and the inability to pitch the car into a turn with the brakes. But also get used to how hard and late you can brake if you are in a car with sophisticated ABS, such as a Porsche or BMW. With these cars, the technique is reasonably simple: Wait to the last possible moment, then hit the brake pedal as hard and fast as you can, and hold it down, letting the system do the work for you. Of course, if you're not used to that style of braking, it will take a little practice.

That's how you apply the brakes with ABS. But how you release them is no different than from non-ABS brakes. Rarely do you snap your foot off the brake pedal. Instead, you trail the brake pressure off.

TRAIL BRAKING

I'm constantly amazed by the number of drivers I talk to who have been told never to trail brake. Then there are others who have been told to always trail brake. Never trail brake, always trail brake? It's a bit confusing, isn't it?

Let's review what trail braking really is because I'm often surprised by the inconsistencies out there among drivers concerning this technique. First, trail braking

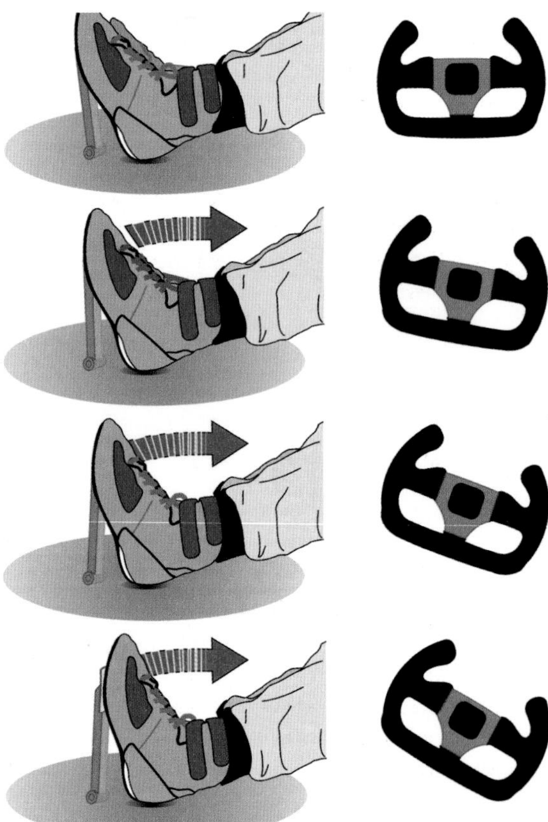

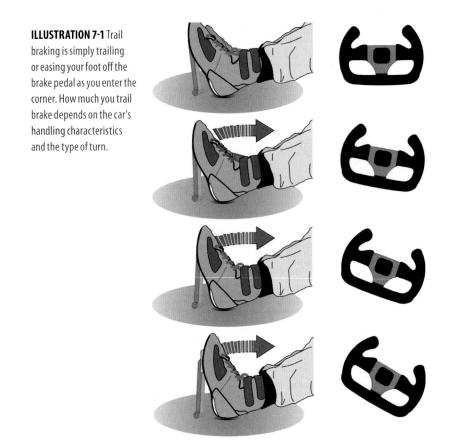

ILLUSTRATION 7-1 Trail braking is simply trailing or easing your foot off the brake pedal as you enter the corner. How much you trail brake depends on the car's handling characteristics and the type of turn.

is not braking against the throttle. In other words, trail braking is not having the brakes applied (even a little amount) while you're on the throttle. Trail braking is not "braking all the way to the apex." Yes, it could be that you are trailing off the brakes all the way to the apex, but it's not necessarily that way. Trail braking is when you gradually release, or trail, your foot off the brake pedal while turning into a corner. If you've completed your braking entirely and your foot is off the pedal at the point you begin to turn into the corner, then you have not trail braked at all. If you have even the slightest amount of brake pedal pressure on while turning into the corner, then you are trail braking. Some times your foot will finally be completely off the brake pedal within a foot or two past the turn-in point, and other times it will not be until you're practically at the apex. Either way, you're trail braking.

So, trail braking is when you trail your foot off the brakes as you release pressure while turning into a corner. And the reason for doing so is twofold. First, if helps keep load on the front tires so the car will turn into the corner better. It will "rotate" better. And second, it helps you use all of the tires' traction throughout the corner. If you get to the turn-in point and suddenly take your foot off the brake pedal as you turn in, there will be a fraction of a moment when you are not using up all of the tires' traction. You could be using more and carrying more speed.

Some drivers claim that they never trail brake. But when I observe them, or analyze the data on their data-acquisition system, it's obvious that they do. Many drivers trail brake more than they realize, more than they are aware of. And other drivers trail brake less than they are aware of. If you're not sure how much you trail brake in each and every corner, you're not aware, and there may just be some speed available by becoming aware.

My suggestion is that you do become aware, that you become aware of where you finally are finished with the brakes. It's important that you focus your eyes on the end-of-braking point, that point in the corner where you are completely finished with the braking and your foot is off the pedal. Be aware of where that is in every corner.

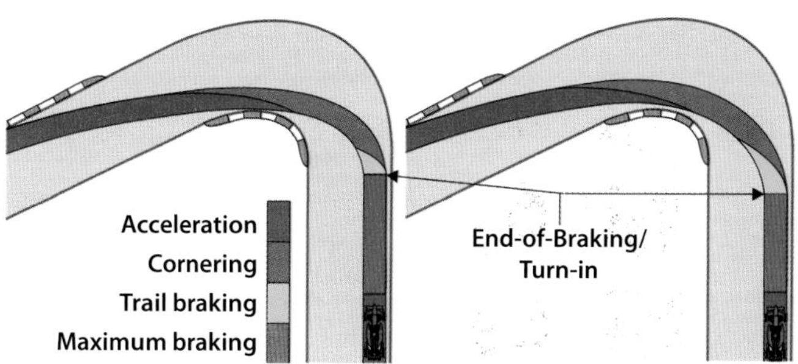

ILLUSTRATION 7-2 Trail braking helps "rotate" the car while entering a corner. In general, the more you trail brake, the earlier you can begin turning into a corner (as in the illustration on the right); the less you trail brake, the later and more abruptly you will have to turn in. The later and more abruptly you turn in, the less speed you will be able to carry into the corner.

SPEED SECRET

How and when you release the brakes will have more effect on your lap times than where you start braking.

Should trail braking be used in every corner? No. There are turns, especially very fast ones, where you want to be squeezing back on the throttle about the time you're turning into the corner, since this helps the car's balance and the overall grip level. As a general rule, the slower and tighter the turn, the more you will use trail braking to help you rotate the car; the faster and more sweeping the turn, the less you will use trail braking.

SPEED SECRET

Speed comes from the timing and rate of release of the brake pedal.

LEFT-FOOT BRAKING

Over the past few years it has become obvious that if you want to win in any form of racing that uses purpose-built race cars (F1, Indy Cars, Indy Lights, Prototype Sports Cars, Formula Ford), you must left-foot brake. Why is that, and what has changed to make that statement so true?

At one time some of these race cars had gearboxes that rewarded the use of the clutch. That is not the case anymore. Many now feature a sequential-shift operation. The point is you do not need to use the clutch to shift. Not only do you not need it, it may actually slow your shifts down if you do.

Of course, the main reason for this change in technique being used

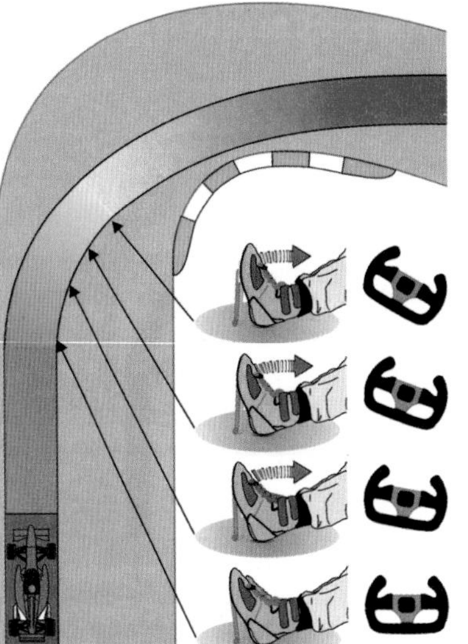

ILLUSTRATION 7-3 As you initiate your turn into the corner, slowly and gently ease, or trail, your foot off the brake pedal. This is the only way of ensuring you use up all 100 percent of the tires' traction entering a corner.

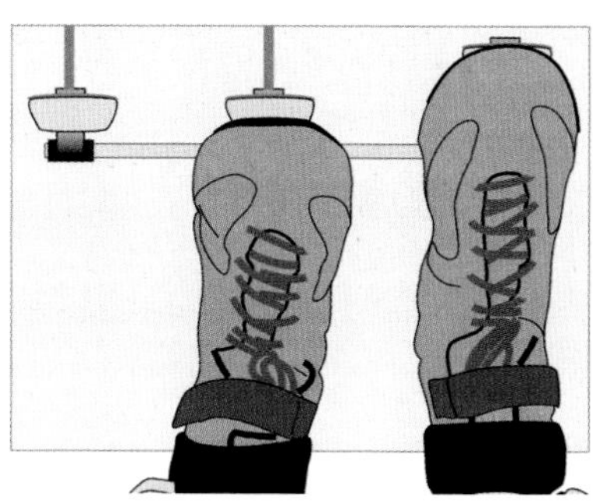

ILLUSTRATION 7-4 Left-foot-braking is one technique that you will need to develop if you want to make it in Formula One, Indy car, NASCAR, and maybe even sports cars.

by today's winning drivers has to do with their backgrounds. Most of the top drivers today have spent many years racing karts. And what do you do with your feet in a kart? It's right foot on the gas, and left foot on the brake pedal. It's the years and years of training the left foot to be sensitive that results in great left-foot braking in race cars later in the career.

If a driver has not spent many years using the left foot for braking in a kart, the driver may never acquire the accuracy and sensitivity to left-foot brake at the level required. Practice may have to take place at an early age to become a well-programmed, subconscious technique. Why? In his book, *Why Michael Couldn't Hit*, neurologist Dr. Harold L. Klawans sheds some light on why Michael Jordan could not hit a baseball well enough to make it in the major leagues during his hiatus from basketball. The bottom line, according to Klawans, is that if a physical technique has not had some programming by the early to midteen years, the brain and body will never be able to produce the psychomotor skill necessary to perform at the highest levels. In other words, Michael Jordan didn't hit enough baseballs as a kid.

The following article written by Matt Bishop in the June 8, 2000, issue of *Autosport* magazine puts the left-foot braking issue into a F1 perspective:

> *If there is one hour in the whole F1 season that sorts the men from the boys, it's qualifying at Monte Carlo. Last Saturday, the front-row stars of this white-knuckle hour were Michael Schumacher and Jarno Trulli. Michael, we know about. But what can we learn from Jarno's dazzling performance? Good question. And we'll come back to it.*
>
> *Three laps into last year's race, Damon Hill exited the tunnel, braked for the chicane, got it wrong, and lightly punted his Jordan into the barrier. I was standing at the swimming pool complex at the time, and 10 minutes later a stern-faced Damon strode into view. He walked 20 yards past me, then stopped. There he stayed, silent, for 10 minutes more, watching the cars speed past.*
>
> *At the next race, I asked him what he had seen. "I stood at Tabac," he told me, "but I couldn't really tell anything because the barriers were too high. So I went to the entrance of the swimming pool, the fast left-right, and you could start to see something. Michael (Schumacher) and Mika (Hakkinen)—but particularly Michael—were going through way quicker.*

"When I walked on to the exit of the swimming pool, the right-left, I could really see Michael was doing something totally different with the car. Really, really, obviously different. But you know, I couldn't really tell you what. I couldn't tell you whether he was understeering or oversteering or how he was braking or what he was doing with the throttle. All I could tell you is that he was different and quicker."

A Ferrari insider later revealed to me that Michael was left-foot braking. But more than that. Because for him, brakes aren't simply a mechanism that slows a car. No. "For Michael, brakes are but one element in an exquisite yet subconscious fusion of techno-dynamic ingredients," said my source.

So as the number 3 Ferrari tore past Damon's baffled gaze that day, inside the cockpit Michael's feet would have been a blur, dancing on the pedals with the agile sensitivity of a Bolivian pickpocket's hands. To be seriously quick these days, my man implied, that's what you've got to be able to do.

Which brings us back to Trulli. A classic karter-turned-racer, Jarno has probably never in his life braked an Fl car with his right foot. Jarno can do the techno-dynamic-fusion bit too.

Fair enough, you say, but most Fl drivers left-foot brake these days. Indeed they do, but for many it's been a thing they've forced themselves to do. Some never really get the knack. In 1998, for example, Hill's Jordan teammate was Ralf Schumacher—like Trulli, a man who began karting as a small child, a man for whom left-foot braking consequently comes as easy as walking. I asked Jordan's then technical director Gary Anderson why Damon wasn't left-foot-braking. "I think it's a case of: 'I know it's quicker, but I'm 38 and I can't quite get on with it,'" the Irishman replied.

That remark sounds ominously like Rubens Barrichello this year. "Michael brakes with his left foot right into the apex of the corner," Rubens admitted at Monaco, "'while I have to take my right foot off the brakes earlier to get back on the power at the same spot as him. I brake with my right foot (for) feel. I've tried using my left foot and it just doesn't feel right. You have to do what you're most comfortable with." Rubens' best lap last Saturday trailed Michael's by a full second, by the way.

Perhaps it's time Rubens put his best foot forward. His left foot. Just ask Damon.

According to this theory, if you have not spent a fair amount of time driving a kart by the time you get your driver's license, you will never be a world champion. While I believe there is a lot of truth in this statement, I certainly don't think that if you haven't grown up driving a kart you should stay away from race-car driving. With work, both physically and mentally, you can adequately train your left foot.

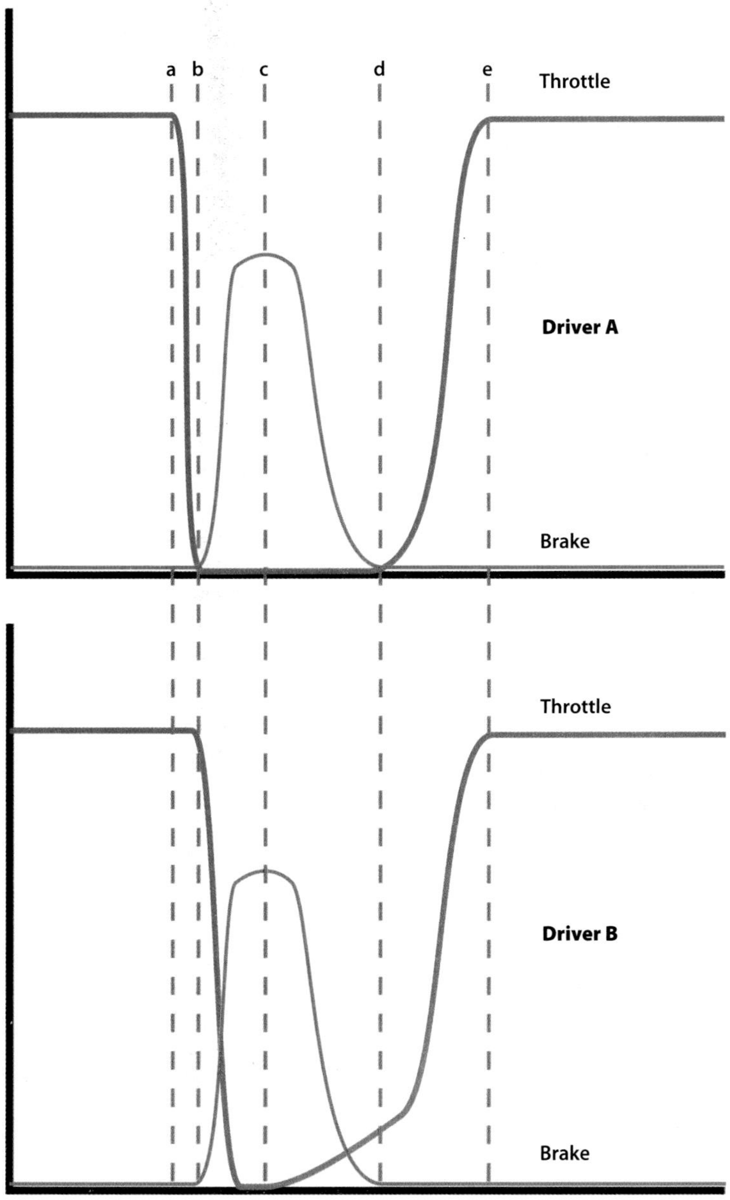

ILLUSTRATION 7-5 The throttle-to-brake graph traces two drivers at the end of a long straightaway. Driver A is a right-foot braker, while Driver B uses his left foot to brake. As you can see, right-foot-braking wastes some time. The amount of time it takes to remove the right foot from the throttle and move it over and onto the brake pedal will cost you (from point "a" to "b" on the graph). With left-foot-braking, there is no lost time; the transition from throttle to braking is immediate.

Of course, the time, focus, and energy spent doing this is time, focus, and energy spent away from working on other areas of your driving. It is still well worth it.

With all this in mind, why exactly is left-foot braking superior to right-foot braking? First, left-foot braking allows you to alter the speed of the car, which is what you use the brakes for, without upsetting the balance as much. It is easier to drive smoothly. Anytime you can drive smoother, upsetting the car's balance less, the higher its traction limits will be. And that means you can drive faster.

Second, it saves time in the transition from throttle to braking to throttle. This is what Barrichello was talking about. With right-foot braking, your right foot must move from the throttle to the brake pedal and back. At the end of a straightaway, that will result in having to brake slightly earlier. In the transition from braking to acceleration, the fraction of a second it takes for your right foot to move from the brake pedal to the throttle is extremely valuable. When using your left foot for braking, that movement or transition doesn't even exist.

In fact, it is possible to actually overlap the end of braking with the beginning of acceleration (squeezing of the throttle), so that there is no time delay at all. If done correctly, the shift in weight balance will be seamless. That saves time and keeps the car better balanced.

With left-foot braking it is also possible to keep your right foot flat to the floor on the throttle and just alter the speed or balance of the car slightly by braking with the left foot at the same time. This is something that I don't recommend be done a lot, as the brakes will overheat eventually. However, if you need to just get the front of the car to "bite" a bit better at turn-in for a fast sweeping corner, sometimes a short, quick squeeze of the brake pedal with the left foot while keeping the throttle flat will do the trick. Or, if you need to take just a little bit of speed off, but don't want to lose the engine's momentum, the same left foot brake application may work.

Can a driver be competitive using his right foot to brake? Certainly. Dario Franchitti, for example, uses his right foot for braking unless driving on an oval track. In the October/November 1999 issue of *Race Tech* magazine, he's quoted as saying, "If the corner requires it, maybe an occasional dab with the left foot, but hardly ever. . . . I am left-foot braking here (the Chicago oval) so I don't get that much pedal feel. I can feel when it is about to lock but my left foot certainly isn't as sensitive as my right foot yet. I am more of a traditionalist." Of course, one wonders what Franchitti could do if he had learned to left-foot brake years ago.

Left foot braking is a technique required when racing on oval tracks. But many drivers never use it in road racing, although it may have some benefits in fast turns where you're not required to downshift prior to the corner. It's also useful when driving turbocharged cars, as it allows you to stay on the throttle with the right foot, keeping the turbo spinning and reducing the throttle lag. As you can imagine though, this is hard on the brakes, so be careful not to overwork them.

With left-foot braking, particularly on oval tracks, some drivers make the common error of having the brakes on slightly while accelerating out of a corner.

This dragging of the brakes wastes time, can overheat the brakes, and is definitely unwanted. Pay attention to this.

BRAKING TECHNIQUE

Okay, okay, you say, that's all fine for drivers who have cars they can left-foot brake, but my car won't allow it. Many cars, in fact, do not allow left-foot braking. Either it is physically impossible to get your left foot in position, or the gearbox does not allow clutchless shifts. If you are using your left foot for the clutch, you can't use it for the brakes, at least not full time.

You can still learn something from what was just said about the technique, though. The basic advantage of left-foot braking is what? It is the ability to make the transition from braking to acceleration as smooth and seamless as possible. Even if you have to use your right foot for braking, that is still your goal.

The speed in which you ease off the brake pedal and the timing of when you come off the brakes as you enter the turns are perhaps the two most important factors in determining the speed you can carry into the corner.

As you know, when you brake to slow the car when approaching a corner, the front of the car dives. Weight has transferred onto the front tires. This dive causes the front springs to compress. When you lift your foot off the brake pedal, the springs expand, popping the front of the car up. Weight is transferring away from the front toward the rear. If you lift your foot off the brake pedal too quickly or abruptly right at the point where you begin to turn the steering into the corner, the front will become unweighted, probably causing it to understeer.

Easing off the brakes slowly (relatively speaking) and gently at just the right speed does three things:

- It keeps the front of the car loaded (weight transferred onto the front tires), which helps the car turn in responsively.
- It often helps you get back to throttle early and with commitment. As the front is loaded at turn-in, the car will rotate better, which means you will not have to wait or modulate the throttle much. Instead, as the car rotates toward the apex, you will be able to get back on the throttle and stay on it.
- It does the obvious. As you ease off the brake pedal, you will carry more speed into the corner. And, due to the two points above, the car will be able to handle the extra speed. If you tried to carry that same speed into the corner by coming off the brakes quickly and immediately, the front would become unloaded and the car would most likely understeer—not rotate toward the apex—reducing the car's speed and delaying the point where you could start to accelerate.

Of course, you can ease off the brakes too slowly or gradually. If you ease off the brake pedal too slowly you will be turning into the corner with too much weight transferred onto the front tires. This may cause the car to oversteer as you turn in. Or it may cause it to understeer by overloading the front tires. You are asking for more than 100 percent from the front tires, so they give up and begin to slide.

How you ease your foot off the brake pedal dictates the balance of the car, and therefore its handling characteristics. Many times a driver will complain that the car understeers too much just after turn-in, and that it results in the driver not being able to get back on the throttle early enough. Perhaps all the driver needs to do is ease the foot off the brake pedal a little more slowly.

One's natural conclusion regarding braking on the approach to a corner is to think that braking as late and hard as possible (at the tires' limit) would be ideal. That it's what you should do for each and every corner. But that's not necessarily true. When you approach a corner with the car "standing on its nose" from braking so hard, how well do you think it will grip the track at corner turn-in?

There are some cars and turns where you need to brake lighter. Ask less than 100 percent from the tires. There are corners that are best approached with the car not standing on its nose. This is something you will need to experiment with. If your car tends to understeer entering a corner, try trail braking more or less to see if either loading the front tires more or less helps. Also, some formula and prototype sports cars are pitch sensitive, meaning the aerodynamics are negatively affected when the car is too far from being level or balanced.

The amount you trail brake also plays a role in the line you drive through a corner. The more you trail brake, the earlier you can turn into the corner. The less you trail brake, the later you have to turn in. The reason is that trail braking helps rotate the car for you, so you don't end up running out of track at the exit of the corner. The advantage to turning slightly earlier is that you can carry more speed. The later you turn in, the sharper you have to turn, meaning you have to be traveling slightly slower. Obviously, these are general rules and not cast in stone, but they do work most of the time.

A common error some drivers make is braking too early. That does something other than the obvious of slowing down too soon. If you brake early, you will arrive at the turn-in point feeling like you have slowed the car enough, which you have. The problem is that often this is prior to the turn-in point. So just before turning in, you come off the brakes, unloading the front of the car. As you turn in, the car will not respond the way you would like, either understeering or just feeling unresponsive. The message your brain receives is "we are at the limit." That message, unless recognized, results in you taking the corner at this (slow) speed over and over again, programming it. A lot of drivers do this, particularly when learning a new track or car, and then it becomes a programmed mistake. And I'm not talking about braking way too early. Even 5 or 10 feet can cause a huge problem.

It is important to be aware of when you are easing off the brakes. If you begin to release the brake pedal prior to initiating your turn-in, you probably began braking too early.

BRAKING PRACTICE

Practice braking when driving on the street. See if you can modulate the brakes so that you can't feel the exact point where the car comes to a complete stop. Work on developing a real feel for the brakes; a sensitive touch is important, especially in poor traction conditions.

Easing off the brakes when coming to a stop when driving your street car teaches your foot the sensitivity and control necessary to ease off the brakes accurately and effectively, with the ideal speed and finesse. If you use your left foot for braking on the racetrack, then you need to practice with it on the street. That's easy if you drive an automatic transmission vehicle and not so easy if you don't.

But even if you drive a manual shift car on the street, you should still practice using your left foot for braking. Any time you approach a corner that does not require downshifting, brake with your left foot. And, although I don't recommend this as a safe street-driving tactic, you may want to shift into neutral on the approach to a stop, and then again use your left foot for braking.

Your main objective is to make your braking as smooth as possible. It doesn't have to be hard braking to do that, either. You want to make squeezing on the brakes (very quickly) and easing off them as natural as breathing.

STEERING WITH YOUR FEET

Most people think you drive a race car primarily with your hands and arms turning the steering. It's the steering wheel that directs the car. Truly fast drivers know that they steer the car with their feet as much, and maybe more, than they do with the steering wheel.

You want to develop the ability to "play" with a car's balance to make it turn more or less, to use your feet to steer the car to rotate it more or less. In other words, use the brakes and throttle to alter the car's balance at various points in a turn to make the car turn.

Ironically, at the very limit, you steer the car more with your feet than you do your hands, and the steering wheel becomes a brake. (Every time you turn it, it scrubs off speed.)

Of course, what you need to do if you want to be fast is control the amount of understeer and oversteer to your advantage, using it to make the car rotate more or less, depending on what you need. What do I mean by rotate? Think of looking directly down on your car from way above, then think of it rotating like the hand of a clock. Driving through a turn, there are times when having the car rotate more is a good thing, like when trying to negotiate a tight turn. There are times when your car can rotate too much, leading to it oversteering so much that you're hanging on for dear life. Obviously, your car's setup plays a role in how much and how quickly it rotates at any one place on the track, but so does how you manage its weight balance.

When you manage the weight just right you can make your car rotate and point through the turn just the way you want—and this is the important part—with little movement of the steering wheel. The less you turn the steering wheel, the more the front tires are pointing straight ahead, and when that's the case, they are scrubbing off less speed. The more you can keep the front wheels pointing straight ahead, and you manage the car's trajectory through a turn with its weight transfer and balance, the faster you'll be. Also, the straighter the steering wheel, the sooner you can commit to throttle, which will give you a higher straight-line speed.

SPEED SECRET

To be fast, you need to be a weight manager.

Let me replay a real life experience with you. In the early 1990s, I was racing an Indy car at Laguna Seca. We were a low-buck operation—so low-buck that I had to sit out a practice session to avoid putting miles on the engine. I wanted to take advantage of what I could learn during that session, so I went out to watch other drivers from as many places around the track as possible.

To this day, the mental image of what I saw is so clear. I stood watching through the fence on the outside of the right-hander, Turn 3. I got to that place about 10 minutes into the session, so the drivers were well up to speed. The first 8 or 10 cars came through the corner pretty much all with the same line and speed. Then, a car approached and turned (relatively) much earlier than any of the others, and yet was carrying as much speed as anyone else. The same thought went through my mind as is probably going through your mind—turn in early, run out of track at the exit, and maybe crash. I was about to step back from the fence to avoid having dirt thrown up in my face from the crash, when the car rotated just before the apex and the driver absolutely stood on the throttle and rocketed out the turn as fast as any other driver. I was shocked! How did that happen, and how did he get away with that kind of "error"? What made it even more shocking was that it was a driver named Mario Andretti. I stood in the exact same place to see how he would recover the next lap. And guess what? He did the exact same thing. Then a young driver named Michael Andretti did something very similar—and he was the quickest driver in the session.

As I continued to study what was going on, I realized that Mario or Michael never put any more steering angle into their cars than any other driver did, but they consistently turned in a little earlier than everyone else. I took segment times and found that they were as quick or quicker than any other driver through that corner, as well. What they were doing, of course, was controlling the car's balance in such a way that it allowed them to turn in a little earlier and yet rotate the car just before the apex to enable them to still get back to power as early as anyone else. And the advantage to this was the initial turn-in was not so abrupt, so they could carry a little more speed into the turn.

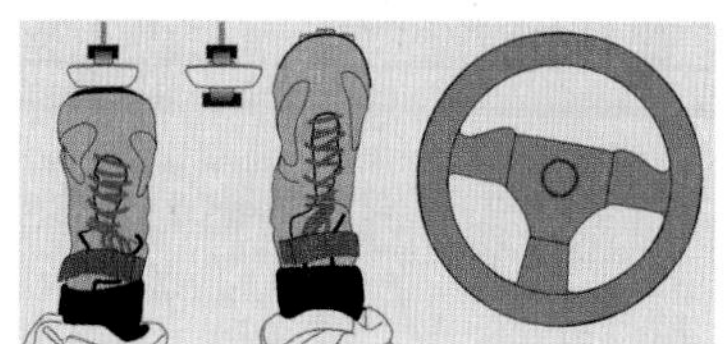

ILLUSTRATION 7-6 When you're driving at the very limit, you steer the car more with your feet. Ironically, when driving at the limit, the steering wheel becomes more of a brake. Being quick requires a seamless tradeoff and balance between feet and hands, using the feet to balance the car, while turning the steering wheel as little as possible.

Over the course of the next couple of seasons, I studied the Andrettis and others from the side of the track and from behind the wheel. While I wasn't driving the most competitive car in the series, every second I spent on the track I focused on learning what I could. Having the opportunity to observe drivers like the Andrettis, Bobby Rahal, Rick Mears, Al Unser Jr., Nigel Mansell, Jimmy Vasser, and Paul Tracy from the cockpit of another Indy car, I was able to see things that few have the luxury of experiencing. I'll never forget the lessons.

What I learned that day watching Mario and Michael was something that I already knew to some extent. I had certainly experienced it myself, otherwise I wouldn't have made it as far as I had in racing. Any driver who wins races knows that he or she can change the direction of the car with the balance of the car. But, from that day on, it became so much more of a focus for me. It's helped me and the drivers I've coached since that time become faster.

By the way, a few years later I was standing in the exact same spot and noticed another driver use the same style as Michael and Mario did, perhaps even carrying it to another level—at least in the speed he was able to carry into the turn and still make it out the other end. His name? Juan Pablo Montoya.

It's an interesting challenge, isn't it? You need to turn the car to get through the corners, but to do that with as little turning of the steering wheel as possible to avoid scrubbing off speed. To do that, you need to alter and manage the weight transfer of the car to change the direction of the car—to make it rotate more or less throughout every single corner on the track to suit your needs.

The key to being fast is the timing and rate of release of the brakes. Get that right—and it is different for every car, every corner, and possibly different lap to lap as conditions change—and the car will rotate or turn with a minimum of steering.

Before we leave this subject, though, I need to make one thing clear. The easiest way to reduce the steering angle you put into the car over the course of a lap, or even through just one corner, is to simply drive slower. If you drive slow enough, you will be able to reduce the amount of steering angle significantly. Obviously, that's not your objective. While you're driving, you need to be aware of whether you're reducing steering angle simply by slowing down. You can—and should—slow down your steering inputs, but without slowing down your corner entry, midcorner, and exit speeds.

SPEED SECRET

The less you turn the steering wheel, the faster you'll be.

8 CORNERING TECHNIQUE

Every racetrack has its own personality. Of course, there are as many shapes and layouts as there are tracks. There are oval tracks (short, long, superspeedways), permanent road racing courses, and temporary circuits (constructed on streets or airport runways) of every length and size. But even two tracks of seemingly identical layout will have a different feel to them.

How well you get to know each track you race on, and how you adapt to them, will play a large role in how successful you are.

As a race driver, your goal in each corner is really quite simple. Well, simple to state here, maybe not so simple to do. You want to:

- Spend as little time in the corner as possible
- Get maximum speed out of the corner by accelerating early to maximize straightaway speed

Often, maximizing one of the above means sacrificing the other. In other words, to achieve the best possible lap times, you may have to compromise corner speed for straightaway speed, or vice versa. It will depend on the specific layout of the track and your car's performance characteristics. The trick is finding the perfect compromise.

While the corners are the biggest challenge, how quickly you get down the straightaways usually matters the most. And the quicker you get down the straight, the sooner you face the challenge of the next corner! *Shutterstock*

THE CORNERING COMPROMISE

As we discussed earlier, becoming a winning race driver requires the ability to drive the race car consistently at the traction limit of the tire and chassis combination (at the limit of the traction circle), and the engine. Having said that, virtually anyone can take a car to its limits on the straightaway, using the engine to its limit. It's driving the car at the limit under braking, cornering, and accelerating out of the corner that separates the winners from the also-rans.

Most races, then, are decided where the cars are moving slowest—in the corners. Yet, it is much easier to pass on the straightaways than it is in the corners. So the faster you are on the straight, the more cars you will pass or gain a time advantage on and the more races you will win. Therefore, the most important goal for the corners is to drive them in such a way as to maximize your straightaway speed.

The skill comes in determining a speed and path through the corners that loses the least amount of time negotiating them, while ensuring maximum acceleration down the following straightaway. This is where true champions shine.

Winning drivers keep their cars at the traction circle limit almost all of the time, though the limit does vary depending on track conditions and the state of the car. For example, as I mentioned in the previous section, aerodynamics are constantly changing the limit. The higher the speed the car is traveling, the more aerodynamic downforce there is, thereby developing more cornering force. At the same time, the acceleration capabilities of the engine are reduced as speeds are increased (at low speed in a low gear the engine has lots of relative power to accelerate at or near the traction limits of the tires; at very high speed, the engine does not have the power to accelerate near the traction limit). Therefore, in reality, the traction circle changes with speed. The higher the speed, the more the top of the circle flattens out and the sides (the cornering forces) expand. Much of the skill you must develop is in being able to read these changing variables, determine from moment to moment where the performance limit lies, and drive the car as close to it as possible. Much of this comes from experience.

Driving the limit of adhesion—the traction circle—through the corners at all times seems like the only thing you have to do to go fast. But, how you drive through the corners can vary. How much time you spend at various points on the traction circle can vary. And how you determine that and the path or "line" through the corner is critical and one of the keys to going fast.

In fact, one of the most important skills to learn is determining the optimum time to spend at the different parts of the traction circle limit. Whereas one driver may spend almost all of the time in the pure cornering region, at almost constant speed, another driver may spend more time braking and accelerating, simply by taking a slightly different line through the corner. Both operate the car at its limit all the time; one may be faster through the individual corner, while the other may be faster down the straightaways.

The trick is to determine which line through the corner results in the best overall lap time, not just what is fastest through each individual corner. Considering the corner and the straightaways on either side as a single problem, rather than just worrying about how to get through the corner itself, is the winner's solution.

To consistently determine that optimum line, you have to take into account the track variables, such as the lengths of the straightaways before and after the corner, the angle of the corner, its inside and outside radii, the track's banking (negative or positive), and the surface's coefficient of friction. And you have to consider the car's variables: its handling characteristics, aerodynamic downforce, acceleration and braking capabilities, and so on. In other words, the optimum solution differs from corner to corner and from car to car on the same corner.

Before I go any further, let's take a look at a few basics.

REFERENCE POINTS

To be consistent in your driving, you should use reference points. They are important for your concentration. The less time and concentration you spend on determining the exact point where you begin your braking for a corner, for instance, the more you can spend on feeling how the car is reacting to your inputs.

These reference points can be anything, such as a crack in the pavement, a point on a curbing, a change in pavement, a marking on a wall on the side of the track, a turn worker station, and so on. Notice though, I do not mention anything that could move during a race, such as a shadow, a turn worker, and so on. Also notice that reference points are not just visual or things that you can see. They're also things that you feel and hear. For example, you may use a bump in a track, and when the engine note reverberating off a wall sounds different as a reference to begin braking.

The three most important reference points are used to help guide you through the corners. They are, in order, the turn-in point, apex, and exit point (see Illustration 8-1). Each point can be dealt with in detail separately, but the

ILLUSTRATION 8-1
A typical 180-degree hairpin turn showing the three most important reference points: the turn-in, apex, and exit.

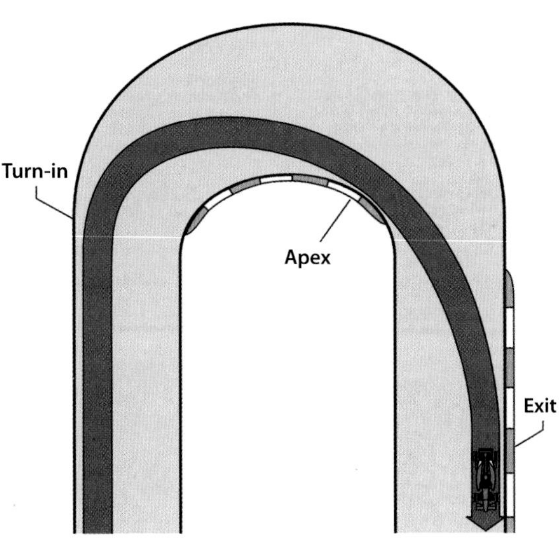

ultimate goal is to combine all three into a smooth, fluid line through a corner by first visually, and then physically, connecting the dots.

The turn-in is probably the most important part of a corner, as this determines how you drive the rest of the corner, where and how fast you apex and exit. As the name suggests, this is the part of a track where you do your initial turn of the steering wheel into the corner. And this turn-in point is determined somewhat by where you want to apex the corner.

The apex of a corner is the point, or area, where the inside wheels run closest to the inside of the road. The apex can also be thought of as the area of a turn where you are no longer driving into the corner but are now driving out. It is sometimes called the "clipping point," as this is where your inside wheels clip past the inside of the roadway.

Where you apex is determined by where and how you entered the turn, and it will affect how you exit it. The ideal apex for a corner can be either early in the turn, in the middle of it, or late in the turn.

Determining whether or not you had the correct apex is simple. If you come out of the corner having to turn more to keep from running off the road, then you had too early an apex. If you chose too late an apex, the car will not be using the entire road on the exit. It will still be too close to the inside of the corner.

If you are doing anything with the steering wheel other than unwinding it after the apex of the corner, you are probably on the wrong line. Most likely, you have turned in and apexed too early. Rarely should you ever be turning the steering wheel tighter once past the apex.

When you hit the apex perfectly, the car will naturally want to follow a path out to the exit point, the point where your car runs closest to the outside edge of the track. In fact, to properly exit the corner you *must* use up *the entire* track. Allow the car to come out wide to the edge of the road. This allows the car to smoothly and gently balance its weight and achieve maximum acceleration. It allows you to "unwind" the car.

I know when I've hit the perfect apex. It's when I'm able to just barely stay on the track at the exit, while accelerating as early and hard as possible. If I have to ease up slightly on the throttle to stay on the track, then I apexed too early. If I wasn't able to unwind the steering after the apex, I apexed too early. But if I still have room left on the exit, then I apexed too late.

THE IDEAL LINE

You can adjust the amount of time spent on each part of the traction circle by taking different lines through a corner. Illustration 8-2 shows two possibilities. The dotted line shows the "geometric line," a constant radius through the corner. This is the fastest way through that particular corner. The solid line shows an altered line in which the driver has started to turn later. The line is tighter than the geometric line at the beginning, but exits in a wider, expanding radius farther down the following straightaway.

This second line, the solid line, is called the "ideal line," and will result in an overall faster lap time. Why?

As I said earlier, you are not just dealing with one particular corner but rather a series of corners connected by straightaways. Considering this, plus the fact that you will spend more time accelerating on a racetrack than you will just cornering, superior exit speed is far more important than cornering speed.

Never forget that the driver who accelerates first out of a corner will arrive first at the other end of the straight and most often the finish line. It doesn't matter how fast you go through the corner. If everyone passes you on the straight, you won't win a race. Drive the corner in such a way as to maximize your straightaway speed.

SPEED SECRET

Races are won on the straightaway, not in the corners.

The driver following the geometric line in Illustration 8-2 spends almost all the time at the limit in the "cornering only" region of the traction circle, keeping the speed almost constant throughout the corner. Remember what the traction circle told us: You cannot begin to accelerate if you are using all the traction for cornering. Therefore, the geometric line does not allow you to accelerate until you've reached the very end of the corner and begun to straighten out the steering.

The ideal line, on the other hand, with its tighter radius at the beginning of the corner forces you to enter slightly slower, but the gentler, expanding radius through the remainder of the corner allows increasingly more acceleration, and therefore higher exit speed. This higher exit speed stays with you all the way down the following straightaway (and even multiplies its effect), more than making up for the slower entrance speed.

SPEED SECRET

Corner exit speed is usually more important than entry speed.

Driving the ideal line, you will spend less time at maximum cornering on the traction circle. You will spend more time at the braking and acceleration limit, however.

Determining how much to alter the path from the geometric line is one of the more complex problems facing a race driver. Altering it too much—turning in too late and probably apexing too late—means the initial part of the corner must be taken so slowly that the time lost there cannot be recouped fully in the following straightaway. It will result in a slower overall lap time. Not altering the line enough (turning in and apexing too early) results in a slow exit and straightaway speed.

As I said, there is no single ideal line for all cars or corners. The same car driven through different corners require different lines. Even for the same corner, different cars will require different lines.

The difference may be subtle, perhaps a couple of inches either way, but it makes all the difference in the world between being a winner or being just a midpack driver.

SPEED SECRET

The more time you spend with the front tires pointed straight ahead, or near straight, and the throttle to the floor, the faster you will be.

In general, the shorter and tighter the corner and the longer the following straightaways on either side, the more the line should be altered from the geometric line. In other words, a later turn-in and apex. Similarly, the greater the acceleration capabilities of the car, the later the turn-in and apex.

Many drivers seem to fall into the habit of driving all corners the same. They fail to adjust their driving appropriately for the different conditions—corners or cars—even though they may drive the car at its traction limit. This may explain why some drivers can be very fast in one type of car or at one track and yet struggle when they get into another car or drive another track. A true champion driver can quickly alter his or her line to suit the track and car, and of course, always drive the limit.

CONTROL PHASES

Breaking the cornering technique down further, there are six activities or phases you go through with your feet on the throttle or brakes in a corner (Illustration 8-3): braking, trail braking, transition, balanced throttle, progressive throttle, and maximum acceleration. The length and timing of each of these phases vary, depending on the car you're driving and the type and shape of corner. When you add the turn-in, apex, and exit reference points to the equation, you have the formula for a successfully completed corner.

Taking a closer look at the braking phase, think of the brakes as a waste of time. Brakes are merely for adjusting speed, not for gaining much. So if you are looking to improve more than one-tenth of a second on an average road racing circuit, don't just look at the brakes. Don't think that by braking later you'll gain a great advantage. You're going to make up more time with the throttle on, not off.

SPEED SECRET

The less time spent braking, the faster you will be.

Racers are always talking about brake reference points. They're always comparing and bragging about how late they begin braking for a corner. But the important reference point is not where you start braking, but actually where you *end* maximum braking. Only use brake reference points as back-up.

Instead, when you begin braking for a corner, focus on the turn-in point to visualize and judge how much braking is necessary to slow the car to the proper speed for entering the turn. Your speed at the start-of-braking may be different due to how well you entered the straight, so that reference point will constantly need slight adjustment. You need to analyze and sense the speed and adapt your braking zone to be at the correct speed at the turn-in reference point so you enter the corner at the ideal speed.

I always wondered why I could never remember where I started braking for a corner, where my braking point was. This was until I realized that I concentrated more on where I needed to have my maximum braking completed, the turn-in point, and the speed I wanted to slow to, instead of the start-of-braking point. Every driver has his or her weak and strong points. One of my strong points has always been in the braking area, which I attribute to my concentration on this "end-of-braking" point.

The most controversial control phase is definitely trail braking. Some "experts" say you should never trail brake. Complete all your braking and be back on the gas by the time you reach the turn-in point. Others say you should trail brake at every corner on every track. Of course, the truth lies somewhere in between. In some corners in some cars, you need to trail brake a lot, and in others use little to no trail braking. It varies depending on the corner and the car you're driving. Your job is to determine what works best and the ability to vary how much you trail brake.

How do you determine how much to trail brake in each corner (and car)? Begin by asking yourself, "Does the car turn in to the corner well?" If not, try trail

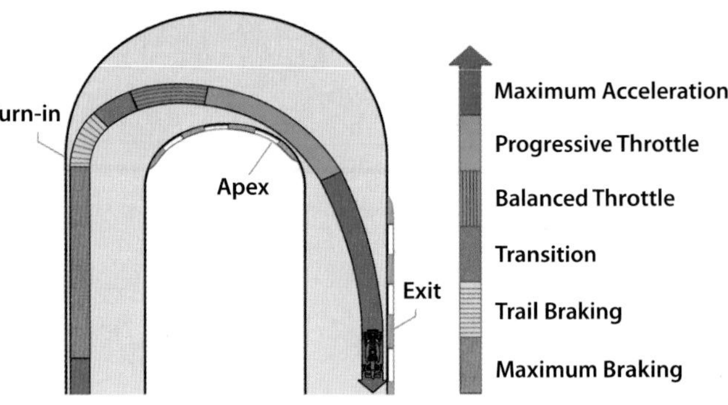

ILLUSTRATION 8-2
The control phases of a corner.

braking a little more, gradually easing (trailing) your foot off the brakes as you turn in. Or, does the car feel unstable or unbalanced going through the turn? If so, try coming off the brakes and getting back on the throttle just as you turn in. In this case, there may not be a trail-braking phase at all.

The transition from braking to acceleration is one area of your technique that may adversely affect the balance of the car most. You should be able release the brakes and begin application of the throttle without feeling the transition whatsoever and as quickly as possible. It should be immediate, as fast as you can possibly move your foot from the brake pedal to the throttle.

Practice this when driving your street car. You should never be able to feel the point where you ease off the brake pedal and begin to squeeze on the throttle.

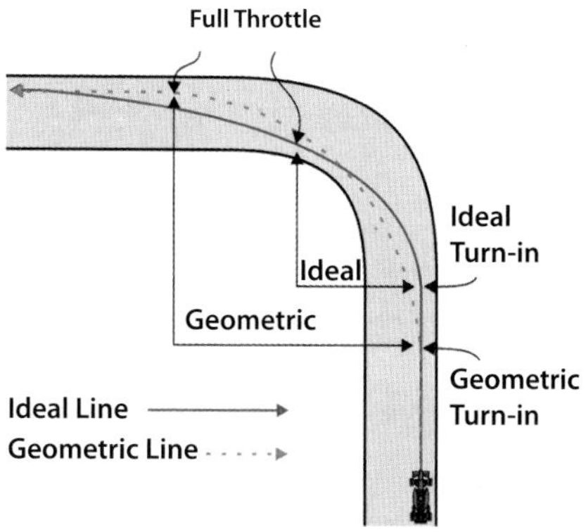

ILLUSTRATION 8-3 The effective length of any corner is from turn-in to the point where you get back to full throttle. This illustration demonstrates how much less time you spend cornering by using a late apex. Note how much more straightaway there is prior to the turn-in point for braking, or how much longer the straightaway is for maximum speed, with the later turn in.

A correctly executed transition from braking to acceleration is paramount. It must be done with perfect smoothness. That's one big reason why one driver can make a car turn into the corner at a slightly higher speed than another driver. Just because you cannot make your car turn into the corner at a specific speed, does not mean Michael Schumacher or Dario Franchitti couldn't. Maybe you are not using the correct technique, not being smooth enough, turning the steering too quickly, unbalancing the car, and so on. Again, how you lift your foot off the brakes is absolutely critical. It has to be eased off the pedal—quickly—so as not to upset the balance of the car. Then, you have to transition over to the throttle so smoothly that you never actually feel the exact point where you have come off the brakes and where you start to apply acceleration.

Remember the traction circle. The relationship between steering position and throttle position is interactive. Steering input must be reduced ("unwound") in order to apply acceleration. Since a tire has a limited amount of traction you cannot use all of it to turn the car and expect it to accelerate at the same time. You have to trade off steering input as you begin to accelerate, otherwise you "pinch" the car into the inside of the corner on the exit, often causing the car to spin and always scrubbing off speed.

And remember, given relatively equal cars, the driver who begins accelerating earliest and hardest will be the fastest on the straightaway. I think that tells you everything you need to know about the progressive throttle and maximum acceleration phases.

9 THE LINE

I want to start the discussion on how to master the line with a reasonably obvious (if not, perhaps, something that you are consciously aware of) piece of physics: Corner speed is proportional to corner radius.

What does this really mean? Simply that the more speed you carry through a corner, the larger the radius of the turn must be. Alternatively, the tighter the radius of the corner, the slower you must drive. Simple enough, right? And, like I said, even if this is something you had not consciously thought of before, I'm sure you knew this at the intuitive or subconscious level.

Of course, I'm talking here about driving at the limit, with the tires at their very limit before breaking loose and beginning to slide the entire way through the turn.

Now, let's take this physics discussion a step further. As you also know at the intuitive level, the tighter the radius or the faster you drive through a corner, the more you feel the g-forces and the more g-forces the car is generating. Again, I know you know what a g-force is from an intuitive point of view, but what does it really mean?

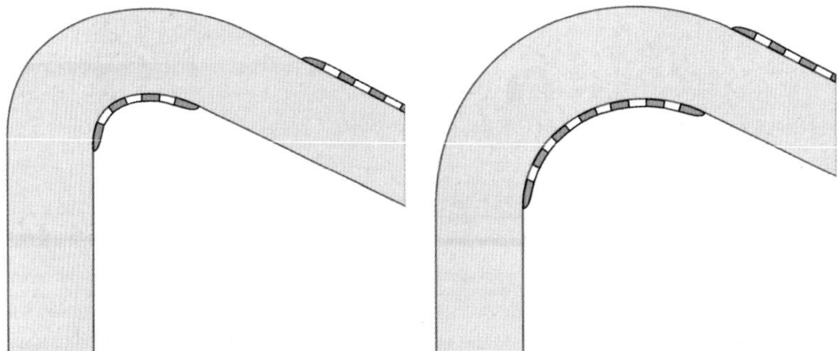

ILLUSTRATION 9-1 Which corner do you think you can drive through the fastest? Right, the one with the largest radius. The same theory applies to the line you drive: The larger the radius or arc you follow, the faster you can drive.

G-force is the lateral force acting on the car and you while going around a corner, with 1.0 g being equal to the force of gravity pushing laterally (sideways) on the car.

Now, let's put these two facts (speed is proportional to corner radius, and the faster you drive or the tighter the radius of the turn, the higher the g-forces) together into one mathematical statement: S = g/R (S represents speed in miles per hour, g is lateral g-force, and R is the radius of the corner in feet). In other words, the speed you can drive through a corner is proportional to the amount of g-force generated, divided by the corner radius.

What does this really mean, and why do you need to know this? You probably don't need to know the actual physics and math that goes along with this. What you do need to know, and what this really means is that the speed you drive through any particular turn is determined by the g-force your car is capable of generating—which is determined by the mechanical and aerodynamic grip the car has, along with your ability to balance the car to maximize the tires' grip—and the radius of the corner.

In terms of your driving, then, there are two areas you can work on to maximize your speed: balancing the car (to maximize tire traction) and increasing the radius of the corner. I'll discuss balancing the car in the sections on corner exit, entry, and midcorner. For now, let's look at the corner radius.

Driving through a turn using the largest possible radius means following what we call the "geometric line." This is the line that you would draw with a compass, using up every inch of the track surface, from outside edge to inside edge and back out to the outside again, on a constant radius. See Illustration 9-3.

If you decrease the radius of a turn by not using all the track, your maximum speed will be significantly reduced. For example, by entering a corner even one foot away from the edge of track, the radius of the turn may be reduced by as much as 1 percent. What's 1 percent worth? As much as half a second on some

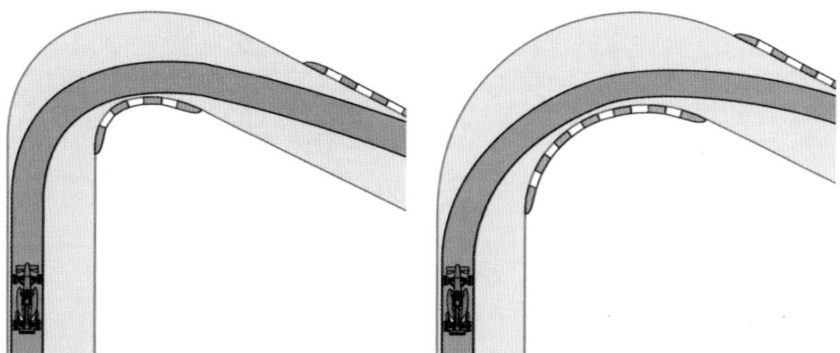

ILLUSTRATION 9-2 Compare the apex point in these corners. The one on the left, with the smaller or tighter radius, uses a later apex than the larger radius turn. As a general rule, the larger the radius of the corner, the earlier the apex. Why? Because you don't have to rely so much on accelerating hard out of the turn, since the larger radius allows you to maintain more speed through the corner.

ILLUSTRATION 9-3 The geometric line may be the fastest way to drive through one corner in isolation, but it isn't necessarily the fastest way around the track.

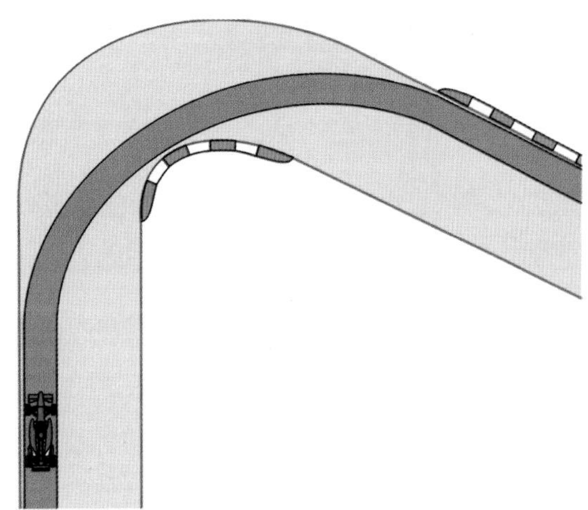

road racing circuits. What's half a second worth to you? From this I'm sure you see how critical using every inch of track surface really is.

Back to our geometric line. Although the geometric line is the fastest way to drive through each particular turn, it is not the fastest way of getting around the entire track. The reason for this has to do with the fact that there is usually something following the turns that is more important: the straightaways. If you have driven at least one race in your life, I'm sure you already know that it is far easier to pass a competitor on the straights than it is in the turns. What may not be so obvious at times is that there is more time to be gained, resulting in lower lap times, by being fast on the straightaways. In other words, it is far more important to be fast on the straights than it is to be fast in the corners.

Obviously, that doesn't mean putt-putting around the turns at a crawl. What it does mean is driving the turns in such a way as to maximize your straightaway speed. And that means altering the line you drive from the geometric line to one that allows for earlier acceleration, one we call the "ideal line." In most cases, that means driving a line with a later turn-in, apex, and exit. See Illustration 9-4 for an example.

The benefits of a line with a later turn-in, apex, and exit are listed here:

- You spend less overall time in the corner. When do you have the most control over your car, when you are on the straight or in the corner? And didn't we already decide that the larger the radius the faster we could drive? Less time in the corner equals more time on the straight—or at least near straight—which is about as large a radius as you can find.
- You are able to begin accelerating earlier. The sooner you get back on the throttle and begin accelerating out of a corner, the faster you will be down the ensuing straightaway.

- You can maintain your speed on the approaching straightaway longer by braking slightly later. Because you are turning into the corner later, the straightaway approaching the turn has effectively become longer, and therefore, you can maintain your speed longer.
- And, you can actually see through the corner better. By turning into the corner a little later, it allows you to see around the turn better. On most road-racing circuits this is not a big benefit, your vision is not blocked anyway; but on street circuits particularly, with the cement walls on the inside of the turn, it can make a big difference.

Along with those benefits comes a negative, however. In this case, due to the later turn-in, your line through the early part of the turn is a tighter radius. You know what that means: lower speed. But this tradeoff is an easy decision. Yes, you have to go a little slower early in the corner, but you more than make up for it down the following straight by beginning to accelerate earlier.

Now that I've convinced you of the benefits of using a late apex in each corner, think about something. Does this analysis of turns and the resulting late apex line work for every corner? Not necessarily. Let me give you a general rule, a very general rule:

SPEED SECRET

The faster the corner, the closer to the geometric line you should drive; the slower the corner, the more you need to alter your line with a later apex.

Let's take a good look at why this is the case.

CHANGE OF SPEED

Change of speed. Remember that phrase. The greater the potential for change of speed from corner entry to corner exit, the straighter the line you want your car pointed to allow for acceleration. In other words, the slower the corner, the later the apex you should use. A corner taken in first or second gear is certainly going to allow a greater change of speed than one taken in fourth or fifth gear, so you would use a later apex in the former than you would in the latter.

Based on what you now know about corner radius versus speed, you could also interpret my general rule as: The tighter the radius, the later the apex; the larger the radius, the earlier the apex, or closer to a geometric line. In simple terms, a slow hairpin will require a later apex than a fast sweeping turn.

Once again, the reason has to do with your change of speed. In a hairpin turn you will be accelerating hard out of the corner. Your change of speed will be relatively high. Your change of speed through a high-speed turn will not be so high, due to your car's reduced ability to accelerate the faster you are traveling.

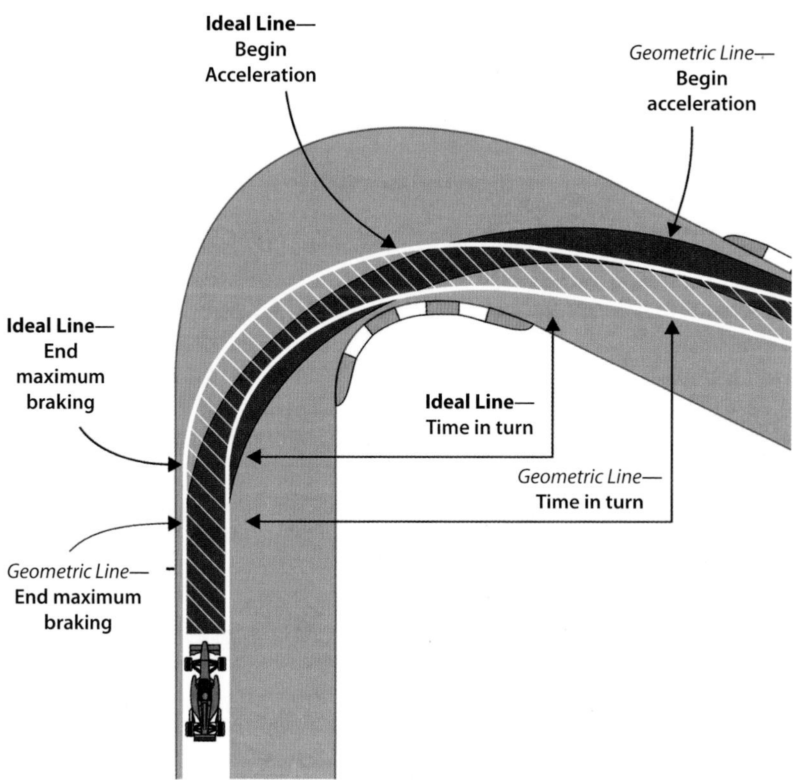

ILLUSTRATION 9-4 Driving the ideal line (the white striped line) does mean you will have to enter the corner a little slower, due to the tighter initial radius, but it allows you to begin accelerating earlier, which will result in faster speeds on the following straight. Also, you spend less overall time cornering, more time braking and accelerating.

If that does not make sense, ask yourself this question, "Can my car accelerate from 100 to 110 miles per hour as quickly as it can from 50 to 60 miles per hour?" The obvious answer is no. Any car, no matter how much engine torque it has, can accelerate quicker in the lower gears than in the higher gears.

So a general theme or objective in slower corners is to turn in and apex late, which allows the car to be driven on an increasing radius (a straighter line) when heavy acceleration is required. In faster corners, where acceleration is limited anyway, you want to use an earlier turn-in and apex, allowing you to maintain or carry more speed into the turn.

The other factor to consider here is how sharply you have to make the initial turn-in. Obviously, to turn in and apex later in the corner, you have to drive a sharper radius early in the turn. And you know what that means. Slower speed. In faster corners, the initial turn-in will be less sharp, allowing you to carry more speed.

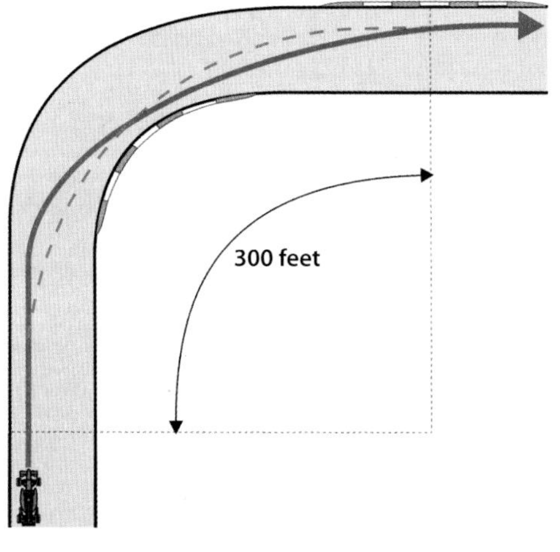

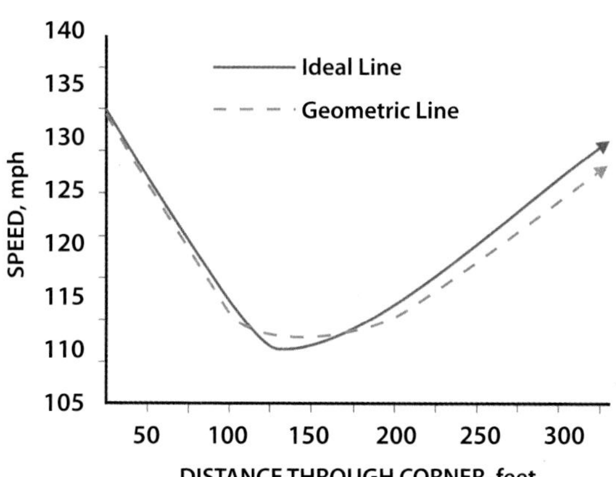

ILLUSTRATION 9-5
The graph shows the speed at various distances throughout the corresponding corner. The red-dotted geometric line is faster through the actual corner; the solid green ideal line is a little slower in the early part of the corner, but allows earlier acceleration, resulting in a much quicker exit speed. The extra speed at the exit will continue, and multiply, all the way down the following straightaway.

Using a later turn-in and apex requires a significantly different speed of turning the steering wheel and a different amount and timing of trail braking than a fast corner's turn-in technique.

G-FORCE JUNKIES

Most people who race or are involved in any type of performance driving—myself included—are what I call "g-force junkies." We are addicted to g-forces, that feeling of being pushed sideways or backwards in the seat, or forced up against the seatbelts. The more g-force, the better. For those who are competing on a racetrack, at least, that is usually a good thing, because as we saw earlier, the more g-force, the faster we are moving. However, that is not always the case. In fact, there are really two ways to feel more g-force:

- Increase your speed, which is one of the objectives of race driving the last time I checked!

- Reduce the radius of the corner; the tighter the line around a corner, the more g-force you will feel. But what's wrong with that? Every time you decrease the radius of the corner, you have to reduce your speed, which certainly isn't the objective.

So to satisfy your g-force addiction you may, without even being consciously aware of it, begin to tighten up the radius of a corner. Or, at least, not drive in such a way as to lessen the g-forces. Where this often shows up is at the exit of a corner, in not unwinding the steering wheel and releasing the car out of the turn.

The ultimate goal is to only increase g-forces by increasing your speed. In fact, one of the goals of driving the ideal line is to minimize g-forces so that you can then increase them by increasing your speed.

Again, we come back to this compromise: Increase the radius of the turn and we can drive faster through the corner, but the more we alter the line toward a late turn-in/apex (a tighter radius early in the corner), the sooner we can begin accelerating and the faster we will be down the straightaway. Finding the ultimate compromise—the ideal line—for each turn on every track you race on is the challenge, one of the biggest challenges of race driving.

10 PRIORITIZING CORNERS

Some corners on a racetrack are more important than others. Fast lap times and winning races come from knowing where to go fast and where to go (relatively) slow. When learning any track, concentrate on learning the most important ones first.

PRIORITIZING CORNERS

There are three important reasons for prioritizing the various corners on a racetrack:

- When you are learning a track, it is difficult (if not impossible) to try to learn every corner all at once. It is much easier to work on one or two corners at a time, moving on to the next when you feel you are doing a good job with the first one(s). When you know which corners are most critical to your overall lap times, you can concentrate on getting them right first.
- There are times when you must compromise the car's setup to suit one corner more than another. In this case, it is best to set up the car for the most important corners.
- Even when you know a racetrack well, you should constantly be trying to find some new or different approach to your driving technique that results in more speed. When doing that, sometimes you may have to compromise one turn's speed for another. If you know which corners are most important, you know which ones can be compromised and which ones can't.

There are two ways of looking at which corners on the track are the most important. The first is which corner is most beneficial to producing the quickest lap time; the second is which is the most challenging.

The old rule was the corner leading onto the longest straight was the most important in terms of your lap time. The new rule is that the most important corner on the track is the fastest one leading on to a long straight. After all, some tracks have straightaways of almost equal length; which, then, is the most important corner? And, even though a 40-mile-per-hour hairpin may lead onto a straight that is a little longer than another corner that can be taken at 80 miles per hour, improvement in the faster corner will result in the greatest gain in speed. It is not as simple as just determining which corner is followed by the longest straight.

SPEED SECRET

Focus on the most important corner first and the least important last.

The corner that is the most challenging or difficult is also a factor in determining the most important turn. Usually, the most difficult corner will give you your greatest improvement in lap time. Why? Simply because that is the corner that you are most likely to be farthest away from having "perfected." And, of course, if it is challenging to you, it probably is to your competitors as well. That means that if you can perfect it, you will have gained the most on the drivers you are racing against.

When you analyze any track you will find that there are only three types of corners:

1. One that leads onto a straightaway
2. One that comes at the end of a straightaway
3. One that connects two other corners

Some people believe the most important corner, in terms of lap speed, is one that leads onto a straightaway; the next most important is one that comes at the end of a straight with little straight after; and the least important is a corner between corners. This way of prioritizing corners was really made popular by Alan Johnson in his 1971 book *Driving in Competition*.

The reasoning here is since it's easier to pass on the straightaway, and on most racetracks you spend more time accelerating than you do cornering, it's most important to maximize your straightaway speed to take advantage of all

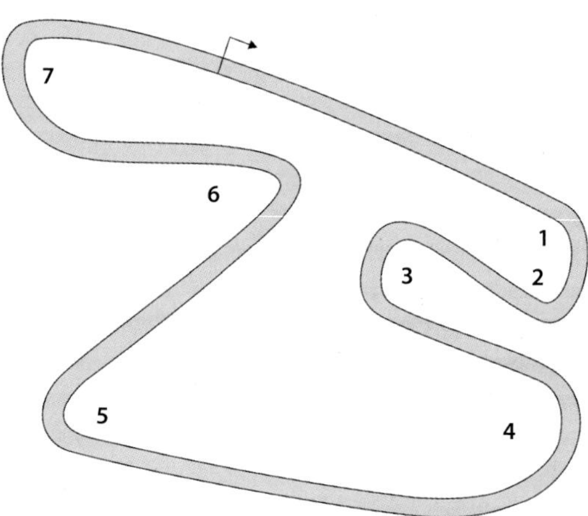

ILLUSTRATION 10-1 Looking at this track map, which turn is the most important corner? Is it Turn 7, the one leading onto the longest straightaway? Or is it Turn 4, the fastest corner leading onto a long straightaway? Right, Turn 4. Which is the least important corner on the track? Probably Turn 2 or 3 because they don't lead onto much of a straightaway.

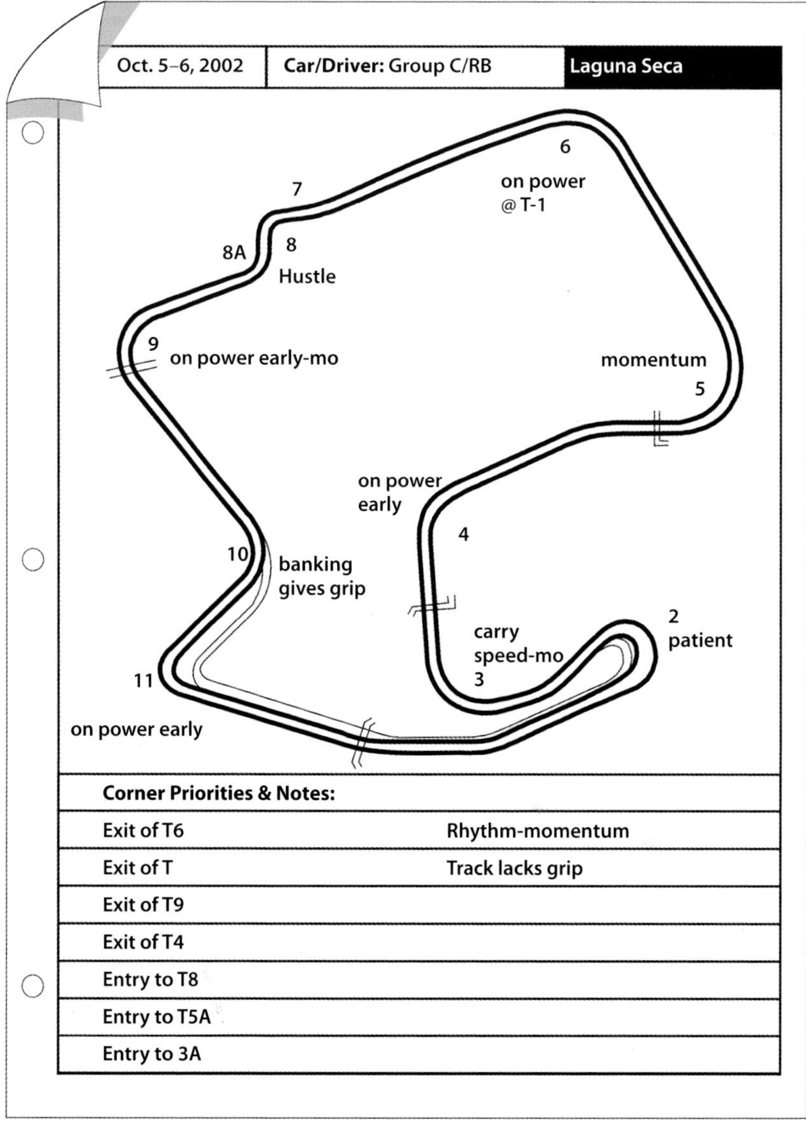

Corner Priorities & Notes:	
Exit of T6	Rhythm-momentum
Exit of T	Track lacks grip
Exit of T9	
Exit of T4	
Entry to T8	
Entry to T5A	
Entry to 3A	

ILLUSTRATION 10-2 Prior to going to a track, sit down and write out what the track's corner priorities are, along with any other notes, thoughts, or ideas you may have about driving the track.

that time spent accelerating. The corner that leads onto a straight will determine your straightaway speed. If you don't begin accelerating early, you will be slow on the straight.

This way of analyzing and prioritizing types of corners is not a bad place to start, but if you want to win, there is more to it than this.

SPEED SECRET

Before you can win, you have to learn where to go fast.

There is far more to gain or lose in a track's fast turns than in the slow turns. In fast corners, since your car has less acceleration capabilities, it is much more difficult to make up for the loss of even 1 mile per hour than it is in slow turns.

Let's look at an example, comparing a slow turn—one usually taken at around 50 miles per hour—and a fast turn taken at around 120 miles per hour. If you make an error and lose 5 miles per hour in the slow turn, it is relatively easy for

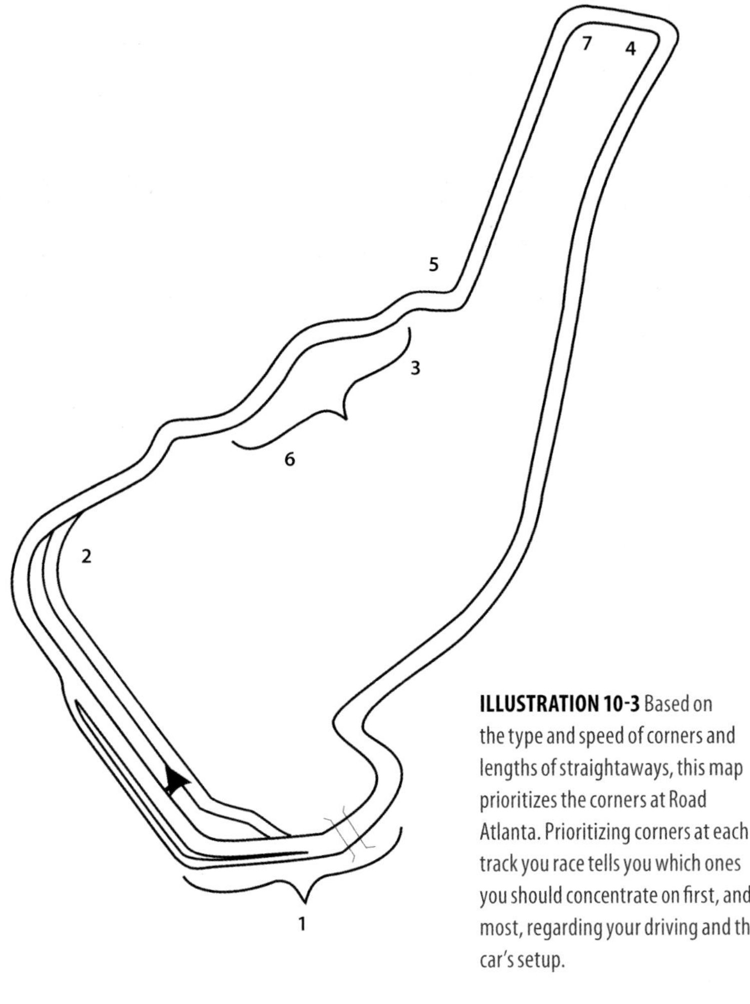

ILLUSTRATION 10-3 Based on the type and speed of corners and lengths of straightaways, this map prioritizes the corners at Road Atlanta. Prioritizing corners at each track you race tells you which ones you should concentrate on first, and most, regarding your driving and the car's setup.

your car to accelerate from 45 to 50 miles per hour. But your car will not accelerate from 115 to 120 miles per hour as quickly.

Another reason the fastest corners are the most important is that many drivers are intimidated by them. Plus, in most cases, the slow corners are the easiest to learn. The sooner you perfect the fast corners, the sooner you will have an advantage over your competition.

So, the most important corner is the *fastest* corner that leads onto a straightaway. The second most important is the next fastest that leads onto a straight, and so on down to the slowest corner that leads onto a straight.

Your next priority is the corners at the end of straightaways, which do not have a useable straight after it. Again, start with the fastest corners and work on down to the slowest.

Finally, concentrate on your speed through the corners that link other corners together.

Analyze where your car works best, as it will handle better in some types of corners than in others. It's a compromise deciding whether or not to change the car to suit a more important corner. Again, the priority should be to work at making the car handle well for the fastest corners leading onto straightaways.

SPEED SECRET

The most important corner is the fastest one leading onto a straightaway.

11 DIFFERENT CORNERS, DIFFERENT LINES

The ideal line in a corner that leads onto a straightaway (Illustration 11-1) is one with a late apex, approximately two-thirds of the way through the corner. This allows you to accelerate very early in the corner.

In any turn leading onto a straight it is best to brake early, get the car well balanced as you turn in and then accelerate hard onto the straight.

When a straight leads into a corner that is not followed by a useable straight—one that is long enough to allow passing or being passed (Illustration 11-2)—an early apex is used. Why? Well, since there is not a lot to be gained on the exit of the turn, you want to maximize the benefit of your entrance straight speed. In other words, sacrifice the corner's exit speed to maintain the straightaway speed for as long as possible. To do this, brake as late as possible, take an early apex, continue braking into the turn and position the car for whatever comes after this corner.

Now, I'm sure you're saying, "Well, lots of corners at the end of a straight also lead onto a straight." You're right. When this is the case, drive them like a corner leading onto a straight, taking a late apex. Again, straightaway speed is of utmost importance. A corner leading onto a straightaway always has priority over one at the end of a straight. That is why you won't experience many of these types of corners. They do exist, however, and it's important to recognize them and know how to deal with them when you do come across them.

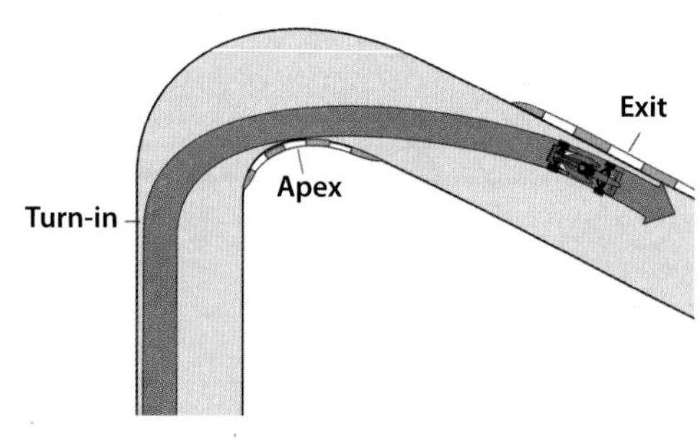

ILLUSTRATION 11-1 To drive a corner that leads onto a straightaway, your priority is the exit speed onto the following straightaway. This means a relatively late turn-in and apex (approximately two-thirds of the way through the turn), early acceleration, and using all the track surface at the exit. Given relatively equal cars, the driver who begins accelerating first will be fastest on the straightaway.

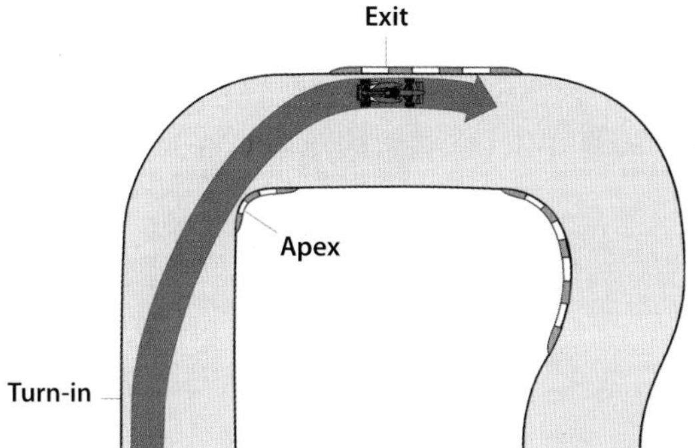

ILLUSTRATION 11-2 Your goal in a corner at the end of a straightaway that does not lead onto another straightaway is to make the preceding straight as long as possible by braking into the turn. Once you're past the apex and you have slowed down the car enough, then tighten the radius and head toward the exit point.

The last type of corner is the compound curve, where two or more turns are linked together, such as esse bends (Illustration 11-3). The rule here is to get set up for the last curve that leads onto a straightaway. Drive this last corner like you would any corner leading onto a straight, with a late apex. The first curves in the series are unimportant and must be used to get set up for the last one. Try to get into a smooth gentle rhythm in this series of turns.

Any time you have a succession of corners, your main concern should be the last turn. Again, concentrate on carrying good speed out onto the straightaway. Drive the corners in such a way as to maximize your performance through the final corner leading to the straight.

CORNERING SPEED

Don't you wish there was a magic formula to figure out what the optimum speed was for each car and corner combination? There is. The tire companies, Formula One, and some Indy-car teams use a sophisticated computer-simulation program to determine tire compounding and construction and chassis setups based on cornering speed. After plugging in hundreds of variables about the car and track, the computer will determine the exact theoretical speed at which the car will be at the limit. What's interesting is that a good driver can usually go faster than the computer says is possible. So it's still up to us to figure out what speed we can drive each particular turn of the track.

Using a very simple mathematical formula, however, and knowing some basic information (radius of the turn and coefficient of friction between the tire and the track surface, and assuming an unbanked track), you can calculate the approximate theoretical maximum cornering speed through a turn. Obviously, this is of little "real world" value. How could you ever drive through a corner while accurately monitoring your speed? What this mathematical exercise can do, though, is point out an important point.

For example, let's look at a 90-degree right-hander that has a theoretical maximum cornering speed of 80 miles per hour. By not using all the track

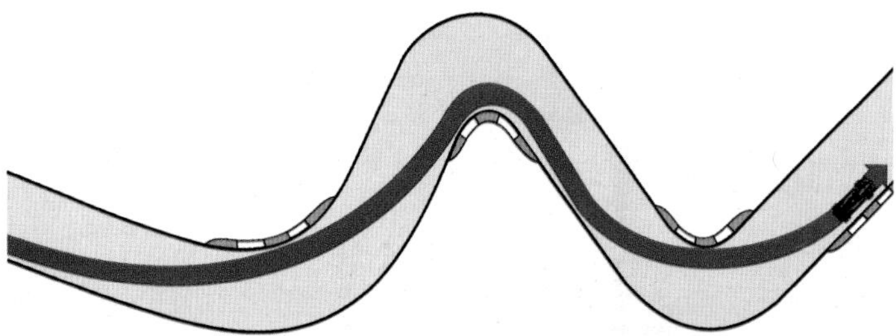

ILLUSTRATION 11-3 When dealing with a corner that links two other corners, your main concern is your exit speed onto the ensuing straight. That often means sacrificing the line through the "linking" turn to maximize speed through the previous and following corners.

surface—entering the corner 1 foot away from the edge of the track, and exiting 1 foot away—you have reduced the radius of the turn to the point where the theoretical maximum speed is now slower, just over 79 miles per hour. Even though that's only a little more than a 1 percent decrease in speed, 1 percent of a 1-minute lap time is more than half a second. That's a lot time to waste by simply not using all the road!

What this demonstrates is the extreme importance of using every little bit of track surface available and just how critical the ideal line is to your cornering speed.

Combined with a precise line through a corner, you must develop a delicate sense of traction and speed, as this is what ultimately determines your cornering speed. But remember: Increasing your radius through a corner effectively increases the speed you can carry, and vice versa.

WINNING PRIORITIES

Looking at what separates the winners from the losers gives us a guideline as to how to approach learning to drive at the limit:

- What separates the winning novice racer from the losing novice? The line, choosing the ideal line on a consistent basis.
- What separates the winning club racer from the losing club racer? The acceleration phase of the corner, how early and hard they get on the power.
- What separates the winning pro racers from the losing pros? Corner-entry speed, how quickly they can make the car enter the turn without delaying the acceleration phase.
- What separates the greats from the rest? Midcorner speed, how much speed they carry through the middle of turn.

Now, before you get any ideas about trying to carry blazing speed through the middle of every corner, realize that the greats became great only after perfecting the line, the acceleration phase, and the corner-entry speed. What this demonstrates is a priority list of how to drive at the limit and hopefully become a great race driver.

12 LEARNING THE TRACK

Before you can consistently drive at the limit, you need to know the track well. That doesn't just mean knowing which direction each corner goes, although that's part of it. It really involves knowing every last detail about the track. With some tracks, that may take a long time. Others are much simpler, taking little time to learn.

When "reading" the track, think about the track surface (asphalt or concrete types, bumps, curbs, and so on), turn radius (decreasing, increasing, constant, tight, large, and so on), road camber (banking: positive, negative, or even), elevation changes (uphill, downhill, and hillcrests), and the length of the straightaways (short or long).

On a track that is new to you, drive all corners with a late apex at first. This will allow you a little extra room on the exit if you find the turn is a little tighter than you thought. Then, with each lap, move the apex earlier and earlier in the turn until you are beginning to run out of track on the exit. Then go back to where you could accelerate out of the corner and still stay on the track. That's the ideal apex.

The banking of a turn may be one of the most critical factors you need to consider. Earlier, I said the radius of a turn determines your speed through it. Well, the radius of the turn may not be as important as its banking in terms of cornering speed.

When driving a positive banked corner, try to get into the banking as soon as possible and stay in it as long as possible. This probably means turning in a little earlier than would be normal if it were not banked. Many drivers underestimate the additional traction resulting from a banked corner. Use the banking to your advantage.

With off-camber (negative-banking) corners, set up so that you are in the off-camber section for as short a time as possible. Also, the banking may vary from the top of the track to the bottom, so look at the track closely. Banking may not be noticed while driving through a corner. That is why it is important to walk a track, making note of the detail changes.

Watch for the uphill and downhill sections of the track. They will have a great effect on the traction limit of the car. You want to use these elevation changes to your advantage and minimize their disadvantages. Just remember, a car going uphill has better traction than one going downhill, as the forward motion of the car tends to push it into the track surface, increasing the vertical load on all four tires. Your goal is to do as much braking, turning, and accelerating as possible on the uphill sections, and as little as possible on the downhill portions.

Make note of pavement changes, especially in the middle of a corner. You may want to alter your line to take advantage, or lessen the disadvantage, of where there is maximum grip. You want to make most of your turn on the grippiest pavement and then run straight on the less-grippy pavement.

After cars have run on a track for any length of time, an accumulation of bits of rubber from the tires, stones, and dust will end up just outside of the ideal line. This is called "the marbles," because it's much like driving on marbles. It can be very slippery. Try to stay out of this area. If, because you moved off line to let another car pass, you had to drive through the marbles, your tires will pick up some of these bits of rubber and stones. They will not have much grip when you get to the next corner. Be careful. Usually, they will clean off once you've driven through one or two turns again.

WALKING THE TRACK

Walking the track is a useful technique in learning to drive it quickly. Many drivers do this, although they make the mistake of turning it into a major social event, walking among a large group of friends. Without wanting to take away some of the fun, I would suggest you will learn and remember the track a lot better if you walk it by yourself, or possibly with one other driver who will give you a few tips or suggestions. Remember, also, to walk the track exactly in line with where you are going to see it from: the driver's seat. Even squat down to see elevation and asphalt changes and how the track looks from the height of your driving position.

Once you've driven a number of tracks, it gets easier. Every time you go to a new track, a corner will remind you of one from another track. You then take that information and apply it to the new corner. This is where experience really pays off.

Having walked tracks for years, I now never walk it until after I've driven it at least for one session. Often, I found that if I walked a track before having a little real experience, I would get false

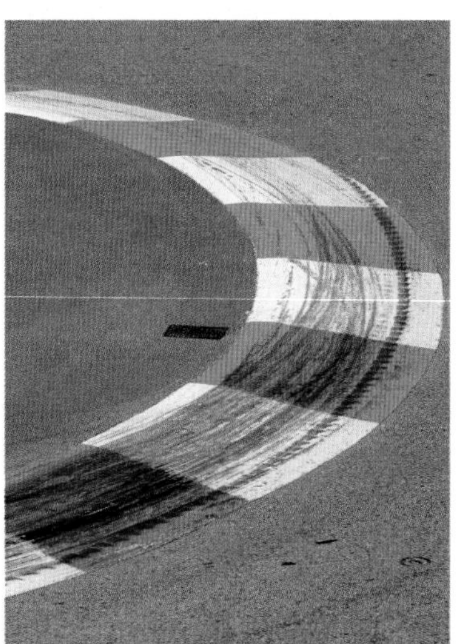

Learning the track means recognizing and placing in your mind every tiny mark, bump, color change, and shape of every inch of the track. The bigger your database of reference points, the better you have learned the track. *Shutterstock*

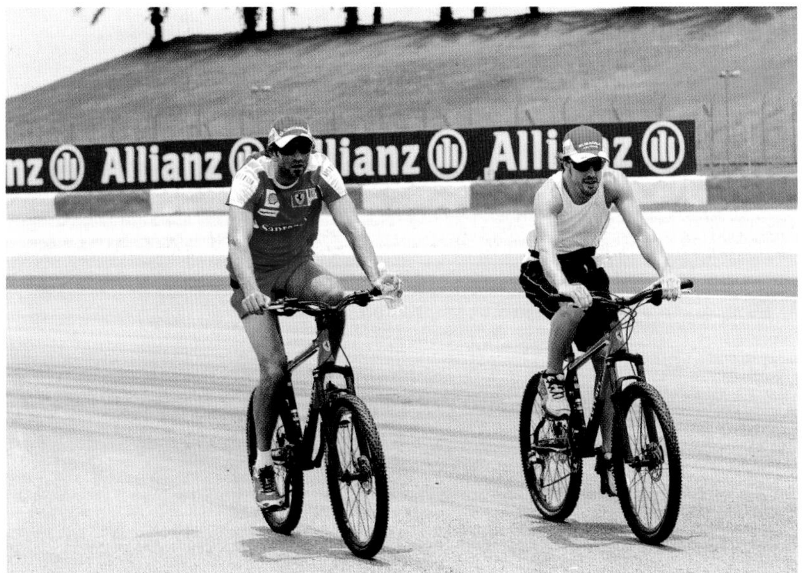

One of the best ways to learn a track is ride it on a bike. You can feel the shape, bumps, and elevation changes better than you can when walking it, while still having the time to take in all the minute details. *Shutterstock*

thoughts and ideas of how to drive it. What may have looked like a third-gear corner while walking the track may really have been a fourth-gear corner. Then, before I could really start to learn the track properly, I first had to unlearn the false thoughts and ideas.

So what I do now is first study a map of the track just to get the direction clear in my head. Then I explore it during the first practice session, trying different gears in the various corners and concentrating on the most important turns first. Then, at the end of the day, I walk it to really work out the details, checking my thoughts and ideas regarding track surfaces, banking, reference points, areas where it is safe to run off the track, and so on. And then, if possible, I try to do a number of laps in a street car at very low speed. This helps program every last detail into my head.

LEARNING A TRACK

When learning a new track, you face two hurdles before you will be able to drive consistently at the limit:

- Discovering and perfecting the ideal line
- Driving the car at its traction limit on that ideal line

It is usually easiest to focus on learning a new track in that order: the line first and then driving it at the limit.

When first learning the ideal line around a track, it is important to use *all* the track surface, even if that means forcing the car toward the edge of the track. At the entrance to the turn—at the turn-in point—it is easy to drive the car to within inches of the edge of the track. At the apex, you need to be right against the inside edge or curb. And, at the exit, drive the car to within a couple of inches of the edge, even use the curb or drop a wheel over the edge to see what it feels like (remember, you're driving relatively slowly at this point).

I hear a lot of drivers talk about how they can always find the right line for a corner simply by following the path of dark black tire marks, "the groove" they call it. They're wrong. The dark black tire marks are a result of drivers trying to tighten their line or make a correction: either feeding in massive amounts of steering, causing understeer, or controlling the back end of the car from a slide (oversteer). When walking a track, follow the path of a really dark tire mark. It usually ends up going off the track to the outside or spinning back across to the inside. The ideal line, or "groove," is usually just inside of the really dark black line through a turn. So, yes, you can use the dark black tire marks as a guide, but don't follow them.

It's important when you first learn a track to force yourself to use every inch of it, to make it a habit, a programmed, subconscious act. As your speed increases, the car will naturally flow or run out to the edge of the track. If you are driving the ideal line, and *you don't hold the car in tight (pinching it)*. Remember to let the car "run free" at the exit. If you hold the car in at the exit, you have greatly increased your chances of spinning, and you are scrubbing off speed or you can't get on the power as early as necessary.

Before moving on to the second part of learning a new track—driving at the limit—the ideal line must be a habit. Driving the line must be a subconscious act. It is difficult to concentrate on two things at once, the line and the amount of traction you have ("traction sensing") to determine whether you are at the limit or if you can carry more speed or accelerate sooner or harder.

After the ideal line becomes a habit, a subconscious act or program, you can begin to work on driving at the limit. The key here is sensing the amount of traction you have (traction sensing). With each lap, begin accelerating a little earlier and harder out of each corner (actually, remember the corner priorities: fastest corner leading onto a straight . . .), sensing the amount of traction available. Keep accelerating earlier until you either begin to run out of track or the car begins to understeer or oversteer *excessively*. Remember, the car must be sliding (understeering, oversteering, or neutral steering) somewhat, or you're not driving at the limit.

ILLUSTRATION 12-1 In some cars, running your inside tires over the curbing can help load the outside tires, providing them with more grip. Of course, this depends on the size and shape of the curb. Additionally, running over the curb usually straightens out the corner, increasing its radius.

Once you feel you're getting close to the limit under acceleration (on the ideal line), then begin to work on your corner-entry speed. Working on the fastest corners first on down to the slowest, carry a little more speed into the turn each lap, until you can't make the car turn in toward the apex the way you would like, until it begins to understeer or oversteer *excessively* in the first one-third to one-half of the corner, or it hurts your ability to get back on the power as early as you could before.

Don't forget that when working on the acceleration or corner-entry phase, that just because you sense you've reached the limit, you can't go faster still. It may be that the technique you are using now results in reaching the limit, but by changing that technique slightly you may be able to accelerate earlier or carry more speed into the corner and raise that limit. For example, you sense the car is beginning to oversteer *too much* under power on the exit of the corner. You've reached the limit with the way you are applying the throttle now. But if you apply the throttle a little smoother, more progressively, the car may stay more balanced and not oversteer as much. Another example: You carry more and more speed into a corner until it begins to understeer as you initiate the turn. You've reached the limit with the technique you're using now; however, if you used a little more braking while you turned (keeping the front tires more heavily loaded), or turned the steering wheel more "crisply," perhaps it wouldn't understeer at all.

The point is not to believe you've reached the ultimate limit just because the car slid a little *one time*. Once you're used to accelerating that early or carrying that much speed into a corner, take a number of laps to see if you can't make the car do what you want by altering your technique slightly.

Corner-entry speed and exit acceleration are related. If your corner-entry speed is too low, you tend to try to make up for that by accelerating hard. The hard acceleration may exceed the rear tires' traction limit, causing oversteer. If your corner-entry speed was a little higher, you wouldn't accelerate so hard and wouldn't notice any oversteer.

Of course, if your corner-entry speed is too high, it may result in getting on the power late. This is going to hurt your straightaway speed.

To recap the strategy for learning a new track:

- The line: At a slightly slower speed (difficult to do in a race weekend practice session with other cars around), drive the ideal line until it becomes habit, a subconscious, programmed act.
- Corner exit acceleration: Working from the fastest corner leading onto a straight down to the slowest, progressively begin accelerating earlier and earlier until you sense the traction limit.
- Corner-entry speed: Working from the fastest corner to the slowest corner, gradually carry more and more speed into the turn until you sense the traction limit.
- Evaluate and alter your technique if required: Try accelerating more progressively or abruptly, trail braking more or less, turning in more crisply or more gently, a slightly different line, whatever it takes to accelerate earlier and carry more speed into the corners.

THE LINE YOU DRIVE

Late turn-in and apex, early turn-in and apex, or mid-turn-in and apex, which one do you use? There is no one answer to that question. Ultimately, it is up to you to figure it out.

If you've been driving on racetracks for some time now, you drive the line you do either because someone told you to, or because it just feels right.

Obviously, if you are driving a line through a corner because of the first reason, let's hope the person who told you to knew what he was talking about. If not, you're in trouble; and you had better find someone else for advice. This is why I suggest you also use the second reason to determine your line. Sure, having someone you trust give you an idea of where you need to be is okay for starters, but you had better move on to what feels right very soon.

Does that mean that you cannot go wrong with the "it feels right" method, that it will always result in the perfect line? Absolutely not! In fact, often a not-so-perfect line will feel good, at least if you rely on only one of your senses for that "feel." For example, many drivers turn in early for corners because it visually "feels right." They look to the apex at the inside of the corner and naturally turn the steering wheel to go there. Visually it looks right, but about the moment the outside tires are dropping off the edge of the track at the exit, kinesthetically it does not feel so good (sometimes referred to as the "pucker factor")!

It is when you use all three sensory inputs to "feel" what's right that the car will tell you what line to drive. And it will do that in an obvious way. It won't be subtle. It will be direct, but only if you pay attention. And by paying attention, I mean being sensitive to what you see, what you feel, and what you hear.

REVIEW

To finish this chapter, let's review a plan for learning a new track quickly:

- **Preparation**—Review any information you can get your hands on, such as track maps, in-car video, computer simulation games, and descriptions of driving a lap written by someone with lots of experience there. The best place to start to find this kind of information is the track's website, then an online search, and then computer games. The main objective during the preparation stage is to become as familiar as possible with the direction the track goes and any references you may be able to use. One thing that is

ILLUSTRATION 12-2 A mistake a lot of drivers make when they first go to a new track is to place too much emphasis on "learning the track," which way the corners go and what the line is through them. If they put that much focus on simply driving the car at its limit, this information about the track would naturally come to them.

not an objective during this stage is to set too much in stone, to program every little detail you pick up from an in-car video, for example. Unless you're driving the exact same car, only use the reference points you see on the video for what they are, reference points. Just because a driver (in a car that may or may not be similar to yours) is using such-in-such as a turn-in point doesn't mean you should. But, and this is the real key to this, you can use the same reference point, but you may have to adjust exactly how you use it (turn in just before it, just after it, or whatever). Again, the objective in preparing to drive a track is to minimize the amount of brain power you have to put into what direction the track goes and what reference points you can use initially.

- **Be a sponge**—During the first few sessions, focus on soaking up as much information about the track as possible, just like a sponge soaks up water. Take in as many reference points as you can, and focus on specific senses to do that. In fact, take a session or two to focus solely on soaking up more visual information, then more kinesthetic (feel, balance, and g-forces), and finally more auditory information.
- **Download**—After each session, make notes on a track map of every detail of what you're doing where on the track (shift points, where you begin braking, where you end braking, where you're back to full throttle, etc.), and all the reference points you've soaked up (every crack in the pavement, the shape and placement of every curb, turn worker stations, signs, bridges, changes in track surface, marks on the surface, etc.).
- **Mental imagery**—After driving the track for a session or more, take all the information you know about the track from your preparation and your actual experience and replay it in your mind. The more repetitions you do during your mental imagery sessions, the more effective it will be. In other words, you will learn the track faster, and be faster.
- **Drive the car, not the track**—Finally, it's time to stop thinking about the track and focus simply on driving the car to its limit. Usually, if you drive the car at its limit, even off line, you will still be faster than if you drove the perfect line but with the car not at its limit. It's at this stage where you really forget about driving the track, and you just trust your mental programming to drive the car. Yes, you need to be consciously aware of how close the car and its tires are to their limits, but you don't want to be consciously thinking about which way the track goes and where your reference points are. By now, if you've done all the previous steps well, the track should be well programmed into your mind, and it will be time for you to simply trigger it and go.

13 THE EXIT PHASE

"If you don't know where you're going, you'll never get there." It's an old saying, but it perfectly describes why I'm discussing the exit phase of corners before the entry phase.

The goal for the exit phase of any corner can be summed up best by the following statement: The driver who begins accelerating first will arrive at the other end of the straightaway, and most times the finish line, first. That is what the exit phase of corners is all about, maximizing your acceleration down the following straightaway. But there are limits.

THE 100 PERCENT TIRE RULE

As I'm sure you already know, a tire's traction can be used for braking, cornering, accelerating, or some combination of these three. In fact, you can use 100 percent of the tires' traction for braking. You can use 100 percent for cornering. Or you can use 100 percent for accelerating. But you cannot use 100 percent for cornering and 100 percent for accelerating at the same time. You can't even use 1 percent for accelerating if you are using 100 percent for cornering. You can only get 100 percent out of the tires, not any more than that.

The key point I want to make here is that overlapping your braking, cornering, and accelerating, without asking for more than 100 percent from the tires, is critical to going fast. It is your ultimate goal.

To drive at the limit, you must use all of the tires' traction throughout the track. As you begin braking for a corner, use 100 percent of the traction for braking. When you reach the turn-in point and begin to turn the steering wheel, you must ease off the brakes, trading

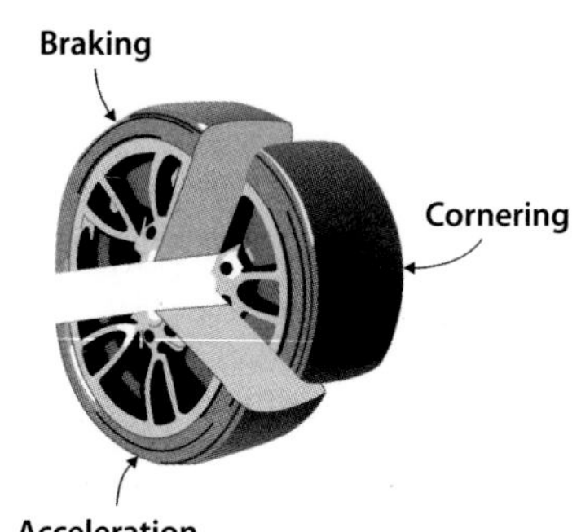

ILLUSTRATION 13-1 A tire can be used to do three things: brake, corner, and accelerate. You can use all the tire's traction for one of these, or combine two of the forces at one time, within limits.

some of the braking traction for cornering traction, going from a combination of 100 percent braking/0 percent cornering, through 50 percent braking/50 percent cornering, to 0 percent braking/100 percent cornering. For some amount of time, ranging from a small fraction of a second to a few seconds, all you are doing is cornering at 100 percent. Then, as you unwind the steering, you trade off cornering traction for acceleration traction.

SPEED SECRET

You can only use 100 percent of the tires' traction, and make sure you do.

That is the key to the exit phase of a corner: unwinding the steering or releasing the car out of the turn, allowing you to use the tires' traction for accelerating. Because, as I said, the sooner you begin accelerating, the faster you will be down the straight.

Remember what I said earlier about being a g-force junkie? Often, to fill your need to feel g-forces, you will subconsciously hold the arc of a turn just a little longer than necessary. In other words, not unwind the steering and allow the car to follow an increasing radius. The longer you hold the car on a tight radius, the more g-forces you will feel, but the slower you will be able to drive.

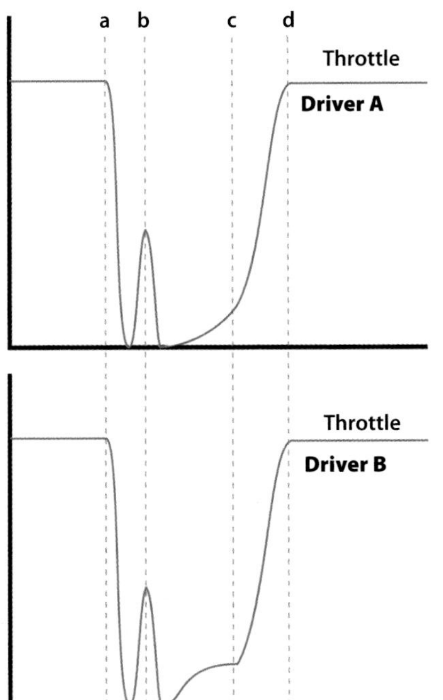

ILLUSTRATION 13-2 Here's the throttle trace graph of two drivers in the same corner. Both drivers come off the gas at the same point at the end of the straightaway, and they both blip the throttle for a downshift at the same spot. But notice the difference in how they got back on the throttle. Driver A in the top graph does a nice job of squeezing back on the throttle; Driver B does something a little different: He picks up the throttle a little earlier, then squeezes fully on. In most cases (but not all), Driver B's throttle application will result in a quicker lap time. Many times, this early touch of the throttle also helps balance the car, as well as improving midcorner speed.

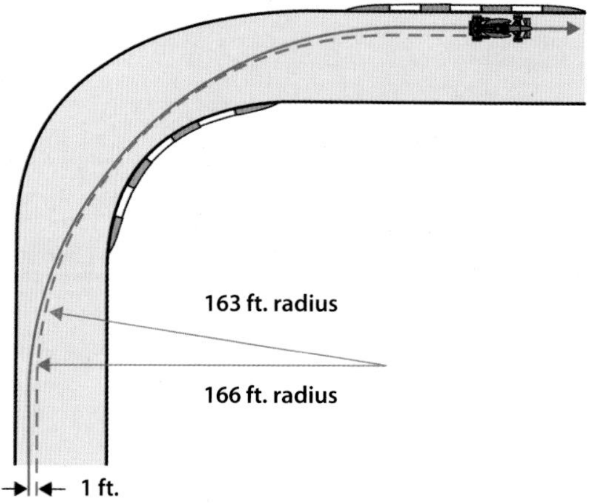

ILLUSTRATION 13-3 If you don't use all the track width, you're giving up speed. By keeping the car even one foot away from the edge of the track at turn-in and exit, you reduce the corner's radius significantly. In this example, the corner radius is reduced by 3 feet, meaning the theoretical maximum speed you can drive through the corner is reduced by more than a half mile per hour. That may not sound like much, but if you do that at every corner on the track it will probably cost you a few tenths of a second; that's giving time away.

So unwind the steering wheel. Let the car run free. If you constantly think about this concept and keep in mind the 100 Percent Tire Rule, you will be more likely to consistently drive at the limit.

A quick, but important point for driving on an oval track, or any road course corner with a concrete wall lining the track, is the more you try to keep the car away from the wall at the exit, the more likely it is you will end up hitting it. Unwind the steering and let the car track out as close to the wall as possible.

When and how you begin accelerating in a corner also plays a critical role in the exit phase. Again, the general rule is the sooner you begin accelerating, the better; however, with some cars you need to be a little patient getting back to power. If you begin to accelerate too soon, all you do is unload the front tires, causing the car to understeer, and then you have to ease back off the throttle to control it.

With other cars, almost from the instant you turn in to the corner, you need to begin squeezing back on the throttle, and quickly. You need to experiment with your car to find out what works in each corner.

One more thing to keep in mind: For every inch of track surface you leave by not letting the car run out to the edge, you are giving speed away. If your outside tires are not at least nibbling at the curbing, riding on top of the curbing, or hanging the outside half an inch of them over the edge of the track, you will not be as fast as you could be. Do you get my hint?

SPEED SECRET

Every inch of track you are not using is costing you speed. You paid for it, so use it all.

14 THE ENTRY PHASE

Compared to the exit phase of a corner, the entry phase is usually much more difficult or challenging. Paul van Valkenburgh once said the skill required to squeeze down on the throttle, keeping the tires at the limit of adhesion while accelerating out of a corner is like walking a tightrope, while the skill required to determine and set the car's speed when entering a turn is like jumping onto a tightrope blindfolded.

While technically the entry phase begins only at the turn-in point, I will also discuss the corner approach, braking and downshifting, in this chapter.

One of the first pieces of advice that new race drivers are given is "Going into the corners slowly and coming out fast is better than the opposite." Although this advice is entirely true, some drivers take this too far. This advice may actually have been the cause of some drivers being slow. Why? Because many drivers do not carry enough speed into the corners.

Ultimately, you want to carry more and more speed into the corner until it begins to negatively affect when you can begin to accelerate. If your corner-entry speed is so high that you have to delay the point where you begin to accelerate, then you need to slow down on the entry.

Let's look at a couple of examples to see what happens when your corner-entry speed is not ideal. First, imagine entering a turn 1 or 2 miles per hour too fast. Although that is not much too fast, it will definitely delay the time when you can get back to throttle and begin accelerating. As I said, if your corner-entry speed negatively affects when you can begin to accelerate, you need to enter slightly slower. Corner-exit speed is usually more important than entry speed.

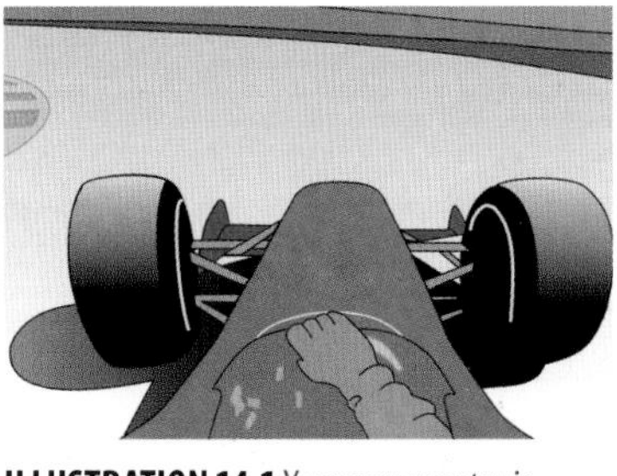

ILLUSTRATION 14-1 Your corner entry is absolutely critical. If you over-slow the car by even 1 mile per hour, you may never be able to make up for that error, no matter how much horsepower your car has.

Having said that, imagine entering a corner 1 or 2 miles per hour *slower* than ideal. What happens then? One of two things. First, and probably the lesser of

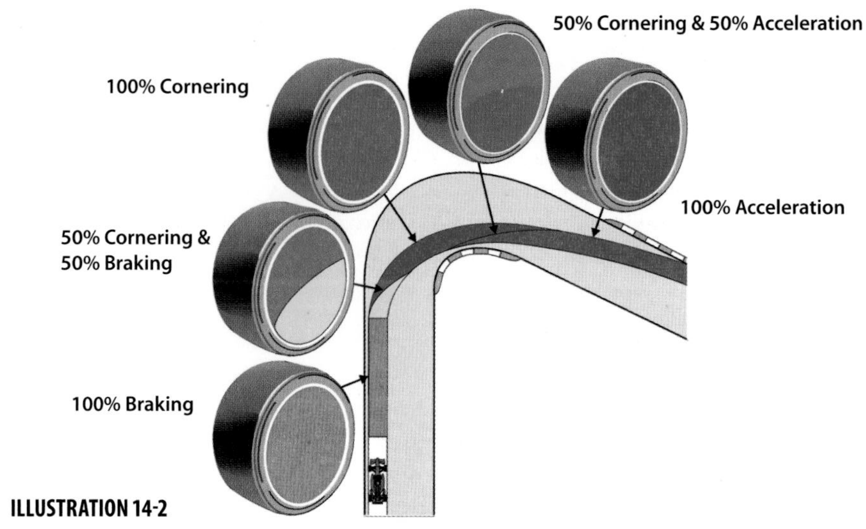

ILLUSTRATION 14-2

the two evils, is you have lost momentum, and momentum is always important. Whether you are driving a 60-horsepower Formula Vee, or a 900-horsepower Champ car, every time you slow down the car, it takes some amount of time to accelerate back up to speed again. If you slow the car more than necessary, it will take time to get that speed back, while in the meantime your competitors are probably pulling away from you.

The second and more damaging, and more difficult to recognize, effect of over-slowing your car for a corner is what I call the change of speed problem (there's the phrase again). I will discuss this in detail a bit later, but the basic idea is that if you over-slow the car on the entry, you will naturally want to accelerate hard to get back up to speed. This acceleration will often result in a form of power oversteer in rear-wheel-drive cars, and power understeer in front-wheel-drivers.

You can see from these examples just how critical getting the exact right speed is when entering a corner. To do this, you need accurate and sensitive traction-sensing skills. It also requires finely tuned speed-sensing skills. In addition to these skills, there are a few techniques that may improve your corner-entry speed.

LATE BRAKING

When asked what strategy might best be used to improve lap times by three or more tenths of a second on a typical road course, what do you think most race drivers' answer would be? If you guessed "brake later," you are absolutely right. I know, I've asked many, many drivers this question.

But is it the best strategy? Start to answer that question by asking yourself the simple question, "Why do I brake for a corner?" The answer is that you have a mental picture or feel for what speed you need to be slowed to by the time you begin to turn in to the corner; you are traveling at "x" speed approaching the corner, and to be slowed to "y" by the turn-in you need to begin braking "now" and "this" hard. In short, you brake to slow the car to a speed you feel is the maximum the car can carry into the turn.

With this in mind, what do you suppose most drivers, and probably you, do when braking later for a corner? Right, you brake later, but harder, for you "know" you need to get the car slowed to "y" speed by the time you turn in. In fact, until you update your mental picture of your corner-entry speed, braking later will only result in you braking harder. Your corner-entry speed will be exactly the same. Of course, what often occurs when braking harder is that you lock up the brakes; at best you begin braking a car-length later and enter the corner at the exact same speed as before. The biggest improvement in lap time you will see from this is no more than a few hundredths of a second.

However, if you update your mental picture of the corner-entry speed, to "y + 2 mph," for example, you will naturally brake a little later and not any harder. This will result in carrying more speed into the corner, and you will see tenths of a second improvement in lap time in one single corner.

SPEED SECRET

Corner-entry speed is more important than late braking.

This is where the big gains in speed come from. So instead of simply using the strategy of braking later, change your mental picture of the corner-entry speed and you will naturally brake later and carry more speed into the corner.

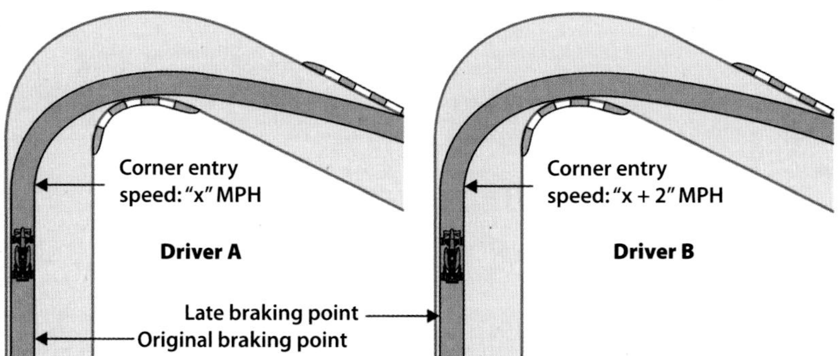

ILLUSTRATION 14-3 A comparison of two approaches to going faster. Driver A brakes later, but enters the corner at the same speed as always ("x" miles per hour) by braking harder. Driver B brakes later, but also carries 2 miles per hour more into the turn by braking as hard as usual. Braking later results in a small gain; carrying a little more speed into the turn results in a big gain.

ROTATION TURNS VERSUS SET TURNS

Another factor that determines how much you should trail brake into a corner is whether it is what I call a "rotation turn" or "set turn." Typically, but not always, "rotation turns" are shorter, tighter, slower corners. "Set turns" are longer and faster.

In many fast-sweeping corners, it is best to be on power, on the throttle from the second you turn in. In other words, no trail braking. Why? Because the car is better balanced this way. The car has taken a set. If you enter the turn while trail braking and then transition to acceleration, the car's weight distribution is changing. The weight is transferring off the front and to the rear, while the car is cornering. In most cases, a car will have more traction or grip—a higher cornering limit—when it is set, when its weight has stopped transferring from one axle to the other.

This is especially important in corners with a long midcorner phase. The reason is since you are spending more time in the corner, then ultimate cornering grip is critical. In shorter, tighter turns whose midcorner phase is almost non-existent, then the ability to rotate or change the direction of the car is more important than overall cornering grip. In this case, you should trail brake more. Trail braking enables you to rotate or turn and change the direction of the car quicker.

Given the choice, if you could accelerate all the way through every corner, rather than having to slow down and rotate the car, you would do that since the car would have more overall traction. You would be able to maintain a higher speed through the corners. But that is not practical or useable in every corner. The main objective for some corners is to change direction. These are "rotation turns," since the main challenge is rotating the car. The main objective in other corners is to maintain as high a speed as possible through the turn. These are "set turns."

SPEED SECRET

The faster and longer the corner, the less trail braking you should use and the earlier you need to be on the power; the slower and tighter the corner, the more trail braking you should use to help rotate the car.

STEERING TECHNIQUE

Obviously, the entry phase of a corner requires you to turn the steering wheel to make the car turn. Through the years I've heard numerous discussions and comments on just how a driver should turn the steering. While some people claim you should pull down on the wheel with the hand on the inside of the corner, others say you should push up with the outside hand. The odd thing is, the people handing out these two conflicting pieces of advice are racing and high-performance driving school instructors, supposed experts on the matter. Both approaches have their pros and cons. Pulling down with one hand usually provides more strength but may be less sensitive and accurate (less smooth). Pushing up with one hand is more accurate, but takes more effort.

Having heard this argument for years, and trying both methods myself and with drivers I've coached, I think the whole discussion of which hand you should primarily use to turn the steering wheel is a complete waste of time. Driving is a two-handed sport! If you are using just one hand to do most of the work, then you are losing out on both strength and accuracy. While one hand is pushing up, the other hand should be pulling down. While one is pulling down, the other is pushing up. That is where smooth-steering inputs come from.

As you approach a corner, there are a number of ways you can turn the steering wheel. You can slowly turn the wheel, or you can quickly crank the wheel. You can start off slowly turning the wheel, and then progressively turn it faster. Or you can do the opposite, quickly initiating the turn and then progressively turn it slower. You can also turn the wheel a little farther than is required to get the car to go where you want, and then quickly unwind it back out. You can arc or bend the car into a turn, or you can make your turn-in crisp. You can have "slow hands" or "quick hands."

So which is the right way of turning the steering wheel? I don't think there is a "right" way. It is all a matter of the type of turn you are approaching, the

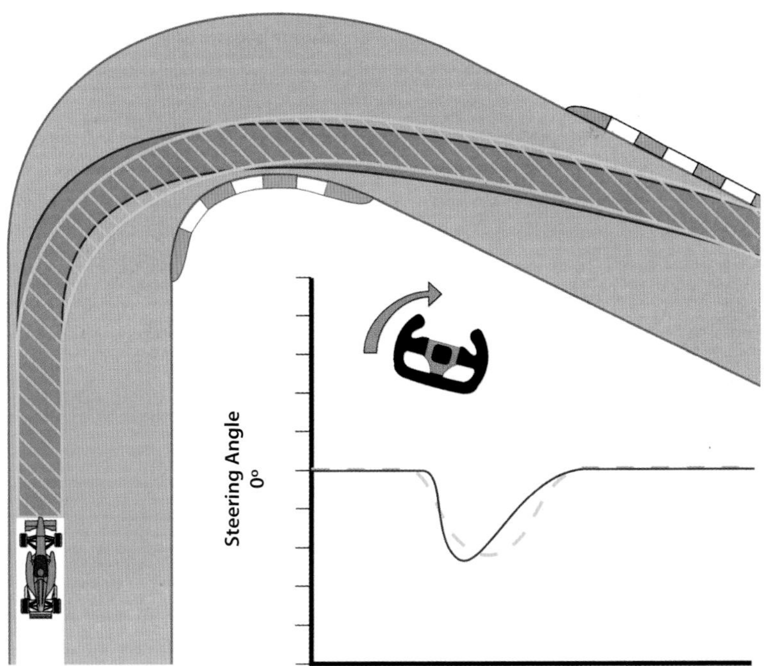

ILLUSTRATION 14-4 A comparison of two corner turn-in techniques. One driver eases, or "bends" the car into the corner, while the other driver uses a crisp, abrupt style. Which one is best? Depends on the car and the corner.

handling characteristics of your car, and your driving style. It is probably some combination of all the ways I mentioned, and maybe even more.

The point is, some corners require a quick, abrupt turn, and others don't. Some are best when you progressively increase the rate at which you turn the wheel, and others are best suited to the opposite approach. The key, then, is to be able to use whatever approach best suits the particular corner and car. To do that, your driving style must be adaptable. Many drivers have one particular way of turning the steering wheel and cannot adapt their style. That is one of the reasons why some drivers are better in fast corners than they are in slow ones, and vice versa.

Time for another general rule.

SPEED SECRET

The slower the corner, the later the apex, the quicker and crisper you need to turn the steering wheel; the faster the corner, the more you need to arc or bend the car into the turn with slow hands.

Former Formula One driver, Johnny Herbert, quoted in *Race Tech* magazine (August/September 1999), talked about this very thing and his steering style when entering a corner:

> *I have a style which, although it is quite smooth in the corner, has quite a hard turn and with these tires I normally break the grip straight away. So I introduce more understeer than I need in slow corners. Fast corners are no problem, because you have the downforce. The slower you go, the worse it gets. So you have got to be very, very smooth. Last year you could still use the front tire as a brake almost, to scrub off a bit of speed. But with these tires, the extra groove doesn't allow that. It just scrubs and doesn't stop; it just understeers. It breaks the traction of the tire.*
>
> *It is finding the right thing to do. It is not a natural thing. Naturally I turn in hard. But if you do that you just introduce much more understeer. But it is very easy to say, "There it is, go and do it..." It is doing it so that it is an advantage. You can't go in slower to stop the understeer, because then you are just too slow anyway. You have to carry the speed or even more speed but be smooth.*

The real key to this situation is being aware of how you turn the steering wheel, and then adapting to what works best, not just saying, "This is my style and I'm sticking to it." Of course, this doesn't mean spending all your conscious thought and awareness on what you are doing with the steering wheel as you enter a turn. If you did that, you would most likely end up stuffed into a wall somewhere on the outside of the corner! No, it is simply an overall, relaxed awareness of what you are doing. Usually, by asking yourself some awareness-building questions before, during, and after driving, you will subconsciously become aware of what you are doing. And most important, you will almost certainly turn the steering wheel in the way most suitable for the corner.

The key to steering correctly is to have a solid mental image of what you feel is ideal, and then be aware of what you are doing now. You can, and should, practice becoming aware of how you turn the steering wheel when driving on the street. If you do it enough on the street, it will become a habit, a program, when you are on the racetrack. As you turn into a corner, ask yourself, "Did I turn the wheel gently and slowly, or did I crank the wheel abruptly?" "Can I turn the wheel more gently?" "Did I turn the wheel slowly at turn-in and progressively turn it faster, or the other way?" "Did I turn it farther than required to get the car to go where I wanted, and then have to unwind it prior to the apex?" "Did I unwind the steering from the apex on out and release the car toward the exit?"

The more positive questions you ask, and the deeper you dig, the more aware you will become. And that awareness will lead to positive, accurate steering techniques.

Since more and more race drivers today are coming from a karting background, it is important for me to point out one difference between driving some types of karts and race cars. In many karts, one of the techniques used to make the front tires grip and turn in well is to crank the steering wheel in and then quickly unwind it to the point required to get the kart to follow the desired line. While this works with the type of geometry and lack of suspension on a kart, I can't recommend it for any race car. It is one of the habits (mental programs) that kart drivers have to change when beginning to race cars. If not, they may never reach their full potential racing cars. So if are you coming into car racing with lots of karting experience, be aware of your steering technique.

CHANGE OF SPEED

Okay, time to tackle the change of speed problem, one of the most common errors I see drivers make.

The key point is the change in speed through a corner may be causing the car to make you believe you are driving at the limit when, in reality, you have created an artificially low limit. Let me use an example to demonstrate what I mean.

Let's assume you can enter Turn 1 of our imaginary track at 80 miles per hour. That is, at the point you initially turn the steering wheel at the turn-in point, you are traveling at 80 miles per hour and you can carry or maintain that speed to around the apex, where you begin to accelerate. Throughout this corner the car is at its limit of traction, one-half of a mile per hour more, and the car would begin to slide too much, causing it to either scrub off speed or start to spin.

Now, what would happen if you entered the corner at less than 80 miles per hour? What often happens is this: You slow the car down to say, 78 miles per hour at the turn-in point. As you enter the corner, your traction sensing tells you that the car is not at the very limit; there is still some traction to be used. So, your right foot pushes down on the throttle pedal, and the car accelerates. Understand that this happens entirely at the subconscious level; you are certainly not consciously thinking about doing this, it is just happening.

Although entering the corner at 78 miles per hour is not driving at the limit, it is not far off. The tires are close to being at their limit of traction, just before they break away and begin to slide too much. So as you begin accelerating, you are now asking the rear tires (in a rear-drive car) to take on a bigger task. Remember, you can only ever get 100 percent out of the tires, nothing more. If you are using 99 percent of the rear tires' traction for cornering at 78 miles per hour, and then begin accelerating, there is a good chance you will ask for more than 1 percent traction for accelerating. In fact, there is a good chance that with your right foot squeezing down on the throttle, you will be asking for more like 5 or 10 percent

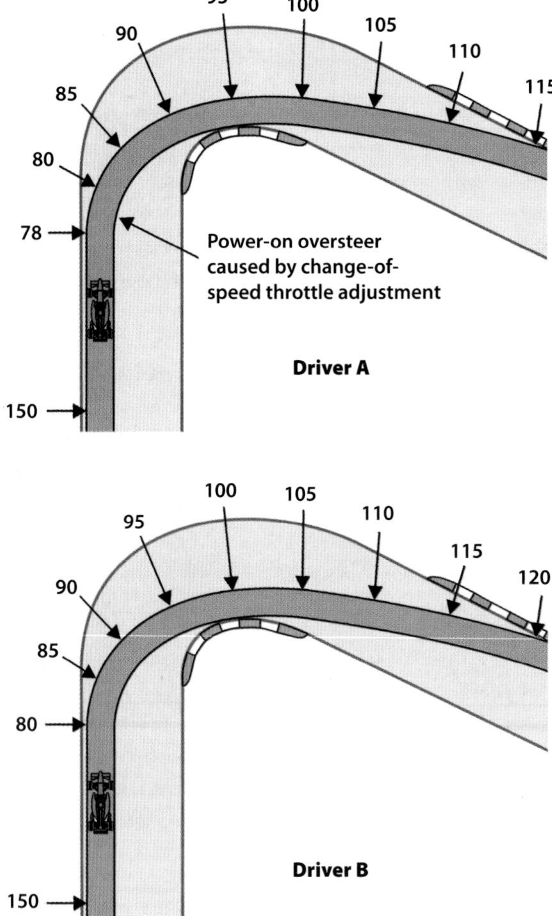

ILLUSTRATION 14-5
Let's look at Driver A and B again. Both approach the corner at the same speed, 150 miles per hour. Driver A slows the car to 78 miles per hour at the turn-in point, turns in, and immediately senses the car is not at the very limit, so he gets on the throttle. This causes a little power-on oversteer, so the driver eases off slightly to correct before getting back on the throttle and then increases the speed through the rest of the corner, hitting 115 miles per hour at the exit. Meanwhile, Driver B enters the turn at 80 miles per hour with the car at the limit, then smoothly increases speed throughout the corner, hitting 120 miles per hour at the exit.

of the traction for accelerating. The result is the rear tires go beyond their limit, and the car begins to oversteer, if even so slightly.

Your read on the situation at this point is that you are driving at, or maybe even slightly beyond, the limit. You are only at somewhere between 78 and 80 miles per hour, which is not as fast as you could be. You think you are at the limit, and you are right to some extent. But you have created this artificially low limit.

You see, it is the change of speed, from 78 to 80 miles per hour that caused the tires to barely exceed the traction limit and create an artificially low limit. If you had entered the corner at 80 miles per hour, your traction sensing would have told you that you are at the limit. You would have squeezed on the throttle appropriately, enabling you to accelerate out of the corner on the limit.

A greater than ideal change in speed also causes excessive weight transfer. It unbalances the car more than necessary. When your traction sensing signals there is more speed required to get the tires to their limit, you tromp on the throttle, and weight is transferred to the rear. Yes, this may be a good thing to help give the rear tires more grip to handle the increased demand for acceleration traction, but it may also cause excessive understeer. The ultimate result is the same: Your traction sensing then feels that the front tires are beyond the limit, and you don't go any faster (maybe even slow down).

This is why your corner-entry speed is so critical. If you enter a corner too slowly (below the limit), and then try to make up for it by accelerating, you may create a limit that is not as high as if you entered the corner at the ideal speed. And your corner-entry speed is one of the reasons your speed-sensing skills are so important. Without them, you will not be able to accurately and consistently gauge, and therefore drive, the correct speed when entering the turns.

Truly great race car drivers have the ability to adjust their entry speeds to within one-half a mile per hour, consistently, of the ideal corner-entry speed for every corner and every lap. Lesser drivers' corner-entry speed may vary from lap to lap anywhere from 1 to 5 or more miles per hour. Until your speed-sensing abilities are finely tuned and consistent, you will never know for sure what technique (or car setup) worked and what didn't work. This is why the speed sensing exercises I suggest are so critical.

SPEED SECRET

The less change in speed through a corner, the faster you will be.

MOMENTUM

A common problem that many drivers have is over-slowing the car entering corners, especially fast ones. The question is, why? There are at least three reasons for this:

- It could be that you have the "in slow, out fast" habit or program too well ingrained. This is one of the first things a race driver is told. It's better to go into a corner slow and come out fast, rather than vice versa. It's great advice for a beginner, who typically does get this wrong. But the problem is that you might never move beyond this advice. You have this technique so ingrained into your mind that you cannot move onto the approach the real fast drivers use: in fast, out faster.
- It could be that you are too focused on the begin-of-braking point, rather than the end-of-braking point. If you are totally focused on where you begin braking, when you get there you're likely to jump on the brakes (especially if you're trying to move the brake point in a little deeper). And if you jump on the brakes, you're likely to over-brake and over-slow the car. If, instead, you focus your eyes into the turn and onto the point where you're planning to release the brakes—the end-of-braking point—it's likely you will not over-slow the car. You will set the ideal corner-entry speed because you're focused on doing just that.
- It could be that you're not looking into the turn far enough. If you do not look far enough down the track and through the turns, you'll drive as if you're connecting dots. You'll drive from one point to the next, using these reference points to find your way around the track, but not planning out where you need to be and how fast you should be going when you get there. Everything is a reaction. There is no planning ahead. In fact, it's impossible to plan ahead if you're not looking far enough ahead.

No matter which one of these reasons is the cause of over-slowing (and it could be a combination of two or even all three), your mental programming is not right. You do not have a mental model of how to set the right speed for corner entry. And the only way you're going to improve it is to change your programming. Typically what happens is you'll make a physical change and then back that up with mental programming (more about this later in the book) to set it in stone. That's natural. Someone helps you learn that you must carry more entry speed to be fast, to focus on the end-of-braking point, or to look farther through the turn. You'll physically practice this enough times that you get it and then begin to form the mental model of that higher entry speed. For some drivers this occurs relatively quickly; for others, it's painfully slow.

Certainly, the technique of setting one's corner-entry speed is universal. Over-slow any car entering a corner and you're going to be in trouble. Too little entry speed is going to lead to too little exit speed. Sure, if you overdo it on the entry and carry too much speed into a corner, you'll be slow exiting it. That's where the "in slow, out fast" advice came from. It's a shame that it's been overemphasized because it is basically good advice, in the right dose.

If you over-slow a car entering a corner, here's what happens. First, you've lost momentum. Slowing a car is a terrible thing to do to it. Every time you slow it down, it's got to work hard to regain that speed, so the less you slow the car, the easier it is to get back up to speed.

Second, over-slowing the car leads to the change of speed problem we've looked at already. If you over-slow a car entering a corner, even by 1 or 2 miles per hour, your natural instinct is going to be that the car has some available traction. Your instinct, then, is going to be to use it. You'll step harder on the throttle, and when you step harder on the throttle, you're going to cause one of two things to happen: understeer or oversteer. The understeer is caused by too much weight transfer to the rear; the oversteer is caused by asking the rear tires (in a rear-wheel-drive car) to do too much, causing power oversteer. So, if you had not slowed the car so much, your instincts would not have encouraged you to step so hard on the throttle, and the change in speed from the minimum to your exit speed will have been less. It's the difference in speed from minimum to exit speed that causes the problem. The less the difference, the better.

15 THE MIDCORNER PHASE

As I mentioned earlier, the midcorner phase is what separates the champions from the truly great drivers. During Michael Schumacher's championship years, the speed he carried through the middle of every turn, without it negatively affecting his corner entry and exit speed, was a major factor in his outright speed advantage. In fact, Schumacher's style, technique, and ability in the midcorner phase may even help the other corner phases.

So what exactly did Schumacher do that enabled him to carry anywhere from 1/2 to 2 miles per hour more through the middle of each corner? Well, I would love to be able to say I know exactly what his "secret" was, but I can't. What I do know is that it has more to do with the way he balanced the car than anything else.

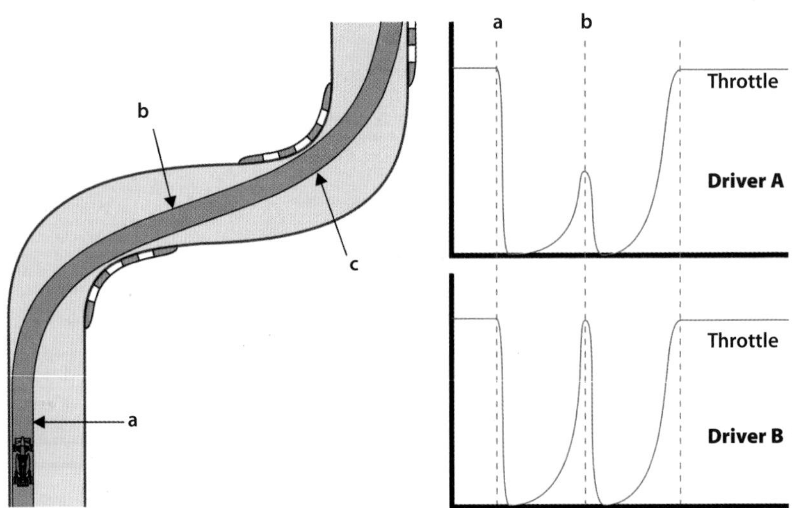

ILLUSTRATION 15-1 Small things can make big improvements. For example, compare these throttle graph traces. Between these two turns, Driver A uses less than half throttle, while Driver B gives a short stab at full throttle. Driver B is "hustling" the car. In these esses, that little burst of throttle could result in being up to three-tenths of a second quicker in this section alone.

Although the line he chose has a little to do with it, that is something that practically every other driver in F1 had figured out, as we have already discussed.

If you ever had the opportunity to really watch Schumacher, you may have noticed what I'm talking about. If you could stand near the edge of the track, as I was fortunate enough to do, you could see that the attitude of his car (the pitch and roll) did not change as much as others. His car stayed better balanced.

So, again, how did he do it? Without being able to get into the cockpit with him, I can only imagine that his footwork was close to perfect. The way he squeezed on the brakes with his left foot, and then eased off of it while beginning to squeeze the throttle with his right foot was seamless, perfectly smooth. My guess is that he had the slightest of overlap between the two, meaning that he had not fully come off the brake before he began to squeeze the throttle. There certainly did not appear to be even a nanosecond of time where he was not either braking or accelerating. I'm sure there was not too much overlap either, otherwise he would be known for being hard on the brakes, which he was not. Having observed closely from the side of the track and seen the bits of data acquisition traces that were shown in magazines of his driving style, that is a well-educated hypotheses. The only way of knowing for sure would to have been in the car with him at the time.

Schumacher's midcorner ability also had something to do with what made him quick everywhere else on the track: his steering technique. He had such light hands on the wheel. It was as if he was hardly gripping it. From the in-car camera, there are times where it looked as though only his fingers were touching the wheel. His palms did not seem to be in contact with the wheel. Of course, to control an F1 car with only the fingers holding the wheel, with the amount of grip and feedback through the steering it had, would require great physical strength. Schumacher had a well-known reputation for being perhaps the fittest driver in the world, maybe even in the history of the sport.

This light, sensitive touch that Schumacher had on the wheel provided him with more feedback from the steering wheel. Any time you grip the steering wheel tightly, tensing the muscles in your arms, the information (vibrations, sense of forces feeding back through the wheel, and so on) from the wheel to your brain will be restricted.

The feedback from the steering wheel provides you with much of your traction sensing. If that is restricted to any degree, you will be less sensitive to how much grip the tires have, and therefore whether you are at the limit or not.

Michael Schumacher's physical strength may just have been one of the keys to his abilities. Due to his strength, his arms and hands may have been more relaxed while controlling the steering wheel. Any time you can relax your muscles, the more feedback will reach your brain. The more feedback your brain has, the better your skills will be. And that leads to the second thing that results from the light touch on the steering wheel: smooth, precise, progressive steering inputs.

SPEED SECRET

Fast midcorner speed comes from great entry speed, car balance, and early exit speed.

BALANCING THE CAR

A car is balanced when there is no weight being transferred forward, as would be the case when you are braking; when there is no weight being transferred to the rear, as is the case when you are accelerating; and when there is no lateral weight transfer, as when you are cornering. This is the car's mechanical balance.

Why is a balanced car so important? Because a balanced car has more traction than an unbalanced car, and the more traction the car has, the faster you can drive.

Aerodynamic balance must also be considered. With some cars the aerodynamic downforce is affected when the car is unbalanced, when the car has a nose-dive, rear squat, or roll attitude. In these cars, when the rake of the underside of the chassis changes in relation to the track, the distribution of front to rear downforce can change dramatically. It can also reduce overall downforce. Once again, the better the car's balance, the more traction it has, and the faster you can drive.

So how do you balance the car better than your competitors? Well, that has to do with using the controls smoothly, not doing anything too abruptly to upset the balance, and then having a great sense of personal balance to become aware of the car's balance. Your footwork, for example, is critical. Without smooth, quick, and seamless transitions from throttle to brake and brake to throttle, the car's balance will suffer.

A great personal sense of balance may be the final reason Schumacher had the ability to carry so much speed through the midcorner phase. He had an amazing sense of balance, not just his own personal balance but the car's balance.

Balancing the car in a way that allows maximum midcorner speed also has much to do with the braking technique used when approaching and entering the turns. Why? If you brake hard, standing the car on its nose during the entry to a corner, the car's balance will not be ideal for the rest of the corner. If you can brake hard on the approach, while being able to rebalance the car as you ease off the brakes during the entry phase, your midcorner speed will be good, without having to be too slow on the approach.

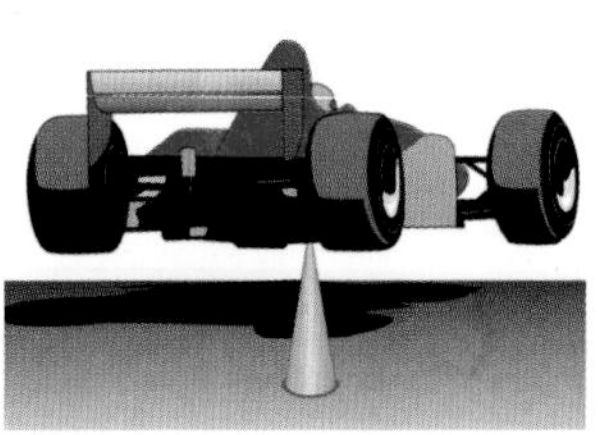

ILLUSTRATION 15-2 Driving at the limit requires keeping the car as balanced as possible, as if it were balancing on top of a single point.

Before you begin to think I consider Michael Schumacher to be some kind of superhero, understand that there are other drivers who are very good at this as well. Alex Zanardi, for example, was extremely good at this when racing Champ and Indy Cars, although he had a difficult time with it in F1. Motorsport journalist Jonathan Ingram made the perfect observation and wrote about this in his "Inside Line" column published in *On Track* (February 17, 2000):

> *It seems the transition from Champ Cars to both grooved tires and carbon brakes must have been a major problem. Zanardi's speed in CART came from not just braking late with cast iron equipment, but* ***braking more lightly entering the corners*** [bold added]. *That's why his moves seemed so unusual to fans, stewards, and fellow drivers—he often carried unexpected speed in the middle of the corners. With less tire on the road and carbon brakes that needed to be leaned on heavily to bring them up to operating temperature, Zanardi's American success could not be translated.*

By the way, I don't believe his struggles in F1 had anything to do with him losing his skills or technique. It was that he could not access them as well. Why? I believe it had more to do with his comfort level within the team and his lack of ability to adapt his behavioral traits (more about this in Chapter 27) to suit the environment within that team. Had he been comfortable in his surroundings, he would have carried the same type of speed in the Williams F1 car that he did in Champ cars.

TRANSITION

If you cannot make the transition from brake to throttle in the corner seamlessly, you will never carry good midcorner speed. This is another area where left-foot braking has an advantage, as it is much easier to make a seamless transition. In fact, when using your left foot on the brake and right on the throttle, there is usually a little bit of overlap, making it much smoother.

A seamless transition from the brakes to the throttle results in a balanced car and fast midcorner speed.

Practice your throttle-brake-throttle transition until you can do it in such a way that if someone were to ride with you blindfolded, they would not be able to tell the exact point where the braking ended and where the acceleration began. That is what I mean by a seamless transition. If you are right-foot braking, the movement of your foot off the brake pedal and onto the throttle should be practiced on the street until it is perfectly smooth.

16 VISION

At least 90 percent of your responses and actions in a race car are a result of the feedback you receive from your eyes and what they report to your brain. Although it's your hands, arms, legs, and feet that you use to control the car, it's what your eyes tell your brain that enables them to do that. Therefore, good vision techniques are critical to driving a race car.

Now, there is a difference between good vision and good eyesight. Eyesight can be measured and corrected with glasses if necessary. Vision is the act of sensing with the eyes. Good vision is something that can be practiced.

This may sound obvious, but look where you want to go, not where you don't want to go. Why? Because your car will go wherever you look, wherever you focus your eyes.

SPEED SECRET

Focus your eyes where you want to go, not where you don't want to go or where you are.

Focus on and visualize the line you wish the car to follow through a corner, constantly trying to see through the turn to the exit. Many drivers spend far too much time (which is any amount of time, even a fraction of a second) focusing on where they don't want to go, such as the curbs, walls, and other things off the edge of the track. And that's where they usually end up.

In fact, this is the key to driving the ideal line. If you want the car to follow a particular line through a corner, then that's where your eyes should be focused. If you don't want the car to go somewhere, like toward a cement wall on the outside of the track, then don't focus there.

Just because the car is pointing in a certain direction does not mean that's where you want to go. For example, when approaching a corner the car is pointing straight ahead. But where you want to go is into the corner, not straight ahead. So look through the corner, then look for the apex and beyond. That probably means consciously turning your head in that direction. The car will follow.

ILLUSTRATION 16-1
As you enter a corner, before you even get to the turn-in point, you should be looking at and through the apex. You have to know where you're going before you can know how much to turn the steering wheel at the turn-in point. Look as far through the corner as possible.

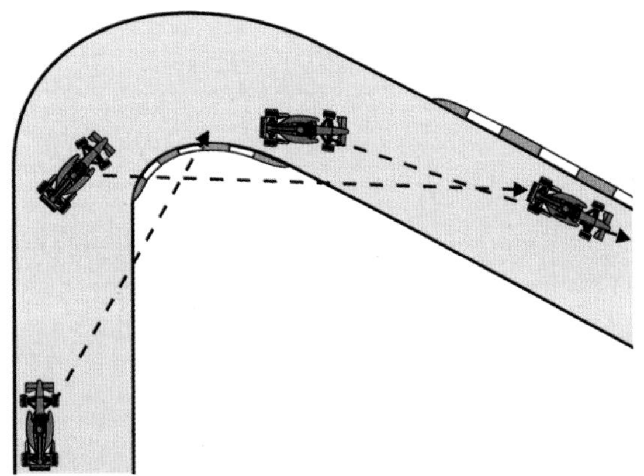

Looking where you want to go is only part of it. I learned this while teaching a student. I told him to look where he wanted to go. He got to the turn-in point and abruptly turned the steering wheel toward the inside of the corner, where he was looking. What I failed to tell him, and what I teach now, is to look where you want to go but have a mental "vision" of the path or line to get there. That's what makes for a smooth arc through a turn.

The better you know the course layout, the better prepared you will be. Always look ahead, and plan your route through the corners. If you mess up a particular turn, forget it, and keep looking ahead to the rest of the track. It really doesn't matter where you currently are, so don't look there. What is happening now on the racetrack was determined by what you did a long time ago. Look now, and plan now, for where you want to go.

SPEED SECRET

Look—and think—as far ahead as possible.

It takes practice to feel comfortable looking farther ahead than you do now, so begin practicing it on the street. You will be amazed at how much it will help and at how far ahead the winners are looking.

As you drive through the corners, keep your head upright. Many drivers wrongly feel they have to lean their head into the corner to be successful. The weight of your head leaning to the inside of the corner is not going to benefit the handling. Watch the best motorcycle racers: Even as they lean their bodies into the corner, their heads are cocked as upright as possible. That's because they realize their brain is used to receiving information from their eyes in the normal upright position, not tipped at an angle. So sit up and keep your head in a normal position. When you turn, move your head from side to side, but do not lean or tip it.

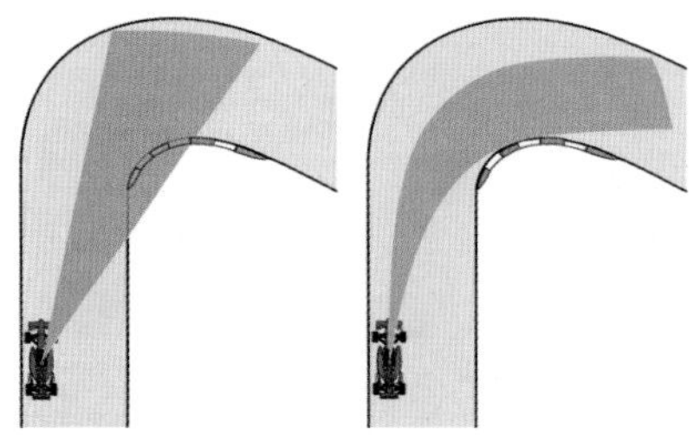

ILLUSTRATION 16-2

As you approach most corners, what you can actually see is often restricted, and the view your eyes give is straight ahead of where they are pointed (left). But you have to see a curved view around the turn in your mind's eye (right); visualize or picture in your mind the path you want the car to follow.

Do not concentrate on just one car in front or behind you. Look well ahead, and watch for anything coming into your overall field of vision. Pay attention all the time. And don't just look farther ahead, *think* farther ahead.

The best race drivers have a tremendous ability to know what's going on around them without having to look. Call it a sixth sense or extraordinary peripheral vision, but it is amazing what a driver notices when driving at speed, with experience. Like a person's field of vision, I call this a driver's "field of awareness," what you are aware of.

Do you remember the first time you drove fast or skied down a mountain? Your field of vision, or awareness, was probably very small, like looking through a scope. But the more you drove quickly or skied, the more your vision expanded, and the more you noticed around you.

Personally, there are times when I notice things to the side or behind me that physically I shouldn't know anything about. But with the adrenaline flowing, my senses are so sharp that I know exactly where a car behind me is, even though the view in the mirror is almost non-existent.

When I first drove an Indy car, my field of awareness was narrowed by the speed at which everything was happening (just as it was when I first drove a Formula Ford, and then a Formula Atlantic car). But as I became more accustomed to the speed, the more my field of vision and awareness expanded once again.

Experience in a fast car, at high speed, will help you become acclimated to this speed and increase your field of awareness. But it is something you can also practice while driving on the street. Work on seeing and being aware of everything around you at all times. Use your mirrors and peripheral vision to keep track of cars behind and beside you, trying to anticipate what they are going to do.

This ability to really know what's going on around you is one of the most important and amazing feats race drivers accomplish. If a driver has to think about it while driving, it won't work. But when it's there, it's not only a great feeling, but also a real key to success. It will come with experience if you "allow" it.

All these things—focusing your eyes on where you want to go, looking far ahead, and using your peripheral vision—are what good vision techniques are all about.

17 RACING IN THE RAIN

Racing in the rain is obviously a little more dangerous than in dry conditions. Driving smoothly and with full concentration is absolutely critical. It cannot be stressed enough. With practice and the right mental attitude, however, you can gain a great advantage over your competitors.

Personally, I love to race in the rain. Having spent many years in cars that were less than competitive, the rain was my equalizer. Since wheelspin is the major limiting factor on a wet track, if a competitor's car had more horsepower, he or she usually couldn't use it, therefore equalizing our cars. Also, having spent many years racing in the Pacific Northwest, I've become accustomed to driving in the rain. Now my mental attitude toward rain is positive, while some of my competitors have negative attitudes toward rain. While I'm loving it, they're hating it, giving me a mental advantage.

THE WET LINE

The general rule in rain driving is to drive where everyone else hasn't. In other words, off the ideal line. The idea is to look for, and use, the grippiest pavement. Through years of cars driving over a particular part of the track, the surface becomes polished smooth and the pores in the pavement are packed with rubber and oil. That is exactly where you don't want to be in the rain. You want to search out the granular, abrasive surface. This can sometimes mean driving around the outside of a corner, or hugging the inside, or even crossing back and forth across the normal line.

Great rain racers were not born that way. They developed the skills to be fast in the rain by practicing sensing the limits, being smooth, and searching out the track surface that provides the most grip. *Shutterstock*

Eventually, of course, you will have to cross the ideal line. When you do, try to have the car pointing as straight as possible, so there is less chance for the car to try to spin.

Since cornering traction is reduced more than acceleration and braking traction in the rain, try driving a line that allows you to drive straight ahead more. That means a later, sharper turn-in and a later apex.

Often, in a race, the rain will stop and the track will begin to dry. Again, watch for, and drive the driest line. This can change dramatically from lap to lap. As the track dries, your rain tires may begin to overheat and tear up. If so, try to drive through puddles on the straights to cool them.

Since water runs downhill, it may be best to drive around the top of a banked corner. Again, search out the pavement that offers better traction. Also, be careful of pavement changes and painted curbing. Often, they are much slicker than the surrounding asphalt.

SPEED SECRET

Look for and drive the grippiest surface.

RAIN TIRES

The optimum slip angle for a tire in the wet is less than in the dry. On dry pavement a tire's optimum slip angle may be in the 6- to 10-degree range; on wet pavement it may be around 3 to 6 degrees. This means you should drive in the rain with the tires slipping less than you would on dry pavement.

This reduced optimum slip angle range also means the line between grip and no grip is a little finer. Plus, once the tires have broken loose and begun to really slide, there is less scrub to slow the car down to a speed where the tires can regain traction. That is why it often feels like a car picks up speed when it spins on a wet track. It's because the rate of deceleration is so little.

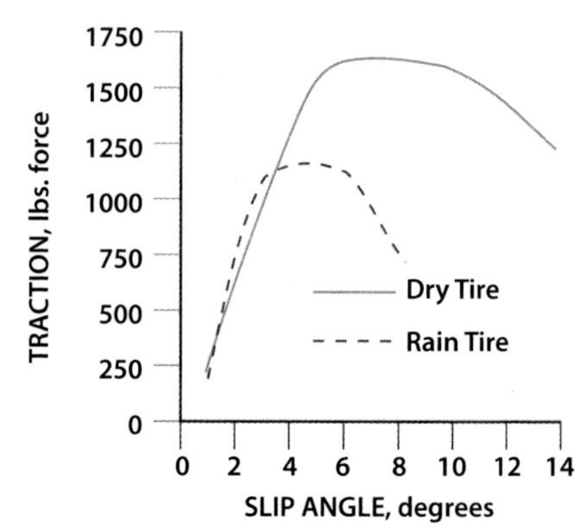

ILLUSTRATION 17-1 This slip angle versus traction graph for a rain tire and a slick racing tire shows that the rain tire is less "progressive." It reaches its limit quicker and lets go quicker. Obviously, its traction limit is lower than the dry tire because the track surface has less grip.

A rain tire is usually less "progressive" than a slick. That is, when the rain tire reaches it maximum traction limit (optimum slip angle) and begins to relax its grip on the road, it does so more quickly than the more progressive dry tire. In other words, the rain tire gives you a little less warning as to when it is going to let go (see Illustration 17-1).

These last two factors, the lack of scrub to slow you down when too much slip occurs and the less progressivity of the rain tire, is why it is critical to make the car slide from the instant you enter a turn in the rain. If you try to drive with no slip, at some point the tires are bound to go beyond the "no slip" range and begin to slide. When that occurs, it is going to take you by surprise. You think you've got lots of control, it's hanging on . . . hanging on . . . and then suddenly it lets go.

SLIDING AND BALANCE

Instead, enter every turn slightly faster than you think possible and make the car understeer, even if that means little or no trail braking at first. Once it is sliding, keep the car's speed up by squeezing on the throttle. If the car is set up right, you can gently make it go from this understeer to a slight oversteer, always keeping the tires slipping. With a little practice, you'll be able to add your trail braking back in (increasing the initial turn-in speed), and make all four tires slip an equal amount all the way through the turn, using the throttle to control the balance of understeer to oversteer, and therefore control the direction of the car by easing off the throttle to rotate the car and vice versa.

By having the car slide all the way through the turn, it will never take you by surprise. You know it's sliding. In fact, the car should be sliding almost all the time. Not too much, mind you, but sliding a smooth, controlled amount.

SPEED SECRET

If the car feels like it's on rails, you are probably driving too slowly.

A car on a wet track takes a set in a turn just like it does in the dry. Recall that "taking a set" is that point when all the weight transfer that is going to take place because cornering force has taken place. In other words, when the car has leaned or rolled in the turn all that it is going to, that is when it has taken a set. This will happen in the rain just as it does when dry, only the overall amount of weight transfer will be less due to the lesser amount of cornering force. It may take a little more sensitivity to feel the car take its set.

Having suggested the car should always be sliding, like anything, gradually work your way up to it. Don't try to put the car in large slides all the way through a corner first time out. But don't drive with the car on rails lap after lap either. With each lap, try entering the corner a little faster, and a little faster, until the slipping feels like it is too much (another tenth of a mile per hour will mean you can't control the amount of slip).

SPEED SECRET

In the rain, initiate slowly, react quickly.

Your initial turning of the steering should be as smooth, slow, and gentle as possible (allowing the tires to gradually build up their cornering forces). But when the car begins to slide, don't wait; catch it quickly with the steering wheel.

As you know, it is critical how you use the throttle and brake pedal in the dry. It is even more important in the rain. Every time you accelerate out of a corner, feed in the throttle by squeezing the pedal down slower than you would in the dry. If you should ever have to lift off the throttle in a turn, "breathe" it, ease out, "feather" it. Do not lift abruptly. That is probably the most common cause of a spin in the rain. Smooth and gentle—finesse—are the keys to driving in the rain.

You must be smoother on the throttle in the rain, squeezing the throttle so you only get just the right amount of wheelspin. Too much and you're either slow (you're not accelerating because of the excessive wheelspin) or you'll spin; too little and you will be slow. Remember the traction limit.

If you get into a little bit of a slide or spin, usually the best advice is, do as little as possible. It's just like driving over an icy bridge in your street car. There is practically no traction anyway, so whatever you do will have no effect, at least no positive effect, although it can often have a negative effect.

Be smooth with your shifts. You may want to try driving one gear higher in the turns than you normally would, using third gear in a corner you normally would use second in. This will lessen the chance of severe wheelspin by reducing the amount of torque available to the driving wheels.

AQUAPLANING

Aquaplaning is one of the trickiest parts of racing in the rain. Basically, it is when the tire cannot "cut" through the buildup of water on the track surface, and it begins to "skim" across the top of the water. Three factors account for this: the amount of water, the depth and effectiveness of the tread on the tires, and the speed the car is traveling. Be prepared for it whenever it rains heavily.

The trick to controlling aquaplaning is to do as little as possible; be gentle. Aquaplaning is much like driving on ice, the less you do, the better your chances of surviving. Do not take your foot completely off the throttle, as the compression braking effect of the engine and forward weight transfer may cause your rear wheels to slip. Under no circumstances should you hit the brakes. This will only cause you to slide even quicker. Nor should you try to accelerate fully through it.

Turning the steering wheel while aquaplaning can also be dangerous. Imagine "skimming" across the top of a puddle with the front wheels turned at an angle (as if trying to corner). When you reach the other side of the puddle, the front tires will now regain traction, while the rears are still on top of the puddle with no traction. The front-end of the car is going to follow the front tires, and the back-end is then going to skid sideways, causing you to spin out. Therefore, whenever you begin to aquaplane, make sure your steering is pointed straight ahead.

RAIN PREPARATIONS

Your chassis and suspension setup may have to be changed for the rain. Generally, you want to run a softer car: softer springs, shocks, and anti-roll bars (in fact, many drivers disconnect the anti-roll bars entirely in the rain). This will help your overall grip while giving you more feel for what the car is doing. If possible, since there will be less forward weight transfer, and therefore braking, by the front wheels, you should adjust the brake bias to the rear. You also may want to add more downforce from the wings and adjust the tire pressures. Use less pressure if there is a little rain and more pressure (causing a slight crown across the tread of the tire) in heavy rain to help avoid aquaplaning.

Perhaps the most difficult and dangerous part of racing in the rain is the lack of visibility. When following other cars, you may need to drive just slightly off to either side, not directly behind, to improve your visibility and to avoid the spray and mist. In fact, do everything possible to make sure you have good visibility. De-fog and clean your windows and helmet visor before driving. There are many anti-fog products on the market today, some that even work.

Driving in the rain can be enjoyable, because it's an extra challenge, as long as you concentrate on the changing conditions and drive smoothly and precisely. I remember, years ago, reading about Niki Lauda's claim that he was born with a natural advantage in avoiding visor fog-up. Because he has buck teeth, when he breathed in his helmet, his breath went downward away from the visor. From that point on, whenever there was a potential for my visor to fog up, I would concentrate on breathing downward. Plus, I always install a brand-new visor on my helmet just prior to driving in the rain. Old visors actually absorb moisture over time and are more susceptible to fogging. It's surprising how much better a new visor is than an old one.

18 RACECRAFT

Passing, being passed, dicing for position. This is what racing is all about. Some drivers can drive fast but can't race. Others can race but aren't particularly fast. To win, obviously, you must be good at both. And the techniques used to be good at both do not always complement each other.

That said, first you must learn to drive fast, then you can begin to race. Many drivers never learn to drive fast because they're too busy racing other drivers. Others are fast, but never learn how to really race, how to pass, defend their position, and so on.

COMPETITORS

I consider other race cars to be part of the track. Therefore, the racetrack is constantly changing as their positioning in relation to you changes. You'll be much more successful in your racing if you concentrate on your *own* performance

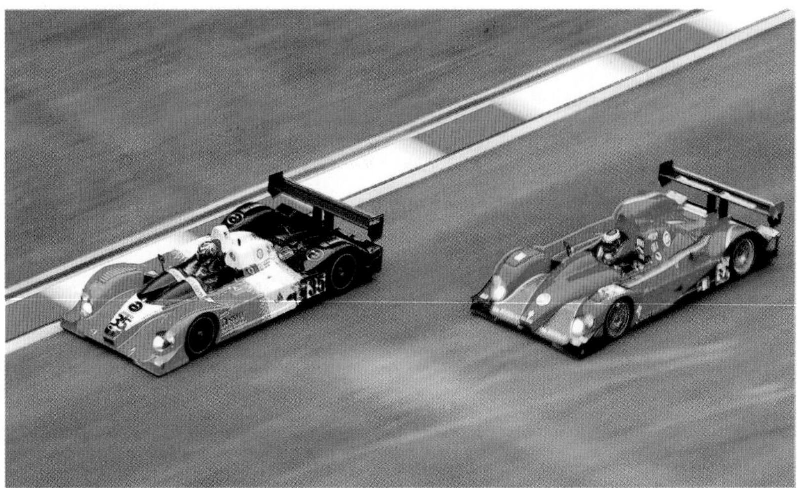

One of the most important parts of passing other cars is to place yourself in a position that you can "present" yourself to take control of the track. Put your car in a position where other drivers can see you easily, and you have the line. *Shutterstock*

rather than on the competition. So if you think of the competitors' cars as simply changes in the track layout, you'll be more relaxed and able to achieve your own peak performance.

It's important to be aware of everything and everyone around you, especially in a pack of cars. Train yourself to be able to be focused and yet able to notice other things around you. Practice this on the street. Concentrate on where you are going, but try to make note of all the other cars around you, especially the ones you can't see directly in the mirrors. This ability can make the difference between being just a fast driver and being a great racer.

PASSING

No matter what, you are going to have to modify your line when passing and being passed. It's part of racing. With any luck, you can do this to your advantage, not your disadvantage. The goal is to deviate from your ideal line as little as possible while passing and being passed.

A good habit to get into during practice sessions is to try driving "passing lines," that is, where you think you may be able to pass competitors in the race. Practice sessions are the time to test the track for grip "off line."

In passing maneuvers, the general racing rule is that the overtaking car is responsible for making a clean, safe pass. If the overtaking car is approximately halfway or more past the slower car and on the inside when entering a turn, it is that car's line. I repeat, though, this is a general rule. The "approximately halfway" is a bit of a gray area.

There are really three ways or places to pass another car:

- By outbraking it while approaching a corner
- By passing on a straightaway (either because your car is faster, you got better acceleration out of the corner leading onto the straight, or by drafting the other car)
- By passing in a corner (by far the most difficult)

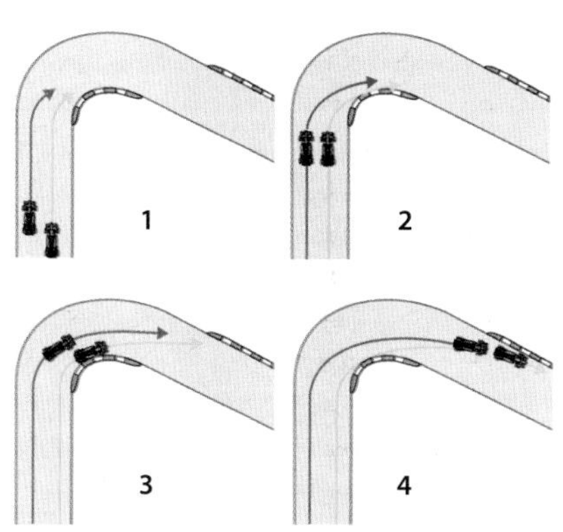

ILLUSTRATION 18-1 The correct way to outbrake a competitor into a corner: As you can see, all you want to do is get beside your competitor by the time you're at the turn-in point. That way, the corner is yours. The only thing your competitor can do is follow.

Probably the most important aspect of passing is to "present" yourself, making sure you get into a position where your competitor can see you. When you go into a corner on the inside of him, it is not necessary to pass him completely (see Illustration 18-1). Often, if you try to go too deep into a corner to get completely by another car, you overdo it, and one of three things

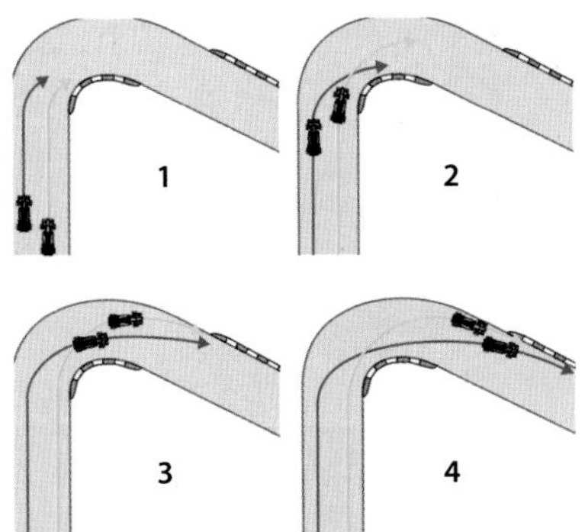

ILLUSTRATION 18-2 The wrong way to outbrake a competitor. If you get too enthusiastic and go too far past your competitor, it opens the door for him to repass you on the exit of the corner. This will probably be easily done, as you have gone too fast into the corner, cannot get back on line to block him, and will not be able to begin accelerating as early as your competitor.

happen: you spin, are unable to make a proper turn-in, or you come out of the corner so wide and with so little speed that the other car re-passes on the straightaway (see Illustration 18-2). All you really have to do is get beside your competitor, and the line through the corner is all yours. Just match your braking with the other driver's braking. There is nothing he or she can do about it at that point.

When outbraking a competitor on the inside approaching a corner, do you turn in at the same turn-in point? No. If you did, it would be much too early. Instead, continue straight down the inside until you intersect, and then blend in with your usual ideal line. That puts you in position to begin accelerating earlier than your competitor.

When following a group of cars into a corner, you most likely will *not* be able to brake as late as you normally do. As each car in front starts to brake, the cars begin to "stack" up in front of you. If you tried to go as deep as usual, you will run into the rear of someone.

When trying to pass another car, sometimes you actually have to hang back a little so you can get a "run" on him at a part of track where it is easier to pass. You often see a driver in a faster car that cannot pass a slower car because it is constantly running into the corner with the nose just barely inside the other car. Of course, the driver of the slower car takes the line through the turn and the faster car then needs to slow down as well, losing all its momentum. That driver would have been better off easing back just a little early for the turn, giving some room between himself or herself and the slower car, then accelerate early, driving the corner hard to gain the momentum down the straightaway, where it is easy to pass.

Remember that anytime you slow slightly while trying to pass another car, you are not at the limit anymore. Therefore, you can probably alter your line to almost anywhere on the track without being concerned about spinning.

SPEED SECRET

When passing, always "present" yourself.

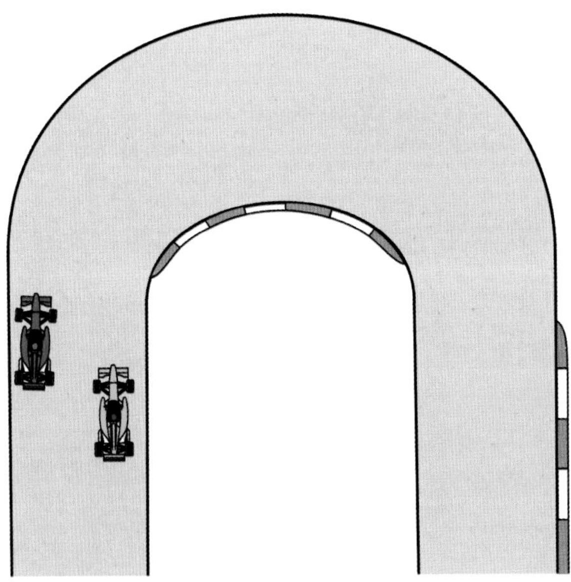

ILLUSTRATION 18-3 The driver of the car on the inside deserves to have the door shut on him by the car on the outside, for two reasons. First, the driver hasn't "presented" himself to the other car by getting far enough alongside him. Second, he is too close to the inside edge of the track—too far away from the car on the outside—making it very difficult for the other driver to see him. Instead, he should have eased off the brakes slightly to get farther alongside the other car and run closer to it as well (a small side benefit of running close to the other car is if the two cars do hit, the impact will be lighter).

If you and another car just in front of you are passing another car, consider that the driver of the car about to be passed probably only sees the first passing car and not you. Be prepared!

If you are obviously slower than the car behind you, try to let that car by. But do so on a straightaway, not in a corner. If you have already entered the corner, you are committed to the line. It is your corner. If you change your line in a corner after you are committed to it, you are going to confuse the faster car behind and possibly put yourself in a dangerous position. Be predictable! Wait until you are out of the corner and on the straight; then point to where you want the car to pass, and let it by. Pointing is important, but make it one or two quick points, then get your hand back on the steering wheel and concentrate on your own driving.

BLOCKING

Blocking is a controversial subject. A general rule is you can defend your position by altering your line, but only once. If you weave down the straight or alter your line two or three times on the approach to a corner, that's called blocking.

I don't think blocking is right. Not only is it dangerous, but if that is what it takes to keep a competitor behind, you don't deserve to be in front. Of course,

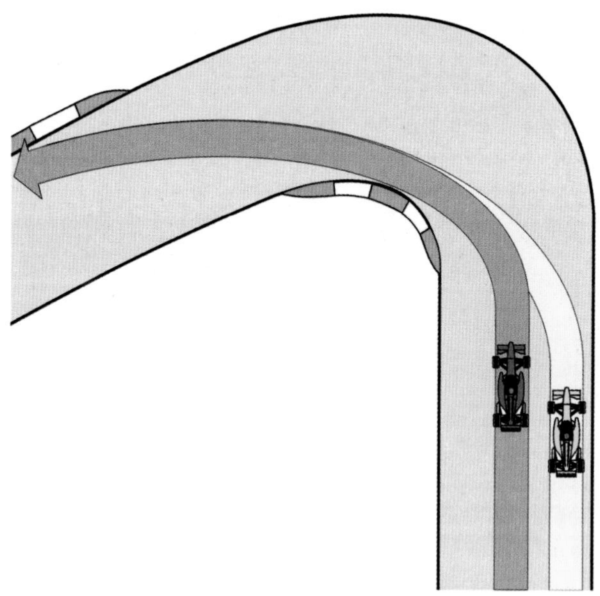

ILLUSTRATION 18-4 When you have modified your line to pass a competitor, simply "blend" back onto the ideal line as soon as possible.

in the last few laps of a race, almost anything goes, as long as you remember that you're not going to win if you crash both of you out of the race. The balance between being a good aggressive racer and being a blocker is a fine one. Having a reputation as a fair but tough driver is great; having the reputation as a "dirty" driver or blocker usually ends up costing you eventually.

You will learn who you can trust to race wheel to wheel. Generally, these drivers will not surprise you by doing something unexpected. They will not suddenly change their line drastically because you're trying to pass. They are predictable. They may change their line slightly to discourage you from trying to pass, but that's to be expected.

Remember, there are no real hard and fast rules regarding passing on the racetrack. And no insurance on a race car (well, you can get it, but it's expensive and you still have to pay the deductible yourself, no matter who's fault it is!). So, it takes respect and courtesy for your fellow competitors for all of us to "play" safe.

19 DIFFERENT CARS, DIFFERENT TECHNIQUES

What about different cars? Are there different techniques required to drive a front-wheel-drive versus rear-wheel-drive car? What about midengine open-wheel cars versus front-engined production-based race cars?

The answer is yes and no. It doesn't matter whether it's rear-wheel-drive, front-wheel-drive, four-wheel-drive, midengined, or front-engined, a race car is a race car. The basic technique is the same. The only difference is in the timing and amount of application of the technique and the slight variations in the ideal line I talked about earlier.

In fact, there may be just as much difference between two rear-wheel-drive cars (a Formula Ford and a GT car, for example) as there is between a front-wheel-drive and rear-wheel-drive.

The biggest difference with a front-wheel-drive car is this: The front tires are doing all the work, steering, accelerating, and most of the braking. Therefore, it's easy to overload or overwork them. If you overwork the front tires, they will overheat and lose even more traction.

With a front-wheel-drive car you have to be careful while accelerating in a corner. If you get on the throttle too hard, you overwork the front tires' traction limit while causing a serious rearward weight transfer, resulting in extreme understeer. Be smooth with the throttle; squeeze the throttle.

Since a front-wheel-drive car has a tendency to understeer (due to all the weight over the front end), it's important to trail brake a little more on the entrance to corners. Left-foot braking is used by many front-wheel-drive racers to help with this trail braking. Additionally, you may have to use "trailing throttle oversteer" to control the understeer in the middle of a long corner. This means quickly easing off, or "trailing" off, the throttle in the middle of the corner to cause forward weight transfer, reducing the understeer.

With rear-wheel-drive, you can "kick" the rear around tight corners with power oversteer by quickly applying lots of throttle, but not with front-wheel-drive. If you try this with a front-wheel-drive car all you'll do is increase the understeer.

Some say you must be more precise, that there's less room for error, when racing a front-wheel-drive car. Definitely, you can't be as harsh with the throttle to help overcome an error, as that will usually overload the front tires.

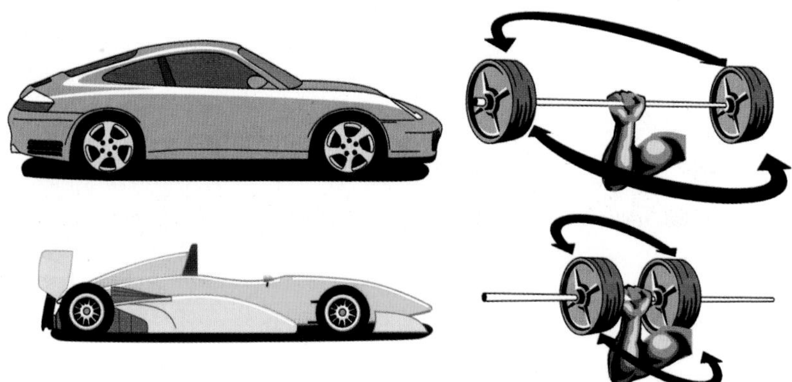

ILLUSTRATION 19-1 A vehicle's "polar moment of inertia" can be compared to a barbell. The more a car's weight is concentrated toward its center, the easier and quicker it will respond to a change in direction. You need to adapt your driving technique to suit your car's polar moment of inertia, by changing—among many things—the timing and motion of your initial turn-in. Usually, the higher your car's polar moment of inertia, the sooner you need to begin turning in, as it will take longer to respond.

You may want to try to straighten the front wheels a little sooner when exiting a corner with a front-wheel-drive, as the limit of how much throttle you can give while the wheels are turned may be less, due to the additional forces on the front tires. Usually, a later apex is required. And you know what to do to drive a later apex.

The key to being a versatile driver is being able to adjust or modify your style or technique to best suit the slight variations of different types of cars. However, there is one rather subtle but important thing you should keep in mind when switching from one type of car to another. Fortunately, it can be summed up in one statement.

What does this mean? Imagine holding a 4-foot-long barbell with a 10-pound weight at each end above your head with one hand. Begin to twist or rotate the barbell in one direction and then back in the other direction. What would happen? It would be difficult to change direction, causing your arm to twist prior to stopping rotation in one direction and going back in the other.

Now imagine sliding those two 10-pound weights in toward the center of the barbell, until they are about an inch from either side of your hand. Rotate or twist the bar again and then reverse direction. Much easier to change direction, right?

The same thing occurs with a car. The farther the mass or weight of the car is distributed from the center of the car (as with a production car), the higher its moment of inertia, and the more difficult it will be to change its direction. The closer the mass of the car is located to the center (as with an open-wheel car), the quicker responding and more maneuverable it will be.

Therefore, when driving a car with a high moment of inertia, it will take longer for it to react to your initial turn-in. To compensate, begin your turn-in slightly earlier and make the turn of the steering wheel more progressive. If you don't do that, you will probably find yourself struggling to get the car tucked right in close against the apex without over-slowing the car.

20 HOW THE DRIVER'S MIND WORKS

The physical act of driving a race car is relatively simple in comparison with the mental aspects. In other words, your results are largely dependent upon your mental performance. Yogi Berra's comments about baseball could be adapted to racing as well: "Racing is 90 percent mental, and the other half is physical."

If you want to win, having an understanding of how your mind works is not only beneficial, it is critical.

My goal here is to give you enough information so that you will buy into the concepts and tools I want you to use. Without this basic understanding, I doubt whether you will believe in the concepts, and therefore, will not use them. With this as a framework, let's dive into the driver's mind.

THE PERFORMANCE MODEL

The Performance Model was developed by my friend, Ronn Langford. It's used to explain and understand how we, as humans, perform practically any activity. The model works like this. Information, primarily from our senses, is input into our brains, which we can look at as operating like a computer. In this "bio-computer," the information is processed based upon our software or programming, resulting in an output. When it comes to driving a race car, this output is some form of action or reaction: using the pedals or steering wheel, looking at something, making a decision, behaving a certain way, having confidence, or literally millions of other actions.

Within your software, or mental programming, are your psychomotor skills (physical actions and movements that you can do without having to think about), state of mind, decisions, behavioral traits, and your belief system.

You could have the latest and greatest super-computer, with the best software or programming available, but if you give it poor quality or little quantity input, you will not get the output you were looking for. Conversely, if you give an old computer with a slow processor lots of great quality input, you still will not get the output you were looking for. In other words, the processing speed of your brain and your software (programming), determines the output as well. And your output is your driving performance.

USING YOUR WHOLE BRAIN

Have you ever had days or times when you feel completely switched on and performing at a high level, and other days when it seems you can't get out of your own way? Part of the reason for this is how well you're using your whole brain. When you're switched on and performing at your best, you're using your whole brain, and you're processing information quickly and efficiently. When you're not performing well, it's as if you're only using half a brain; you're not processing information fast.

Surprising to many drivers is the fact that you can actually use some exercises to speed up your brain's ability to process information and therefore drive faster and smarter. I'll look into brain integration and how you can improve your brain's functioning in the next chapter on brain integration.

SENSORY INFORMATION

Anyone familiar with computers will have heard of the slogan, GIGO, which means garbage in, garbage out. The same thing applies to our minds: If we input garbage, the output will be garbage. Of course, the opposite is also true: quality in, quality out.

So where do you get the information that is then input into your brain? From two main sources: sensory input and thoughts. Sensory inputs can be broken down into three categories: visual, kinesthetic, and auditory. Since your sense of smell is only used when driving to deal with problems (overheated brakes or engine, for example), and not to improve your performance, I'll not deal with it in this book. And, of course, we don't use our sense of taste while driving on the track, at least we shouldn't!

It's obvious that most of the information that you put into your brain when driving comes visually. What is not so obvious is what exactly visual means. To many people, having 20-20 vision means having good visual input. While central vision acuity, which is what the 20-20 measurement relates to, is important, it is not the most important part of the visual input. For example, visual-spatial awareness, peripheral vision, depth perception, and the ability to change focal points rapidly are much more critical to race-car driving. This is why some drivers with 20/20 vision do not "see" as much as others who have lesser vision.

ILLUSTRATION 20-1 Information from the driver's senses (visual, kinesthetic, and auditory) and his thoughts are input into his brain, which operates like a computer. Based on the software or programming in the brain, a psychomotor skill is triggered. This is an action. Then the loop begins all over again in reaction to the action.

ILLUSTRATION 20-2 A driver receives sensory input from his vision, his kinesthetic (feeling, motion, balance, g-forces), and auditory. The better the quality, and the more quantity of sensory input the brain has to work with, the better the quality the output, the better he will drive.

The kinesthetic sense involves much more than just the sense of touch. It also includes your proprioceptive system (the ability to sense forces acting against your body) and your vestibular system (sense of balance). Is your sense of balance important to driving a race car? Is your ability to sense the g-forces against your body important? Is your ability to feel the vibrations and feedback through the steering wheel, pedals, and seat important? You bet!

Some people seem to think that auditory input is not that important when it comes to driving race cars. Boy, are they wrong! Great drivers receive a lot of input from their hearing. The drivers sense when the tires are at their limit of traction to a great extent by the sounds they make. They sense and set the corner-entry speed by the sound of the air rushing past their helmet or car. They use the sound of the engine to tell them a lot about steering angle, shift points, traction, and so much more.

The overall message you should be getting from this is that anything you can do to improve the quantity and quality of sensory information going into your brain, the better your performance will be. I'll get into the details of how exactly to do this in the chapter on Sensory Input.

YOUR SOFTWARE

Everything you do behind the wheel of the car, and outside of the car, is a result of the programming in your brain. What do I mean by programming? Each time you do something, anything, the synapses in your brain that relate to that activity fire off bio-electrical current from one to another. This pathway now becomes the program for doing this act. The more often the act is completed, the deeper the programming becomes.

It is much like the pathway flowing water makes in dirt. The first time the water begins to flow, it seeks out a pathway. The more it flows, the deeper and stronger the pathway becomes. The same is true of the neural pathways in your brain. The more you practice anything, the stronger and deeper the programming becomes.

Let me make one thing clear. A race car must be driven at the subconscious level, not the conscious level. Why? Because a race car is much too fast to be driven effectively at the conscious level. A driver cannot think through each skill and technique as he or she drives the car. If, at the end of a straightaway, a driver thought, "Start braking now by moving my foot onto the brake pedal and squeezing it down, depress the clutch pedal with my left foot, move my right hand onto the shifter, move the shifter forward, blip the throttle, turn the steering wheel," where do you think the car would be? At best, at the back of the pack, but more realistically, crashed into a barrier.

To emphasize the importance of driving at the subconscious level, consider this fact: Your conscious mind processes information at a rate of 2,000 bits of data per second, while your subconscious mind processes at a rate of 4 billion bits of data per second! Is there any wonder the subconscious is better at driving something as fast as a race car?

You must rely on, and trust, your subconscious programming to drive the car. Where does that programming come from? Mostly from experience, physical programming. But it can come from mental programming as well. This is most often referred to as visualization or mental imagery.

Most drivers will tell you that they use visualization, when a majority of them really just close their eyes and think about what they want to achieve. Effective mental programming is more than just that. Mental imagery is really "actualization," where a person not only uses his or her visual sense, but all the senses. He or she not only imagines what a scenario looks like, but how it feels and how it sounds. The more senses a person uses in this mental imagery, and the more real he or she can make it, the more effective a tool it will be.

THE THREE KEYS TO IMPROVING DRIVING PERFORMANCE

Based on the Performance Model, you can see there are three keys to improving your driving performance:

- Faster processing: The faster and more efficiently you process the information in your brain, the better your performance.
- Quality input: The better the quality, and the more quantity of input from your senses, the better the output and the better your performance will be.
- Quality programming: The better your mental programming—your software—is, the better your performance will be.

Because these are so critical to your performance, I'm devoting the next three chapters to just these key factors. I'll then discuss in more detail focus, state of mind, decision making, behavioral traits, and belief system.

21 BRAIN INTEGRATION

You are most likely aware that your brain is made up of two halves or hemispheres. Each hemisphere has its own primary responsibilities: the left hemisphere for logic, math, language, and details; the right for creativity, intuition, art, and the big picture.

How would you describe yourself? Are you a left-brain–dominant person, meaning you're logical, factual, and detail-oriented? Or are you more right-brain dominant: creative, intuitive, and able to see the big picture?

Which do you suppose is the ideal for a race driver? If you answered "both" you are correct. You must be able to see the details and the big picture, be logical and creative, factual and intuitive. You must be what is called "integrated," where both hemispheres of your brain are working at their peak and with the hemispheres working together.

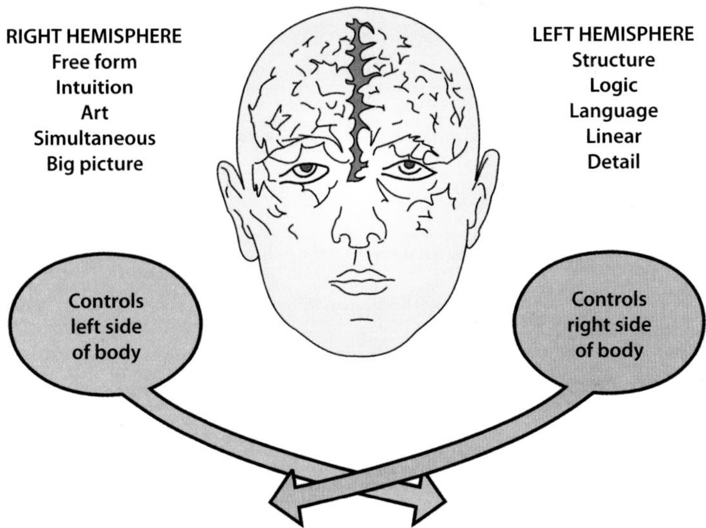

ILLUSTRATION 21-1 Your brain is made up of two halves, or hemispheres, each with its own duties. When you're performing at your best, at your peak, both sides of your brain are operating together, fully integrated.

In fact, sports researchers have shown that one of the most important factors that lead to an athlete performing "in the zone" or "in the flow" is having a fully integrated mind. Between the two hemispheres of your brain is a bundle of nerve fibers called the corpus callosum. This acts as a communications link, transferring bio-electrical current between the hemispheres. It acts like a cable between a computer and a printer. It is as if there is a dimmer switch in this communications link, one that can dial up or down the amount of bio-electrical communication between the hemispheres. When the communication is restricted, you act either more left-brained, or right-brained. When the communication is turned up, your are integrated. That is what leads to great performances; that's when you drive at your best.

Also, the left hemisphere controls the right side of the body, and the right hemisphere the left side of the body. At least, that is the way it should be. Some people, and thus some race drivers—people often referred to as being uncoordinated—do not operate completely in this manner. Instead, their right hemisphere controls the right side of their body, and vice versa, at least partially.

When you are fully integrated, you will think more whole-brained and perform in a more coordinated fashion.

BRAIN INTEGRATION EXERCISES

There are three exercises that will help you improve your level of integration.

Cross Crawl

As I mentioned earlier, the right hemisphere of a driver's brain controls the left side of the body, and vice versa. There is, or at least should be, cross-lateral communication from one side of the body to the opposite side of the brain. This occurs at a high level when you are integrated, and not so much when you are less integrated (this is referred to as disintegrated).

Almost any physical movement that connects one side of the body with the other will help your level of brain integration. However, the simple cross-crawl exercise may be the most effective. Here's how it works.

While standing, raise your right leg, bending it at the knee, and bring your left arm over and touch the right knee. Return to standing. Then raise your left leg and touch the knee with your right hand. Return to standing, and then continue, alternating sides. You will find yourself marching in one place while alternately touching your knees with your opposite hand.

At first, do this at what is a comfortable rate for you, then slow it down to as slow a pace as you can. Doing it at a slow pace puts more stress on your sense of balance, improving it over time. Then speed it up until you are almost running on the spot while touching the opposite knee with your hands. At speed, this is a great exercise to get your body warmed up prior to getting in the car.

There is a reason this exercise is called a cross crawl. When babies first begin to crawl, they most often do it in a unilateral motion. That is, they move their right hand and leg forward, then their left hand and leg, and so on. One side moves, and then the other. After a week or more for most babies, they change to a cross-lateral crawling movement, where they move the right hand with the left leg, the left hand and right leg, and so on. This cross-crawling movement is the first step in the integration process of brain development.

Children who do not do enough cross crawling (often because they go almost directly from unilateral crawling to walking) may miss out on becoming fully integrated at an early age. In many cases, this leads to the child being slightly uncoordinated, or even having what some people call learning disabilities. By simply using the cross-crawl exercise, many children have been able to "recover" from learning difficulties and have become far more physically coordinated. This exercise is extremely powerful.

Do cross crawls for about 30 seconds to two minutes each morning, in the evening, and especially just prior to getting into the race car. After doing this for a few weeks, you will begin to be aware of when you need to become more integrated by doing more cross crawls. You will just feel better—more in the zone—when you're integrated.

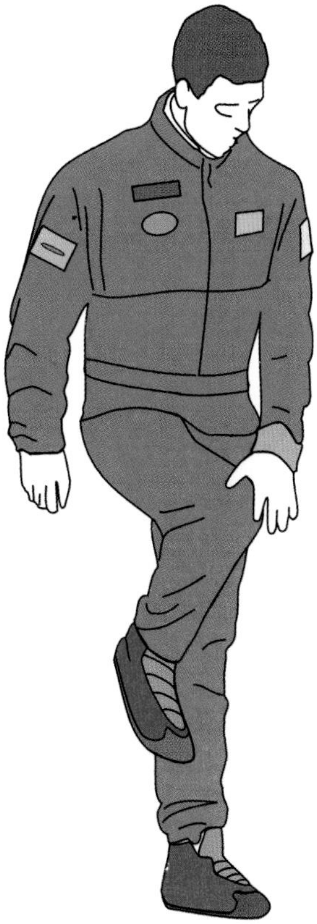

ILLUSTRATION 21-2 Doing cross crawls helps "switch on," or integrate your mind, increasing its ability to process information quickly. Do a few cross crawls just prior to getting into the car and heading onto the track.

Lazy 8s

The second integration exercise is especially effective in helping integrate your vision. Just as there is a cross-lateral connection between your brain and body, there is a similar connection between your brain and eyes. In this case, the information coming into your right eye is sent primarily to the left hemisphere of your brain, and the information from your left eye is sent primarily to your right hemisphere. Once it is in your brain, the information is processed and constructed into what you "see."

If that communication from your eyes to brain, and from hemisphere to hemisphere is restricted in any way, you'll miss a piece of the picture. At the speed you're traveling in the race car, missing just the tiniest piece of information may be catastrophic. And believe me when I say that a large percentage of drivers, even at the highest levels of professional racing, have visual processing problems resulting in incomplete visual pictures. Is it any wonder that many drivers make the wrong decision when trying to cut between a couple of cars in an overtaking maneuver, or that they make a small error (turning in too early for

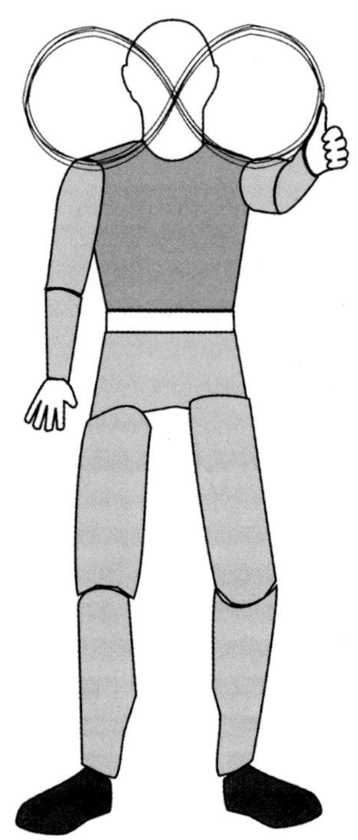

ILLUSTRATION 21-3 Lazy 8s will improve your visual processing, providing your brain with better quality and more visual information. Do Lazy 8s just prior to getting into the car.

a corner, hitting a curb, etc.) that results in a slow lap time or a spin? Either of these may be the ultimate result of a visual-processing problem that can be cured through the use of the lazy 8s exercise.

Here's how it works. Stand with one arm stretched directly out in front of you, with a slight bend in your elbow and your hand in the thumbs-up position. While you keep your head steady, trace an imaginary figure eight laying on its side with his thumb, with your eyes following your thumb. Therefore, your eyes will be tracking this lazy 8 figure.

Do this exercise for about 20 to 30 seconds with each hand, and then with both hands. When doing it with both hands, make two fists, place the knuckles from each hand together, and make a cross with your two thumbs. While flexing your arms and shoulders, trace the lazy 8 while focusing on the cross of your two thumbs. Again, make sure your head stays steady.

At first, have someone watch your eyes closely while you do this. Do they move smoothly or are they notchy? Do they jump ahead in certain areas, skipping part of the figure eight? If so, you may be missing information in that area of your visual field. Do your eyes move congruently (together)?

If your eyes have some notchiness, jumping, or incongruency in your tracking, doing some lazy 8s for even 30 seconds to a minute will probably begin to make some improvement. And even if you didn't notice any problems with the way your eyes track, this exercise will benefit you. Again, it helps with brain integration, and specifically visual integration.

You should do this exercise at least twice a day, and especially just prior to getting into the car. Many drivers report an immediate effect after doing these exercises. They say it helps them become more aware of what is going on around them and much more perceptive. This, obviously, helps improve the quality of visual information being input into his brain.

Most people seem to think that good vision is something you are either born with or not, and that it is something that just goes away with age. And yet, they will agree that if a person does some form of physical exercise that their body will be and stay healthy for a longer period of time. Well, the same thing applies to a person's vision. If you exercise it, it will improve and maintain its health and performance level longer.

Centering

How important is it to the overall performance of the race car that it be well balanced? Critically important, right? But even if the car is perfectly balanced, if your personal sense of balance is not near-perfect, are you going to be able to drive the car to its limit? Or if the car is not perfectly balanced, and neither are you, how effective are you going to be at reading exactly what the car needs?

The point is, of course, your sense of balance is as critical as the car's, perhaps more so. Can it be improved? Yes. How? One way is by centering.

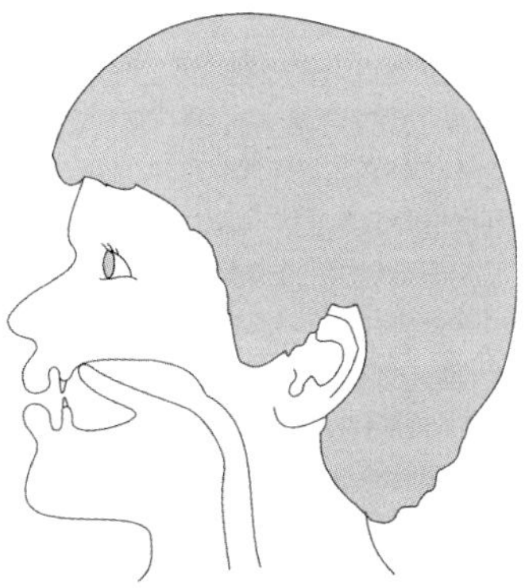

ILLUSTRATION 21-4 To center and calm yourself, as well as improve your sense of balance, gently touch your tongue to the acupressure point on the roof of your mouth, just behind your upper front teeth.

Centering is a technique used in martial arts. All you need do is lightly press the tip of your tongue to the roof of your mouth, toward the front behind your upper teeth, where peanut butter sticks. This area in a person's mouth is a strong acupressure point, which triggers brain integration and an improved sense of balance. To make this fully effective, you should press a couple of fingers from one hand on a point just below your navel, and focus all your energy upon this center point of the body. In martial arts, this point is called the "chi."

Obviously, you cannot drive the car while pressing your navel. But you can place the tip of your tongue on the roof of your mouth, particularly in high stress areas of the track. For example, as you approach the fastest turns on the track, or when trying to brake late for a corner, center (placing your tongue on the roof of your mouth) and breathe. This way, when you most need to be integrated, sensitive to what the car is telling you, and balanced, you are.

This centering technique also has a stress-relieving or relaxing effect. An uptight, stressed-out driver will rarely perform at his peak. By centering, you will be more relaxed, learn at a quicker rate, and perform at your best more consistently.

If you practice centering on a regular basis, you will notice a difference in and out of the car. Just before you leave pit lane to go onto the track, center. Over a short period of time, this will trigger a calm, focused, integrated brain.

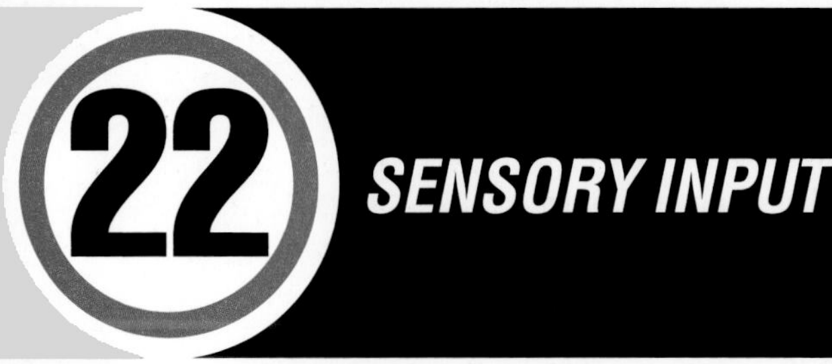

22 SENSORY INPUT

SPEED SECRET

The more quantity and the higher the quality of sensory information going into your brain, the better the quality the output will be and the better you will drive.

Think of it this way. Every little piece of information entering your brain from your senses (primarily vision, feel, and hearing when it comes to race-car driving), results in a decision or physical movement. And just like having more information about the financial performance of a company can help you make a better investment decision, the more information you have about the position of the car in a corner, where other cars are around you, the amount of traction the tires have, the precise speed the car is traveling, the g-forces and vibrations, the sounds from the engine and tires, the better your decisions and physical actions will be.

Most people would agree that hand-eye coordination is a very important part of driving a race car, but few could give you a clue as to how to improve it. Basically, here is what hand-eye coordination is. Information is fed to your brain through your eyes, where it is processed, and then your hand (or any other part of your body) is instructed to perform the appropriate action. From this simple explanation, it is easy to see why any improvement in the quantity or quality of the information going from your eyes to your brain should result in a more "coordinated" action.

In reality, we also rely on hand-ear coordination, where information from your hearing is processed and the appropriate action is then performed by your body, as well as what could be called hand-hand coordination, kinesthetic or feel information providing the input to the brain.

Imagine trying to drive a race car along the ideal cornering line if your vision was restricted 90 percent. Or if your body was completely isolated from the car so that you could not feel any of the vibrations, g-forces, or chassis roll or pitch. Or what if you were deaf? You could not hear any sound whatsoever from the car. Would that affect your ability to drive the car at its limit? Absolutely!

VISUAL INPUT

Have you ever wondered whether what you see is what other people see? Have you wondered whether what you see, or perceive to be the color red for example, is the same as what other people see or perceive as red?

Have you ever wondered whether other drivers see as much or more than you do? Why is it that some drivers seem to be all seeing, all aware, all knowing of everything going on around them, when other drivers seem to have blinders on?

It is a fact that what you see is primarily what your brain constructs. In other words, your eyes send to your brain a small amount of data, where it is turned into a lot of useful information. Most people think it is your eyes that provide you with what you see, where in reality it is more your brain. Vision researchers have proved this point. And that is why some people with 20/20 vision "see" more than others with 20/20 vision. Some older drivers whose eyesight may not be as good as younger drivers', for example, see and are aware of more.

Some drivers see a tiny flash of something in their mirrors and know exactly what it is. For others, that same amount of data from the eyes results in little to no visual construction in the mind. No information is assimilated. That, of course, is why some drivers seem to be able to stay out of trouble; whereas, others seem to be attracted to it. It's just that these drivers are not able to make any useful sense out of a miniscule amount of data being sent to the brain by the eyes.

Think of a driver who has a reputation for making a lot of bad decisions and crashing a lot. To most observers, it is simply a shame that he or she makes so many mistakes, since the driver is so fast, so talented otherwise; the driver is written off as a serious championship contender.

For a driver who makes a lot of mistakes, the root cause of the problem is often a lack of quality sensory input, especially visual input. For instance, where most drivers entering a corner in a pack of cars may recognize that there is not enough room to make a pass, our "crasher" may see it as an opening. The reason is that for a fraction of a second, the driver is not seeing the whole picture. For a variety of reasons, the driver's visual input is restricted. As you know, it would only take a small piece of the puzzle to be missing for an error to occur at racing speeds.

ILLUSTRATION 22-1 Think of your brain as a sponge: Its job is to soak up information about the track and what the car is doing. The more information it has, the better the communication to your body will be. In other words, the more quality information your brain has, the better your performance will be. Be a sponge.

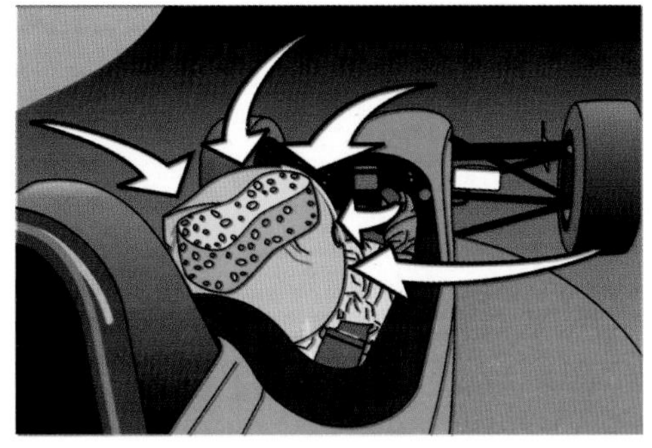

When you think about it with this in mind, you can see why some drivers make more mistakes than others, why they seem to be attracted to problems, and why they seem to make more than their fair share of bad decisions. I have personally witnessed race drivers who have 20/20 vision but whose visual input is significantly restricted.

Fortunately, visual processing is something that can be developed. How? First, by using the Lazy 8s exercise I recommended in the previous chapter. The improvement I've witnessed in drivers who use these exercises on a regular basis is nothing short of amazing. And second, through a short period of sensory deprivation, which can lead to an increase in sensitivity.

Consider for a moment a blind person. Despite their lack of vision, are their other senses (feel, hearing, taste, smell) usually more or less sensitive than people with sight? Better, right? Why is that? Because they have been forced to develop the senses they have.

By restricting some of your senses for a short period of time, you are forced to develop your other senses. Of course, it is not something you consciously set about doing. It is something your mind automatically does on its own. I have, at times, joked about going onto the racetrack blindfolded and how that would improve your other senses, if you survived!

The same type of thing can be used to develop the visual construction process. If you restrict the amount of information your eyes can send to your brain but ask your brain for just as much information output, it is up to your brain to make up the difference. In other words, send your brain a little bit of data and expect your brain to output a lot of information. Now, going on to a racetrack with your vision even slightly restricted is dangerous. But what if you practiced taking the same amount of data from your eyes as you usually do, but asked your brain to output more information for that little data? What you are doing, in fact, is practicing being more aware, practicing using your brain's vision construction abilities, practicing being sensitive to visual data.

It is, in fact, something you can and should practice not only on the racetrack. It is something you should definitely practice while driving on the street, and also in all other activities in your life. For example, while driving down the highway and using your vision as you normally do, ask your brain for as much information as possible. Ask it to be aware of everything along the side of the roadway. Make note of the ground, the grass, and the trees in great detail as you pass it by. But not just the amount of them. Note the colors, the type and amount of leaves on the trees, the condition of the bark, whether the ground is made up of mostly dirt or of rocks, and the speed at which they pass by.

When doing this, though, don't look directly at the ground, grass, and trees. Look down the road like you normally would, but allow your brain to take in more information (actually, construct more information from the data your eyes are supplying to it). Ask your brain to provide more information. There is a physical limit to how much your eyes can take in. It is practically limitless what your brain can do with that information.

What you are doing is practicing becoming more aware of everything around you, using the same amount of visual data supplied by your eyes. Practicing this in your everyday world will greatly enhance your performance on the racetrack.

While driving in traffic on the street, practice being aware of every car, truck, pedestrian, and anything else. The more you practice this, the more aware you will be of other cars around you on the track, without having to put much, if any, concentration on it. The less focus and concentration you put into noticing where your competitors are, the more you have to put into more important things like track conditions, reference points, and speed and traction sensing.

One of the more difficult visual challenges you face is seeing around corners. Often, your view through the turns is restricted. What you need to do is see *around* the corner.

SPEED SECRET

Practice—every day—being aware.

A number of years ago (in the early to mid-1990s), when Al Unser Jr. was at the top of his game, I noticed how his head was turned and cocked to the side on the approach to a corner when racing on street circuits. It was as if he was trying to stretch his neck to peek around the cement wall on the inside of the turn. I wasn't sure if this was something Unser did on purpose, or whether he did it at all, but it sure looked like it. Now, I think it was something he did without realizing while trying to stretch *his vision*, not his head or neck, around the corners. I wonder if this is one of the reasons Unser was so dominant on street circuits for so many years.

If you consciously try to stretch your vision—try looking as far around the corner as possible, even if it means using your imagination—over and over again when practicing, eventually it will become a habit, or a mental program. Then, it will be something that you do without any conscious thought, like Al Jr. did. It is as if you are building a mental picture to fill in the holes in the visual picture.

KINESTHETIC INPUT

Feel and auditory input is similar to visual input in that most of the information is constructed in your brain. If you were to practice feeling things with your hands over and over, do they become more sensitive? Yes and no. Actually, your hands themselves do not become more sensitive, but your brain becomes better at constructing the feelings from the same amount of data sent to it. So, in the end, yes you do become more sensitive, but that's because your brain has become more sensitive.

I had a dramatic demonstration of this while conducting an Inner Speed Secrets seminar. As part of a demonstration of the importance of sensory input, and for a little lighthearted fun, I ask two participants to compete in a race. Not a car race, though. The competition is to see who can pull on a pair of women's pantyhose in the least amount of time, while blindfolded and wearing thick ski gloves. As you can imagine, with no visual input, and little kinesthetic input, this is a real challenge (and a few laughs for the rest of us).

I had become quite accustomed from previous seminars to how long it takes a person to complete the competition. Then, in this one particular seminar, a participant completed it in less than half the time it normally takes. It seemed as though he was pulling on the nylons without the gloves. It wasn't until the end of our little race that the participant told us that he was a dental surgeon and worked all day long at delicate maneuvers while wearing gloves and not being able to see well. To him, even through the thick ski gloves, he had some sensitivity. That sensitivity had been developed through years of working with gloves with restricted visibility.

If you were to practice driving with thick driving gloves, and then switch to ones that allowed more sensitivity when it came time to really perform (such as in qualifying or the race), your kinesthetic sensitivity would be enhanced. Therefore, your performance would improve.

The real point is, again, that your sensory input can be improved and developed; and that the more it is developed, the more sensitivity you will have to control your race car at the limit. The key is to practice being aware. Many people go through life without really being aware of what is going on around them, what they can really see, feel, hear, smell, and taste.

AUDITORY INPUT

The same is true of your auditory sense. Try practicing while using heavy-duty earplugs, greatly restricting your auditory input. Then, go back to your regular earplugs and notice how much more auditory input you receive.

Imagine driving your race car with overly effective earplugs, ones that blocked out almost all the sound. You are driving the track, shifting up and down through the gears, the engine revving, the tires growling, the brakes grinding. But you can barely hear all this auditory activity. You strain to hear the engine, having to rely more on the tachometer than you have in a long time to determine when to shift. There is auditory data going into your brain, just not as much as usual. You strain your hearing again to take in as much as possible.

By the end of the session you've got your driving rhythm back, you've learned to adapt to the lack of auditory input. The fact is, your brain is extremely adaptable. In that short session, it has learned to perform nearly at the same level as before you restricted your auditory input. It has learned to be more sensitive.

Now, back in the car for the next session but this time with your regular earplugs. These restrict the noise just enough to protect your hearing from damage but still allow an abundance of auditory input. In fact, you may not have even realized how much sensory input you took in through your hearing before. But now you do. You are hearing the crispness of the engine's throttle response like never before. You had never noticed that sound from the tires before as you cross the concrete patch through the middle of the corner. What does that tell you about the tires' grip level? It changed, didn't it?

Wow! What a session! What a performance! You were magic in the car. It was as if things just happened; you didn't need to try to go fast. It was easy. That's what happens when you crank up the boost on just one of your sensory inputs.

Again, what you are doing is forcing your brain to work with restricted sensory input. Then, when it has gotten used to constructing the information with little data input, give it back all the sensory input you can when it really counts.

A word of warning here, a serious one. Driving a race car, or just hanging around the track, without adequate hearing protection is a big mistake. In just a short amount of time you can permanently damage your hearing. And you should now know just how much your driving performance will be negatively affected if your auditory input is reduced. Before you get any ideas of heading out on to the track with little or no hearing protection, don't!

SENSORY INPUT SESSION

There is a relatively simply way of improving the quality and quantity of sensory input you take in when driving. I call it Sensory Input Sessions. Of all the "tools" I've used when coaching drivers, this is definitely one of, if not the, most effective. How does it work?

First, go out onto the track for a session with the sole objective of taking in more sensory input. The best way of doing this is to determine how much time you have to do this exercise, then split that up into three sessions. These sessions should be at least 10 minutes in length, but not much more than 15.

In the first session, simply focus on everything you can hear. Focus on the engine note, the sound coming from the tires, noise from the brakes, wind noise, and so on. Take in everything auditorially, whatever you can hear.

For the second session, focus on kinesthetic input, everything you can feel. You should notice all the vibrations through the steering wheel, pedals, and seat; the amount the car pitches, rolls, and squats; if the steering wheel gets lighter or heavier as the tires reach their limits; any vibration or chattering of the tires as they corner at the limit; and the g-forces working against your body.

The third session has you taking in everything visually. You should focus on what you can see, on being more visually aware of everything. Focus on discovering track surface irregularities; what you see on the horizon; to notice any vibrations and movements of the steering wheel and other parts of the car; to expand your view to take in more in your peripheral vision; and, if driving an open-wheel car, any changes in the surface of the front tires.

To make this most effective, come into the pits after each session and debrief. Ideally, describe to someone what you heard, what you felt, and what you took in visually. Prod yourself for as much information and feedback as possible by asking yourself questions. If you do not have the time to come in to debrief after each session, use some form of communication to know when to change from auditory, to kinesthetic, and to visual. This can be done either by use of a radio, or by a signal from a lap board. Either way, sit and write down what you heard, felt, and saw at the end of the sessions.

CHAPTER 22

ILLUSTRATION 22-2 Sensory Input Sessions are one of the best tools you can use to improve your ability to sense the limit, and therefore, drive at the limit. During these short sessions, you have one simple objective: to soak up as much sensory input as possible. Break it down into three sessions: one for visual information, a kinesthetic session, and one for soaking up more auditory information.

This is not a one-time deal. This should be done often, especially after switching to a new car or setup. It should definitely be a part of your routine for learning a new track. The ultimate goal is to become more sensitive to all the sensory inputs. This will help you learn a new track quicker and become better at sensing when you're driving at the limit, and it will provide you with much more feedback for developing the car's setup.

Ultimately, there are three main benefits for Sensory Input Sessions.

First, as I mentioned earlier, the better the quality and the greater the quantity of sensory information, the better the quality of your performance. Any time you focus on one specific sensory input, the more sensitive you will become. It is much like a person who loses his sight. By focusing on and isolating the other senses, they become much more sensitive.

The second benefit of Sensory Input Sessions is that they stop you from trying to drive fast and from thinking too much. Trying to go fast never works. Race cars are way too fast to drive at the conscious, trying level. They must be driven at the subconscious level, with the conscious mind observing and being aware.

Often, what you may need is a way to "distract" your conscious mind from trying to drive fast. And what better distraction than having the conscious mind focused on providing the brain with more quality sensory input?

I was once coaching a driver on an oval track who was running lap times in practice that were about four-tenths of a second off where he had been the day before. Worse, this was after making a number of changes to the car to make it better. With the engineer on the radio telling him how many tenths off the quickest car he was, and the team owner telling him to carry more speed into turn three, the driver was trying hard to go fast. But he wasn't. Finally, as he came out of turn two, I got on the radio and asked him to simply focus on what the car felt like for the next four laps. Within two laps he was back down to the times he had done the previous day and was providing great feedback on the car that the engineer could really use to develop it.

I probably could have asked the driver to tell me what he ate last night, and it would have had much the same effect. No, it would not have provided his brain with more quality sensory input, but it would have gotten his conscious mind focused on something other than trying to drive fast. If you can learn to recognize when your driver is trying too hard, and all drivers do at some time, you can use this technique to great effect (though I do recommend asking your drivers to focus on sensory input, not on what they ate last night!).

Third, Sensory Input Sessions reduce the number and the extent of errors, both short-term and long-term. How?

Can you think of a driver who has a reputation for making "bad decisions" in the car? Often, the reason a driver makes poor decisions is because he lacks the information on which to base the decision. It's like trying to make the decision to invest in a stock without having any past financial statements or annual reports.

If you dive down the inside of two other cars on the entry to a turn, with no hope in you-know-where of making it (and crashing), you and others may say you made a bad decision. You may wonder, "What was I thinking?"

If you really want to know why you crashed, you have to dig to the core of the problem. You may think the core is that you just made a poor decision. But the reason, the cause, of the poor decision may be a lack of good information, a lack of quality sensory input.

In this example, what you saw as a large enough gap to make the pass was not. You didn't have all the information. With more quality sensory input, your decision making will improve, whether you currently make good decisions or not.

Sensory Input Sessions can also minimize the effects of errors. Do you think experienced, champion race drivers make any fewer errors than inexperienced drivers do? I don't think so. The only difference is the experienced driver is better at minimizing the effects of them. I have definitely witnessed and experienced this myself.

When the experienced driver makes an error, such as turning into a corner too soon, he or she recognizes it immediately, makes a small subtle correction, and makes the best of the situation. When a less-experienced driver makes the same error, the driver may not recognize it until he or she is passing by the apex. At that point the correction is going to have to be much bigger, sometimes causing a further problem, or at least having a drastic negative effect on the lap time.

So how does the experienced driver minimize the effects sooner? By recognizing the error sooner. How? By having more reference points. Most drivers have three reference points for each corner: turn-in, apex, and exit. Great race drivers, whether they are conscious of it or not, have dozens of reference points between each of those three. To become a great driver, you need to practice sucking up more information about the track, so that you see much more than just three reference points for each turn. You need an almost continual path of reference points. And these need to be in your mind at the subconscious level. That way, if you turn in too late for a corner, you recognize this just a foot or so later, at the subconscious level, rather than when you are all the way to the apex. The sooner you recognize it, the more subtle and effective the correction will be, and the less negative impact it will have. At that level, many drivers are not even aware that they made an error.

As you already know, the better the quality and the more quantity of sensory information you put into you brain, the better your performance will be. As I said earlier, this is similar to but opposite of the computer slogan, GIGO—garbage in, garbage out. In this case, it is quality in, quality out.

When should you use the Sensory Input Sessions? Often. When I've suggested to drivers to use these exercises, they sometimes claim that they do not have time. After all, they only have one practice session and then qualifying, and they certainly don't want to "waste" that time just taking in sensory input. Wrong. That is exactly the time to focus on sensory input. The goal is to learn as quickly as possible, and this is one of the best ways of doing that.

SPEED SENSING

One of the most amazing things a race driver does is determine the speed the car needs to be traveling when entering a turn, and then slowing it to that exact speed. We all do this type of thing every time we come to a stop at a red traffic light. We look ahead and make the decision to begin braking *now*, with *this* amount of pressure, and therefore we will stop at *that* point up ahead. No one tells us when to begin braking; there are no brake reference points on the street that I've seen.

This is even more difficult when we are not coming to a full stop; we're slowing to a specific speed, one that only your "gut feel" can tell you is at or near the limit of traction. The great drivers do that within a fraction of a mile per hour, consistently keeping the car at the very limit. This wouldn't be so difficult, or amazing, if the driver had the time to look at a speedometer while driving into the corners, but the driver doesn't.

If you used a radar gun to measure what speed a group of drivers entered one particular turn at, you might be surprised. As I said, the great drivers would consistently enter the turn at a speed that did not vary much more than a mile per hour. The not-so-greats' entry speed would vary by as much as 5 miles per hour or more!

Of course, I'm talking about slowing and setting the entry speed accurately and consistently so that the car is at or very near the limit. Almost anyone can consistently set the entry speed at something 10 or 20 miles per hour below the limit.

Until you can consistently set your entry speed within a mile per hour or two, you will never be able to begin shaving the last few tenths or hundredths of a second off your lap times. So, your ability to sense speed is critical. Developing this without miles and miles of track time is not an easy thing to do; however, there are a couple of exercises you can use to fine-tune your speed-sensing abilities.

Speed sensing, particularly as it applies to the entry phase of a corner, covers a couple of areas. First, it is having the innate ability to accurately determine the ideal speed to slow the car to when entering a turn. Now, understand this does not mean knowing the car has to be traveling at 88.3 miles per hour when you reach the turn-in point. Obviously that wouldn't do you any good, as it is not possible to look at a speedometer when you are just about to enter a turn. That is why I say it has to be an innate sense.

The second area is the ability to consistently adjust the car's speed to the appropriate level for entering a turn. Just knowing deep down inside how fast you should be entering the corner doesn't help if you cannot tell the difference between 88.3 and 82.1 miles per hour. Great race drivers can sense the difference in speed within 1 mile per hour. The superstars are far more sensitive than that. And they can consistently dial the car into that speed.

Developing your speed-sensing abilities is not an easy thing to do without miles and miles of track time. However, there are a couple of exercises I've come up with that will enhance and speed up the process of developing them.

The first is done in your street car on the street. All you are going to do is practice estimating speed, based simply on sensory input, not the speedometer. Cut out a piece of cardboard that you can easily slip in place to cover the speedometer, then go for a drive. As you are driving along at say, 55 miles per hour, slip the cardboard cover in place. Then change your speed a few times by speeding up and

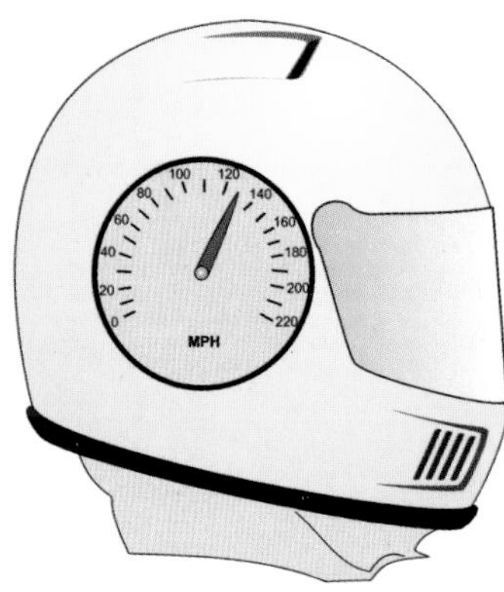

ILLUSTRATION 22-3 You must have an innate sense of the speed you are traveling and how much it needs to be altered in preparation for each and every corner.

slowing down, and finally by trying to put the car back to 55 miles per hour again. Pull the cardboard cover off and check to see how accurate you are. Do it again and again.

An alternative method is to simply leave the cardboard cover in place and pick a speed you want to travel at. Then, accelerate to what you feel is that speed, and pull the cover off and check how well you did.

If you do these exercises over and over again, you will become very accurate, and most important, consistent at judging and establishing a specified speed. And no, it doesn't really matter that you are not at the same speed you will be at on the racetrack. The main objective is that you can consistently get so that you can set the speed of the car the same over and over again, within 1 mile per hour or so, simply using sensory input as your guide. That is accurate and consistent speed sensing.

Another technique to improve your speed-sensing abilities requires a radar gun, someone to operate it, and you driving your race car. Choose the most important corner on the track and have your assistant with the radar gun positioned so that he or she can check your speed just as you turn into the corner (using a pylon or pavement marking as a reference point). Take a couple of laps to warm up, and then drive 10 laps with the main goal of entering the turn at exactly the same speed. Of course, it does no good to drive slowly during this exercise; you should be within a couple of tenths of your best lap times. Have your assistant radio to you the speed you were traveling at just as you turned in to the corner.

The goal, of course, is to consistently be at the same speed as you turn in to the corner. If your corner-entry speed varies more than a mile per hour, you really need to practice this more. Ultimately, you should be able to enter every corner on a racetrack at the same speed for at least 10 laps in a row, within 1 mile per hour.

Your assistant should then ask you to increase your corner-entry speed by 2 miles per hour. See if you know what that small increase feels like. Try 1 mile per hour less. How does that feel? The objective is to calibrate your speed sensing with reality. If you determine that increasing your Turn 3 corner-entry speed by 1 mile per hour is desirable, now you will have a better idea of what that feels like. You will have a better chance of going and doing that, not increasing your entry speed by 4 miles per hour, but by the 1 mile per hour you wanted.

Of course, this can also be done with a data-acquisition system, although the feedback is delayed. The lack of instant feedback is certainly a drawback. With instant feedback, your mind learns more quickly. The real-time feedback from an assistant is more effective.

TRACTION SENSING

Traction-sensing skills are one of the key differences between a truly great driver and all the rest.

To be able to drive at the limit, and to use every bit, but not any more, of the tires' traction, you must be able to feel or sense how much traction the tires have. I know that sounds pretty obvious, but that is what traction sensing is: the ability to sense at any and all points around the racetrack exactly how much traction the car has. Put another way, it is the ability to sense if and when the car is at the limit, the traction limit.

The one question I'm asked more than just about any other by new and relatively new race drivers is, "How can I tell exactly when I'm driving at the limit?" It is perhaps the most difficult question to answer, for knowing precisely when you are driving at the limit is, besides being the key to driving at the limit, an innate feel that one develops. I don't believe it is something that a person is either born with or not. Yes, some drivers seem to have a more natural feel or instinct for it, but with any driver, it can and must be developed.

Where does this ability to sense how much traction the tires have come from? Primarily from your senses, and specifically your senses of feel, vision, and hearing.

By simply being aware of the tires' traction all the time, including when you are driving on the street, your traction-sensing skills will improve. In addition, there are a few specific exercises you can use to develop your traction-sensing skills.

Perhaps the best all-around exercise for developing your raw traction-sensing skills is still a skid pad. For all the money that drivers and teams spend on practicing and testing, it seems ridiculous that little, if any, is spent on something as simple and effective as skid pad training.

As part of my development program for a Formula Atlantic driver I worked with one year, we did a skid-pad session. Although it was rather short, it was one of the most beneficial bits of training we did. The driver's understanding of how to control understeer and oversteer was really enhanced, as was his traction-sensing sensitivity. You may be thinking that you fully understand how to control understeer and oversteer already, and that may be true. So did this driver. But, it is not until you physically practice over and over adjusting the throttle and steering input that you truly understand it. Overall, I would estimate that his car-control skills improved by at least 50 percent after that one skid-pad session.

You do not need a full-blown skid pad to do this type of training. As I did with the Atlantic driver, all you need is a large, smooth, paved parking lot, some way of wetting it down (we hired a water tank truck to intermittently spray the area), and some cones. Set the cones up to form a circle at least 50 feet across. Then, drive your race car around the circle faster and faster until either the front or rear tires begin to lose traction. On a skid pad like this, you should be able to hold the car in a steady-state understeer or oversteer slide for at least three or four laps of the circle. In other words, you should be able to keep the car in an oversteer drift, with the tail hanging out and you controlling it with throttle and steering lap after lap. The same is true with understeer.

To make this exercise most effective, it may be necessary to fiddle with the car a bit. I've gone as far as running rain tires on the front and slicks on the rear, and vice versa. Usually though, just adjusting or disconnecting anti-roll bars are enough. Your objective is to be able to exaggerate the car's ability to understeer and oversteer.

I am a big believer in using street driving to develop your race driving skills. And, you don't need to be driving at anywhere near racetrack speeds to do this. In fact, driving fast often defeats the purpose of what you are trying to accomplish. Unless you are driving at speed, at the limit, the correlation between street and track is not there. And only an idiot would drive that fast on the street. What you are trying to do is program specific skills and techniques in a relaxed, unhurried atmosphere. That way, when you are on the track, these skills and techniques come naturally, without any conscious thought whatsoever.

SPEED SECRET

Use your street driving to make you a better track driver.

One of the first things you can do while driving on the street to enhance your traction-sensing skills is simply to make note of the tires' traction. Do this by paying attention to the noise coming from them and by the feeling through the steering wheel. Notice how both these factors change when going from a straight line to cornering. Yes, on the street, the noise and feel will be minute, subtle things. But if you can read the tire traction at this level, sensing it at racetrack speeds will be easy.

I'd like you to try an experiment for me. While driving on the street, try holding the steering wheel with a tight grip, with your whole hands wrapping around the wheel rim so that your palm is in contact with the wheel. Notice the vibrations back through the wheel. Next, hold the steering wheel with just your fingers, with a light, relaxed touch. Now notice the vibrations through the wheel. Which provides the most feedback? In which way do you feel the most vibrations? With the light touch of the fingers on the wheel, right?

Does this tell you something about how you should hold the steering wheel? I hope so. If you practice holding the steering wheel in your street car with your fingers, with a light touch, that will become a habit, a program. Yes, I know that some race cars require more of a grip of the wheel than what just your fingers can apply. But if you make a light touch a habit, you will apply the lightest touch possible on your race car steering wheel. And that will lead to increased sensitivity and increased traction-sensing abilities.

In all the coaching I've done, there is one exercise that has made the single biggest improvement with the drivers I work with: Traction Sensing Sessions. All you do with this exercise is dedicate part or all of a practice or test session to simply

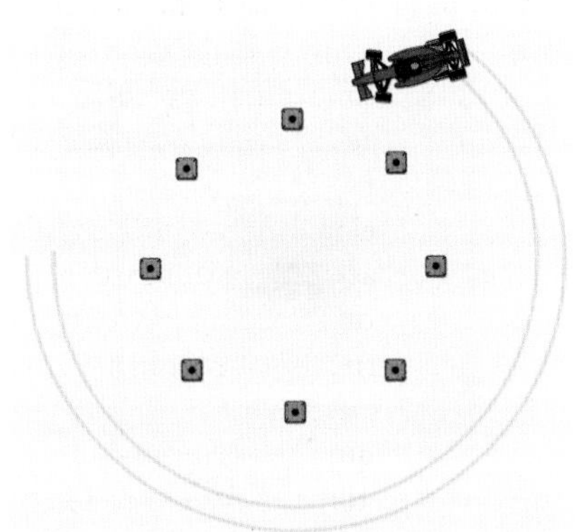

ILLUSTRATION 22-4 You can set up a makeshift skid pad in a large paved parking lot to develop your traction-sensing and car-control skills. Use eight or more cones to describe a circle at least 50 feet in diameter, then add water, and go out and play; I mean practice.

focusing on sensing the tires' traction. While driving on the racetrack, make note of the vibrations and feedback through the steering wheel. Does the steering get lighter or heavier as the tires slide more? Make note of the sound coming from the tires. Do they make more or less noise as they slide more? Overall, how does the car feel as the tires begin to slide more and more? How much warning do the tires give before they start to slide too much? Forgetting practically everything else—and especially lap times—practice reading how much traction the tires have around every inch of the track.

You may even want to put a 1 to 10 rating scale on it, with 10 being the limit of traction just before the tires start to let go, and 1 being the grip they have while going down a straightaway. Then, as you drive around the track, you can actually call out to yourself the amount of traction the tires have.

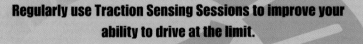

SPEED SECRET

Regularly use Traction Sensing Sessions to improve your ability to drive at the limit.

If you use these techniques on regularly, I guarantee your traction-sensing abilities will improve, and that will lead to your ability to drive more consistently at the limit.

23 MENTAL PROGRAMMING

Everything you do while driving is a result of either the mental programming you have, or a lack of mental programming to do something. The same can be said of everything we do in our lives.

An example is throwing a ball. At an early age, you observed someone throwing a ball; then, maybe one of your parents tossed a ball to you and asked you to throw it back. Rather crudely and without coordination, you managed to toss the ball in some direction. At that point, a neural pathway formed in your brain, representing the physical act of throwing. You threw the ball again and the pathway became a little stronger; you threw again and the pathway grew stronger yet again, and so on.

The first few times you threw the ball, you had to consciously think about how to do it. At some point, when the neural programming became strong enough, you no longer had to think about it. You just automatically, subconsciously, ran the mental program and threw the ball.

The same is true of the techniques required to drive a race car. At first, while you are learning or programming the technique, you are consciously thinking about how to do it. Then, with repetition, your brain forms neural pathways or programs, allowing you to head out on the track and simply execute the appropriate program at the appropriate time.

Think of it this way. Imagine taking a cup of water and pouring it on top of a big mound of dirt, letting the water run down the hill. The first time you do this, the water will try to follow the path of least resistance and begin to make a shallow pathway. This is much like the neural pathway in your brain after doing something for the first time. It's there but not well established. The second time you pour a cup of water on top of the hill, it may follow the same path, or it may find an easier, even more natural pathway. If it follows the same path, that pathway will become deeper and more ingrained, just like the neural pathway in your brain after doing something twice. If it takes another path, then it begins the path-building process all over again.

Now, imagine pouring that same cup of water on the top of the hill a few thousand times a year, for more than 20 years. The pathway would be extremely well routed. It would almost be impossible for the water to follow any other pathway. This is what my own personal neural pathways were like for the mental program that operated my right foot when upshifting through the gears. After more than a quarter of a million repetitions of quickly lifting and then planting my foot back down on the throttle to make the upshift, I had to change that program when I first drove a race car with a "no lift shift" electronic engine-management system. Instead of briefly easing off the throttle to make the shift, as I'd always done before, I had to keep my foot flat to the floor and just pull back on the sequential shift lever.

Needless to say, I found it difficult to not lift my foot at first, and for good reason, right? After all, with that amount of repetition, that strong a mental program, that deep a neural pathway, it was almost as natural a movement as breathing.

The good news is that for years that program was so well developed that I never had to give it even a fraction of a second of conscious thought. That left my conscious mind open to being used for more important things, like considering what a change in my cornering line might do, how a shock absorber adjustment might help the car's handling, or where my competitors were in relationship to me. That is why it is so critical for the basic driving techniques to become habit or mental programs, to allow your mind to concentrate on far more important things.

SPEED SECRET

Drive in your mind before driving on the track.

Now, the bad news. Any technique that has been programmed into your brain can be difficult to change, as I discovered when having to learn to upshift without lifting off the throttle. Do race cars change? Do track conditions change, requiring different techniques? Do all race cars react the same and require the same driving technique? The answers to these questions are, of course, yes, yes, and no, meaning you have to be able to change or alter your mental programs quickly and efficiently.

More good news: mental programming can be changed. How do you do that? Through the deliberate use of what most people refer to as visualization, but what is really mental imagery. Why the distinction between visualization and mental imagery? Because visualization, by the very definition of the word, uses only one sense, vision, in your imagined experience. Mental imagery uses imagined visual, auditory, and kinesthetic sensory input.

ILLUSTRATION 23-1 Everything you do behind the wheel of a race car is a result of one of the countless number of programs in your mind. The key is selecting and fine-tuning the right program for the task.

Jacques Villeneuve had an interesting remark when asked to comment on then-20-year-old Jenson Button's signing to the Williams-BMW F1 team for the 2000 season (*On Track*, February 17, 2000):

> *F1 is 10 times more physical to start with, and then there's the speed. The first time you drive cars that quick, everything happens so fast. Your heartbeat goes up 20–30 bpm just because of it. It takes time to adjust.* You spend more time thinking about what to do rather than just doing it. . . . *You have to be able to adapt right away, but for it to become natural you need mileage. You can do quick laps, but* unless it's natural *you can't work properly on the setup and you can't do a whole race.*

By "unless it's natural," Villeneuve is referring to driving at the subconscious level. And until the act of driving at the limit becomes a subconscious action, part of your conscious mind will be used for thinking about what you are doing, rather than being aware of more important things.

DRIVING SUBCONSCIOUSLY

It is not possible to drive a race car effectively (read: fast, at the limit) by consciously thinking about each movement, maneuver, and technique. A race car is much too fast to allow you the time to think through each function. Your conscious mind cannot react and respond quickly enough to physically operate the controls of a race car at speed. It must be a subconscious act.

To do this, you have to program your mind, just like a computer. How? By practicing, both mentally and physically. At first, it is a conscious act. Your conscious mind tells your right foot to move from the throttle to the brake pedal, your arms to begin arcing the steering wheel into a corner, and so on. But, after doing this particular function over and over, it becomes programmed into your subconscious mind. Then, when required, it just happens automatically, without actually thinking about it.

It's the same as going to the refrigerator for a drink. You don't have to consciously think to stand up, move your left leg in front of your right leg, the right leg in front of the left, and so on. You've done it so often, it's a subconscious act.

When you drive at a subconscious level, it allows your conscious mind to "watch" what you are doing, to see if there is anything you can do to improve technique-wise, or sense what the car is doing handling-wise. As you drive subconsciously, by your "program," your conscious mind watches, senses, and interprets what you and the car are doing, and then makes changes to the "program" (subconscious) to improve. There is no point in continuously driving subconsciously if your "program" doesn't have you driving at the limit. Your conscious mind must always be working at reprogramming or updating your mind's "program," your subconscious.

That is why it is important to start off slowly when learning a new track or car, gradually building up your speed. It allows the conscious mind to keep up to the speed of the car, while it programs your subconscious.

There are times when I'm out on the track and I'm not really even thinking about what I'm doing; I'm just driving. I come in and I can't actually remember what I did. Obviously, the car has to be working well or I'll be thinking too much about it, but I will have more concentration on what the car is doing and therefore be more sensitive to what the car is doing.

Again, as I mentioned earlier, this programming can be done with actual physical practice, or by visualizing it. But it does take some time.

MENTAL PRACTICE

Your brain does not distinguish between real and imagined occurrences. Fortunately for you, it sees and accepts all images as if they were real. Therefore, it makes sense to "visualize," imagine, or mentally practice driving. Not only is it free, but it may be the only place where you can really drive a perfect lap.

In your mind's eye, see yourself repeatedly driving exactly the way you want: driving the perfect line, balancing the car smoothly at the very limit, making a well-executed pass.

Mentally drive the race car, but do it successfully. It's amazing how often an error in a driver's mental visualization of a lap actually happens. So visualize yourself doing it right.

Visualization, or mental practice, is so effective for a number of reasons. First, it's perfectly safe. You can never hurt either the car or yourself. Second, you can visualize anywhere. You don't need a racetrack or a car. And because of that, it's free. I don't need to remind you how important this is.

Next, there's no fear of failure. You always drive perfectly, always as you wish. You can even win every time out, if you wish.

You can visualize in slow motion. This gives you time to be aware of each minute detail of the technique, perfecting it before heading out on the track.

You can mentally prepare for things that may only happen once a season or so. But when it does, you're ready for it, and you can respond in the best way possible. For example, you can visualize the start of a race, "seeing" different scenarios: someone spinning in front of you and you reacting to it; a driver moving to the inside of a corner to block you from passing and you setting up to accelerate early and pass him on the exit of the corner.

When I raced Formula Ford, a fellow competitor and I were good friends. We battled really hard with each other on the track, maybe even harder than against other competitors because we could trust each other. Then, after the race was over and before the next one, we would spend hours talking about the various passing moves we made, others made, and what we could have done if the situation had been different. We didn't realize it at the time, but we were helping each other visualize racing strategy and techniques. We literally practiced thousands of passes. We drove hundreds of races that season in our minds. The result was, when we were in a race, we made quick, aggressive, decisive passes. And they were easy because we had practiced them so many times. We won a lot of races.

And finally, by visualizing prior to heading out onto the track, it automatically forces you to focus and concentrate.

ILLUSTRATION 23-2 If you can't do something in your mind, you'll never be able to do it physically. Prior to hitting the track, take a few minutes to imagine every detail you can about what you want to do.

I like to use a stopwatch to time my visualization laps. If I knew the track well, my mental lap times would be within a second of my real lap times. That told me I was visualizing accurately, which meant I was probably going to be very fast.

Of course, before visualizing, you must have some kind of "feel" for what you are doing. There's no point in visualizing yourself driving a car or track you've never actually seen before. Without some prior knowledge, some background information, you may just be practicing something the wrong way. Remember, visualizing an error is practicing an error. Practicing an error is a sure way of ensuring you will repeat it.

As you turn into a corner, have a mental picture of where you want to be at the exit. You can't get somewhere if you don't know where it is you're going. In fact, as I mentioned earlier, one of the most common errors—turning in early to a corner—is usually caused by not knowing where you want to be at the exit.

Visualization is programming, programming your mind just like you would a computer. And programming allows you to drive using your subconscious mind instead of your conscious mind.

MENTAL IMAGERY

Building a mental model through mental imagery is something that practically every superstar of every sport does. Do you? Do you want to be a superstar? Do you just want to improve your abilities and have more fun? Either way, mental imagery can provide you with a model of how to do things, when to do things, and even why to do things.

Mental imagery is an extremely powerful technique that results in the development of mental programs. These mental programs, then, allow you to do things without "consciously thinking" about them. You do them "automatically," by habit. Just like launching a software program on your computer, a mental program can be launched or "triggered" when you need it.

Just as you have a mental program for walking, and therefore do not have to think about how to walk, you can develop a mental program for the act of driving a car and then rely on that program to drive the car. In fact, that's the goal: to get to the point where the act of driving or being a race driver is a program, where you no longer have to consciously think about what needs to be done, you just do it. And the reason you "just do it" is because you've developed a mental program that resides in your subconscious, and you've triggered or launched it.

Of course, you can develop mental programs to do something through physical practice. That's what seat time does. It develops the habits or programming to do things without having to think about them. However, there are a few problems with relying *only* on physical practice to build your mental programming:

- You'll spend a lot of money.
- You'll use (and maybe waste) a lot of time.
- You'll make mistakes part of your programming. Every time you make a mistake, you've made that mistake part of your mental programming. Remember, practice does not make perfect; only perfect practice makes perfect. If you practice making mistakes, you'll only get better at making mistakes.
- You'll find it difficult to physically change your driving. If you're trying to do something that you've never done before, it can be almost impossible to do it physically. As proof, have you ever known that a particular corner can be taken at full throttle, but you can't get your right foot to go along with what your mind knows? The only way you can change or develop the program to take the corner at full throttle is in your mind, not on the track.
- It can be risky. Experimenting with a new technique could lead to a miscalculation and result as a big "off."

Practicing the same techniques mentally, through the use of mental imagery, costs nothing. It doesn't take much time, and you can practice everything perfectly, improving your ability to do it perfectly on the track.

Many people will ask, "But if I've never done it before, how can I even imagine doing it, much less create mental imagery of it?" That's a good point. Without some idea of what doing "it" looks, feels, and sounds like, it's hard to imagine it. That's one of the reasons I've written this book: to help give you the look, feel, and sound of doing things perfectly.

But how effective is mental imagery compared to actually, physically doing something? As an example of the power of mental imagery, I'd like for you to read the following italicized narrative at least three times. After you've read it, close your eyes, breathe deeply and slowly, relax, and then imagine the scenario that you've just read.

To begin, make yourself comfortable, with your hands resting in front of you. Close your eyes. Breathe deeply, taking nice, slow breaths. Relax your body. Allow your muscles to relax. Feel your body sink into the chair. Feel your body get heavy and relaxed. Hear your heartbeat slow down. Continue to breathe slowly and deeply. If you should feel yourself start to drift off to sleep, just take two or three quick, deep breaths and that will bring you back to a relaxed but awake state. Breathe slowly. Relax your muscles.

Breathe. Relax.

Imagine a bright yellow lemon sitting on a table in front of you—a bright, shiny, yellow lemon.

Now, imagine picking that bright yellow lemon up with both hands. Feel the texture of the skin and shape of the lemon. Notice how bright the yellow is.

Imagine placing the lemon back on the table in front of you. There is a knife sitting on the table. Pick it up. Place the blade on the lemon and slice it in half, hearing the sound of the blade slicing through the lemon.

Notice the juices dripping onto the table. See the lemon juice on the blade of the knife. See the lemon in two halves, with juice on the table around it.

Pick up one half of the lemon and give it a squeeze. As you feel the lemon squish, notice the juices on the face of the lemon, dripping back onto the table.

Bring it up to your nose and smell the lemon. Breathe deeply as you smell the scent of the lemon.

Now, bring the lemon to your mouth, stick out your tongue, and slowly lick the juices off the face of the lemon. Taste the juice.

Continue to taste the lemon juice in your mouth.

Okay. When you're comfortable, slowly open your eyes as you mentally come back into the room.

Again, once you've read this narrative three times, close your eyes and imagine going through it in your imagination. Try to imagine as many of the details that you read as you can. See, feel, hear, smell, and taste the scenario.

What happened? What did you experience? Did you experience anything, like your mouth puckering up? Did you have saliva build up in your mouth? Yes? If you're like most people your mouth began to salivate. Why? Because your brain can't tell the difference between a real and an imagined event. Because your brain thought there was real lemon juice coming into your mouth and your brain then triggered saliva to water down the citric acid of the lemon.

This is a simple example of the power of mental imagery. It's why superstar athletes (in fact, anyone who depends on performing at a high level) use it. It's why, if you want to make a change in your behavior, or improve or develop a skill, using mental imagery is a critical step, perhaps the most important.

Let's look at some background on mental imagery. Of the many research studies and examples of the impact of mental imagery, I've selected three:

- Hunter College basketball players study: A group of basketball players were asked to shoot free-throws, and the success percentage was measured. They were then split into three groups. The first group was told not to practice whatsoever, physically or mentally. The second group was asked to practice daily by actually shooting free-throws. The third group was asked to not physically touch a basketball, only to do mental imagery of shooting perfect free-throws each day. When the players' free-throw percentage was checked a week later, the results were interesting. The first "no practice" group showed no improvement whatsoever, which was no surprise. The second group, the ones that had physically practiced every day shooting free-throws, improved their shooting accuracy 23 percent. And the last group, the ones who did not touch a basketball but did mental imagery each day? Well, they improved their free-throw shooting 22 percent. Without touching a basketball, they improved 22 percent, essentially the same amount as the group that physically practiced each day.

- Soviet Union Olympic study: During the 1980s, the then-Soviet Union's Olympic team tested various training procedures. Athletes from various sports were split into four groups. The first group trained entirely with physical practice 100 percent of the time. The second group used physical training 75 percent of the time and mental imagery 25 percent of the time. The third group split its training 50-50, and the fourth group spent 25 percent of the time training physically and 75 percent of the time mentally. The mental training required them to spend time every day practicing their sport in their minds using mental imagery. At the end of the study, the group that had the made the biggest gains or improvement was the fourth group, the one that had spent only 25 percent of the time training physically and 75 percent of the time using mental imagery. Interesting, the group that trained 100 percent of the time physically actually improved the least amount.

- While this is not a formal research study, the following true story provides a great example of the power of mental imagery. An American prisoner of war who was held captive for five years loved golf. It was his passion; it was his favorite past time, prior to the war, of course. During the time he was held captive, he mentally played a couple of rounds of golf every single day. And he played them perfectly. He saw the green of the grass and the shots he hit. He felt his swing, the connection with the ball, and he even felt the way the grass felt underneath his shoes as he walked the course. He heard the sounds of the birds in the trees, the wind, the sound of his club hitting the ball, and the sound of it soaring down the fairways. He imagined every last detail of what his perfect round of golf would look, feel, and sound like. When released from prison, one of the first things he did was hit the links. Despite not having touched an actual golf club in more than five years and only having played golf mentally through that time, he shot the best round of golf he ever had.

For many drivers, visualization is used to familiarize themselves with a track and possibly to prepractice a specific technique before hitting the track. For these drivers, they are missing out on some of mental imagery's best uses. Overall, mental imagery can and should be used for the following:

- **To see success**: You can develop your belief system by recalling past success and preplaying success in future events. Your belief about your abilities may be the most important key to success, more important than natural or developed skill, and it can be improved upon with mental imagery.

- **To motivate**: By recalling the emotional feelings of past successes and imagining them for future events, you can remind yourself of what you enjoy most about racing. When things are not going well (as they often do for every driver at some point in their driving career), focusing on what you truly get out of racing can lead to a superior performance.

- **To perfect skills**: If you vividly imagine the look, feel, and sound of performing a skill, the likelihood of you doing it on the track is improved. These skills may be a specific driving technique, an interpersonal skill, or just about any other skill, technique, or act you need.

- **To familiarize**: You can use mental imagery to drive thousands of laps of a track to learn it for the first time, refresh your memory of it, or fine-tune the details of it. It can also be used to preplay a meeting, a speech, a media interview, or any other activity, helping you to feel more at ease when the real situation occurs.

- **To trigger a performance state of mind**: By vividly recalling your feelings of a past success, it's almost impossible not to get into a great state of mind. Over time, and by building in a "trigger" word or action, you can literally say a word and get into the ideal state to perform at your peak.

- **To program behavior**: You need to behave in different ways, in different situations. By preplaying these situations and adapting your behavior, you improve your ability to act in the ideal manner, that is, more aggressively, more patiently, or in a more outgoing manner when the need arises. Mental imagery programs the ability to adapt your behavior to suit the situation.

- **To preplan**: Although there are infinite possibilities that can happen in a race, preplanning for as many of them as you can will allow you to act more quickly, more accurately, more confidently, and with more ease. For example, preplaying a number of scenarios that could occur at the start of a race will help you develop the attitude that "it doesn't matter what happens; I'm ready."

- **To refocus**: If you form a mental image of yourself dealing with problems on the track, especially the problem of losing your concentration and then immediately refocusing and continuing, you will develop a program for doing this. When it happens on the track, it will be much easier to regain your focus.

Sports psychologists define two different types of mental imagery: cognitive and motivational.

Cognitive is essentially focused around forming mental images of techniques and strategies; for example, the line, braking zones, trail braking, sensing the limits of the car, racecraft, and so on. Motivational is focused on your belief system, state of mind, ability to focus during a race, pushing hard when laps are down on the field, mental toughness, control or use of emotions, and the "rewards" that come from performing the skill or technique well. Each is equally important. In fact, every time you do mental imagery, there should be a balance between cognitive (technique-specific) and motivational (relaxed, balanced, confident, enjoyment-specific).

Ultimately, these two types are further broken down into the following:

- Cognitive general
- Cognitive specific
- Motivational general
- Motivational specific

To give you an idea of how you would use each type of mental imagery, take a look at this chart.

	Motivational	**Cognitive**
Specific	Goals and goal attainment: setting mental objectives and goals for an event or session and having them be more than just something you've thought about at the conscious level; they've become a part of your mental programming.	Rehearse specific skills: the line, braking points, throttle application, applying what you've felt from the car to a chassis setup change, and so on.
General	Arousal control, self-confidence, mental toughness: where you develop mental programming of your beliefs (confidence), your state of mind, your behavioral traits, as well as feeling the rewards of a job well done.	Rehearse race strategies: how you handle the start, problems with handling throughout the race, pit stops, and so on.

Mental imagery can also be "associated" or "dissociated." Associated means you see, feel, and hear yourself in the very act; that is, you're behind the wheel and your view is from this vantage point. Dissociated is as if you're seeing yourself from above or from a camera view, not from behind the wheel. For some reason, some people naturally do mental imagery from an associated perspective, while others do it from a dissociated perspective.

Which is best, associated or dissociated? Some people will tell you that neither one is better than the other; however, I disagree. While there is nothing wrong with dissociated mental imagery, it's best to program your mind from the perspective that you'll experience when driving, from behind the wheel. Having said that, doing some mental imagery from a dissociated perspective, as if you're watching yourself perform from a TV camera's viewpoint is also valuable. The key is to make your mental imagery as real as possible by seeing the view from the cockpit, by feeling the car and motions from the seat, and hearing everything from behind the wheel. A message you'll hear over and over from me is that the more realistic you can make your mental imagery, the more effective this whole process will be.

The more senses you include in your mental imagery, the more effective it will be. Notice in the lemon example that you used all five senses: you "saw" the lemon (visual sense), you "felt" it (kinesthetic), you "heard" the knife cutting the lemon (auditory), you "smelled" the lemon juice (olfactory), and you "tasted" the lemon (taste). But did you really see, feel, hear, smell, and taste the lemon? Only in your mind, right? Only in your imagination. By involving all five senses, you made the experience real in your imagination. If you had only "seen" the lemon, using only your visual sense, it's likely that your mouth would not have salivated because you would not have made it real enough to your mind.

Obviously, your smell and taste have relatively little to do with driving a car, but certainly visual, kinesthetic, and auditory have a lot to do with it. Most people who claim to do mental imagery really only visualize. That is, they only imagine the visual scene in their minds; they do not imagine what they feel and hear. And that's the difference between mental imagery and visualization; mental imagery involves more than just your visual sense, and visualization doesn't.

In addition to your visual, kinesthetic, and auditory senses, experience all the emotions and feelings that you possibly can. The more you tie your emotions and feelings to each mental imagery session, the more real it will become to your mind. More important, the easier it will be to trigger the specific mental program in the future, on and off the track. This motivational piece of mental imagery is critical to your success.

Prior to beginning a mental imagery session, know exactly what you want to accomplish. This is a good case for writing out a narrative, even if it's a simple bullet-pointed list. By doing this before beginning, it's more likely that you will stay focused throughout the imagery session. This book is full of examples of narratives. You can make them as detailed as I've made them here, or you can make a few short bullet points to help you remember the key points you want to program.

Many drivers talk to me about not being able stay focused on the specific scenario that they want to imagine and program for long. Some drivers have a difficult time doing a specific mental imagery scenario for much longer than a minute or two. If that's you, know that you're human. Everyone I've ever talked to about imagery admits that they are similar. The ones that continued to practice imagery got better and better, to the point where they may be able to stay focused on imaging a scenario for up to an hour or even more. The drivers who get frustrated that they can't do it perfectly the first time tend to quit and, of course, never improve.

Emerson Fittipaldi, the year he won the Indianapolis 500, spent more than three hours sitting in his car in the garage the day before the race. He claimed that if he couldn't imagine sitting in the car for that long there was no way he was going to be able to drive the entire race without losing his focus.

Mental imagery takes practice. Don't expect instant results. It's best to perform mental imagery at least twice a day, once in the morning and once at night, in addition to the specific sessions you do at the track. Also, have a plan for the type of skill or technique you're going to work on. Mondays could be for programming specific driving skills, Tuesdays for working on your beliefs about your abilities, Wednesdays for race strategy, Thursdays for your overall state of mind, and so on. Like any skill, and mental imagery is a skill, be patient and your ability will improve. It's an acquired skill.

Which brings me to a major and common misconception. Thinking about driving the car is *not* mental programming; it's not mental imagery. Thinking about something is done at the conscious level, which is not the most effective state for your mind to be in to truly learn something, for it to become part of your mental programming. Until you allow your brain to get into the right mental state, you're not using mental imagery to program your subconscious.

So what is the ideal mental state to be in to program your mind? Let's start with a little background or theory. Doctors and researchers define four brainwave states, as measured by an electroencephalograph (EEG). By attaching a few probes to your head, the EEG can "read" the bio-electrical activity going on inside your head and therefore measure the brainwaves. These brainwaves are broken down into four levels or states:

- Beta is where your brain is primarily producing brainwaves in the 13- to 25-hertz (cycles per second) range. When you're in a conscious, thinking, active state, like when you're reading this book, your mind is in the beta range.

- Alpha is where your brain is primarily producing brainwaves in the 7- to 13-hertz range. Alpha is when you close your eyes, relax, and begin slowing down your mind. While getting to this state, you should feel your body relax, your muscles letting go, and your body sinking into its chair or seat.

- Theta is where your brain is primarily producing brainwaves in the 4- to 7-hertz range. Just before you fall asleep you pass through a state where you can feel yourself drifting off, but you're barely aware that you're doing this. You may also have flashes of odd images in your mind. This is theta.

- Delta is where your brain is primarily producing brainwaves in the 0- to 4-hertz range. When you're asleep, your mind is producing mostly delta waves.

Notice that I never said that your brain is only producing one level of brainwave at a time. No, it's always producing some amount of all four, but it's the concentration that matters. When consciously awake and conversing, thinking, reading, and driving a car, your mind is producing mostly beta waves, some amount less of alpha, less theta, and much less delta. When asleep, you're mostly producing delta, less theta, less alpha, and even less beta. In other words, your brain is practically always producing some amount of all four levels, but the concentration of the various levels change depending on what state you're in.

What does this all mean?

For your brain to be in the most receptive, effective programming state, you want to relax your mind enough so that it is in what is called an alpha-theta state, where your brain is primarily producing waves in the 6- to 12-hertz range. The best way to describe this state is that your mind is not busy, it's relaxed, you haven't drifted off to sleep yet, but if you let your mind relax and slow down much more it would.

If you simply close your eyes and consciously think about something, your mind will be in a beta state, and the effectiveness of your mental imagery will be much, much reduced. By allowing yourself to get to an alpha-theta state by relaxing your mind and body, everything that you imagine will become imprinted within your brain much more deeply and strongly. This is how you program your mind.

Prior to beginning a mental imagery session, take a few minutes to allow your mind to relax and slow down, to get into an alpha-theta state. If you do, whatever you focus on will seem more real to your subconscious mind and therefore become more deeply programmed in your mind. From there, your ability to reproduce it on the track is much greater.

Of course, getting your mind into an alpha-theta state at home lying on your bed is relatively easy. But getting there while in your trailer or sitting in your race car at the track is more difficult. Why? Because of the distractions and the stress at the track. Fortunately, like most things in life, the more you practice, the easier it will get.

Have you ever installed a piece of software on a computer and then couldn't locate the icon to click on to the launch the program? Imagine the frustration! How useful would a computer program be without an icon or trigger to launch it? Not very, right? The same thing applies to a mental program.

Imagine if you did mental imagery over and over again and developed the programming to flat the one fast corner that you've never been able to take at full throttle before. You can see it, feel, and hear it so clearly in your head. Then, on track, you can't locate that mental program in the vast hard drive in your head. Imagine the frustration. It's the same thing, isn't it? As important as mental imagery is, it's just as important to develop a trigger to launch this program. As an example, you could do this by simply saying the word "flat" in your mind as you imagine doing it, over and over again. The more you do this, the more your mind will associate the word "flat" with the program, eventually getting to the point where you almost can't help but hold your foot at full throttle when you say it. It's just like Pavlov's dog.

Even when preplaying a success to develop your confidence or motivation, focus your mental imagery on the act or performance. Sure, see yourself be successful, but focus on what led to the success: the way you felt, the way you performed, the state of mind you were in, and the actual skills and techniques it took to get there.

One last point about mental imagery: Don't expect it to compensate for a lack of knowledge or for hard work and practice. While it can and will make a huge difference to your performance, it can't perform miracles.

TRIGGERS

Imagine installing the latest, most powerful software package into your computer, but it doesn't have an icon on the desktop or start menu to access it. That is exactly what it would be like if you had the ideal program for driving, but not a "trigger." A trigger is an action or word that allows you to access, or activate, a mental program. Without it, you could have all the mental programming in the world but will never activate it. A trigger word or action works just like a gun's trigger. Once used, it fires the appropriate program.

Trigger words or actions should have some special meaning, or generate a vivid mental image. As an example, to trigger a program to suck up sensory input, you may use the word "sponge" to see yourself as a sponge, sucking up all there is to know. Other trigger words I've used either myself or with other drivers are "car dancing," "watch this," "party time," "play time," "time to kill," "crank it up," and "step it up." Trigger actions could include giving the steering wheel a quick squeeze, looking at a specific sign or message on the dash, or a hand signal from your crew. You may need to spend some time discovering just the right word, phrase, or action to use as a trigger word.

Then, when you are using mental imagery to program something, initiate it with a specific trigger word or action. That way, when you are in the heat of the battle, the second you say the trigger word or use the action, the program will kick in.

THE PLAN

Okay, with that background, here's the plan I'd suggest you follow.

Seven days a week (you can take a day off if you want, although I bet some of your competitors won't and that means if you take a day off, they are getting an advantage on you) you need to spend a minimum of 20 minutes, twice a day doing a mental-imagery session. You can do one in the morning and one just before bed at night, one in the afternoon and one before bed, one after dinner and one before bed, or whenever. But there should be at least one hour in between the two sessions. You should also determine one consistent place to do them. Preferably this is not lying on your bed, as there is too much of a tendency to fall asleep while doing it (if you feel yourself drifting off to sleep, take two or three deep, quick breaths). Sitting in a chair is good, and in your car is even better. You want to make sure you are comfortable and relaxed, it is quiet, and you will not be disturbed.

If you can do it in your race car, it's even better if the car is up on jack stands, as you can then turn the steering wheel freely. Remember, the more you use your body, the more the session will program muscle memory to do things right when you're on the track. If you don't use your car, use as many "props" as you can to make it as realistic as possible. You can wear your helmet, and you can hold a real steering wheel. You can ask someone to sit on the floor in front of you and use their feet as pedals. This is a very effective technique, as you can even have the other person provide feedback to you on the amount of pressure you're applying and the progressiveness of the application. Sitting in one of the latest driving simulators is fantastic since you can begin by driving a track and then go into your mental imagery session. Of course, you have a steering wheel and pedals to use with a simulator, as well. Again, the more realistic you can make your imagery session, the more effective it will be.

When you're ready to begin a mental imagery session, read over the narratives that you've prepared. The reason for writing out a plan for your narratives is to make sure you stay on task. By that, I mean staying focused on what you're working on: being the perfect driver, dealing with big moments, driving at the limit, working on a specific track or technique, or whatever. The only way for this programming to be really effective is to repeat it enough times. Repetition is critical. If you begin to stray off of what you've outlined in your narratives, you will not be repeating the session enough times.

After you've read through the narrative and you've got it memorized enough to be able to follow it over and over again for 20 minutes, begin to prepare for your imagery session. Get yourself in position. Close your eyes, and breathe deep and slow. Relax. Feel yourself relaxing. Feel your muscles begin to let go. Feel yourself begin to sink into your chair or seat. Listen to your heartbeat slow down. Notice your breathing slow down, become even more relaxed. Notice the images of you as a relaxed person sinking into your chair or seat. Breathe. Relax your muscles. Breathe.

Your objective is to get into that alpha-theta state. It's where your mind is slowed down and receptive. You may notice some odd and unrelated images flashing through your mind. It's near that stage just before you fall asleep, but you're still awake enough to be aware of what's going on around you. It can take anywhere from two to five minutes to get to this state (it will probably take less and less time with practice).

And to stress the importance again, the more senses you involve in mental imagery, the more effective the programming will be. Use your hands; hold the steering wheel (even if it's an imaginary one). Move your feet; feel the pedals. Hear the engine, the wind noise, the brakes, the tires.

Breathe. Relax. As you do each of these imagery sessions, continue to breathe and be relaxed. Part of what you're programming is the ability to feel relaxed and breathe normally throughout these scenarios.

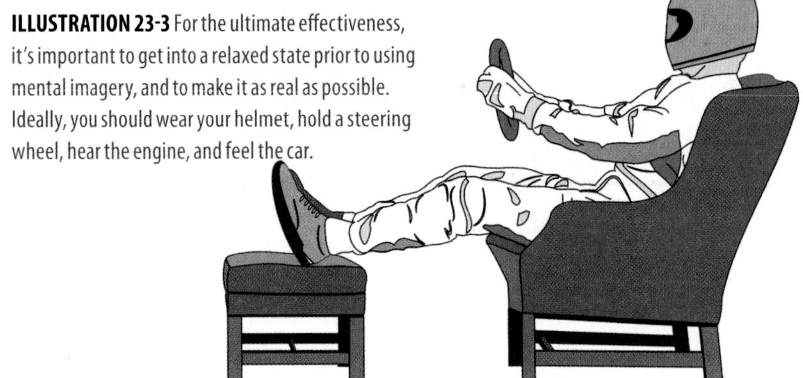

ILLUSTRATION 23-3 For the ultimate effectiveness, it's important to get into a relaxed state prior to using mental imagery, and to make it as real as possible. Ideally, you should wear your helmet, hold a steering wheel, hear the engine, and feel the car.

Breathe. Relax.

You're now ready to begin a mental imagery session.

COMPUTER SIMULATION

Years ago, Jacques Villeneuve may have done more for the computer game and simulation industry than most people realize. You may have read that in his first season in Formula One he used computer games extensively to help learn the tracks that were new to him. Prior to going to Spa, which is generally accepted as the most difficult Grand Prix track, for the first time, he practiced by driving lap after lap on the computer. What happened when he got there for the race? With a limited number of practice laps, he qualified on the pole. At that moment, race drivers around the world ran out and bought more computer games.

I have to admit to not having spent much time with computer games, although I would like to do more, especially with the latest stuff that's out there. In fact, these computer games, and more realistically, simulations, can be a useful adjunct to mental imagery.

I believe these simulations can be valuable in helping a driver develop a virtual-reality visualization, although there are a couple of limitations. First, of the three sensory inputs a race driver relies on (visual, auditory, and kinesthetic), simulations do a good job with two (visual and auditory), and limited, if any, job of the third (kinesthetic). And second, how much does driving an F1 car at Spa help a Formula Ford driver at Thunderhill? Mental rehearsal of driving one type of car on a specific track does not necessarily apply to another.

Having said that, there are things that a "sim driver" can practice that will help once on a real track. He can practice the ability to focus concentration for a period of time. He can also develop a fine sensitivity and control of the steering wheel. And finally, he can practice learning. As I said earlier, it doesn't matter whether you are driving a 900-horsepower Champ car, a 6-horsepower kart on an indoor track, or a computer simulation, the process of determining what works, what doesn't, what effect a change in technique has, or whether a change is necessary is one of the things that separates the great race drivers from the rest.

Earlier, I talked about how a driver must be able to adapt his personality and behavioral traits to suit various situations, to control and trigger the ideal state of mind, and make quick, appropriate decisions. If

ILLUSTRATION 23-4 Computer games and simulations are becoming more and more realistic, and therefore they are useful training tools. You can use them to develop and fine-tune your reactions, learn new tracks, refresh your memory of a track, practice techniques, develop your ability to focus for long periods of time, and build mental programming.

used correctly, a racing simulation can certainly help a driver mentally program and develop these abilities.

Overall, I think simulation is another tool that race drivers can and should use to develop and maximize their on-track performance. Just as only using physical practice on a racetrack, simply studying track maps, relying solely on mental imagery, or practicing only in go-karts will limit your ability to learn, only using simulations will not result in you becoming the next world champion. But combined with all the other tools a race driver has today, simulations are valuable.

EXPECTATIONS AND POSSIBILITIES

Expectations can be dangerous things. Possibilities and potentialities can be marvelous things.

Expectations do not have any direction. It is like me saying, "I expect to be in New York City." That certainly doesn't get me to New York City, does it? No, there is no plan, strategy, or direction. If I say, "My goal is to get to New York City," it naturally leads me to developing a plan to get there.

If you expect a particular result, and it does not occur, you become frustrated or disappointed. Neither feelings help get you any closer to the result you were looking for. If you focus on your goals, such as performing at your very best, you then have a direction to follow, one that will more likely lead to the result you were looking for.

Expectations can also be limitations. Going into qualifying, for example, you think that if you turn a 1:20.5, that will put you on the front row. You head onto the track and do a 1:20.8, a 1:20.6, and then a 1:20.5. What are the odds of you going much quicker? Not good. After all, you matched your expectations. Consciously, you may not be satisfied with the time, but if you put a time into your subconscious, your mind will do what it takes to match it, and not go beyond it. But what if the track conditions changed, for the better? Some tracks change significantly throughout a race weekend, becoming faster with each session. Perhaps a 1:20.5 would have put you on the front row based on a previous session, but only put you on the fourth or fifth row in qualifying.

Karl Wendlinger, who was severely hurt driving a Sauber F1 car at the 1994 Monaco Grand Prix, provides a perfect example of this in Christopher Hilton's book *Inside the Mind of the Grand Prix Driver.* In this case, he is referring to when he went back to testing for the Sauber team in 1995, having been away from the cockpit for almost a year.

> *Because I'd only done a little driving I had time to prepare my body, to do training. I did a lot of concentration exercises. Then I arrived at Mugello. It was a two-and-a-half-day test and, the evening before, I thought, "OK, one minute 30.4 would be a good time." I concentrated, I closed my eyes and as I crossed the line into the lap I started my stopwatch. I did a whole lap in my brain and looked. The stopwatch said 1:30.4. The next day on the track I did 1:30.4.*

Then I said to myself, "It was too easy, tomorrow you have to do 1:29.3." The best lap that Heinz-Harald (Frentzen, Wendlinger's teammate) did in Mugello all year was 1:29.0 and because I had done so little driving I thought 1:29.3 was competitive. I sat in the hotel again, closed my eyes, and started the watch. I "drove" the lap and looked. 1:29.3. Next day I did 1:29.3. You know the best thing was—and this fascinates me about what you can do with your brain—I only did the 1:29.3 because *I made a mistake and lost three-tenths. If I hadn't made that mistake I would have done 1:29.0, but the evening before I had fixed in my brain 1:29.3 and not zero, and that's what happened. If I had fixed 1:29.0 maybe I wouldn't have made the mistake and I'd done the time easily.*

Like I said, expectations are limitations, and you rarely exceed your expectations. Wendlinger proved that, along with the power of mental imagery. Expectations program results into your mind, and your mind is very efficient at running those programs, sometimes too efficient! In Wendlinger's case, it sounds as if had he expected to turn a 1:29.0, he would not have made the error that ultimately cost him three-tenths of a second. That's the (negative) power of expectations at work.

SPEED SECRET

Delete your expectations.
Focus on your possibilities.

THOUGHTS

Behind the wheel of your race car, have you ever thought to yourself, "that was a dumb move to make" or "why did I turn in so early for that corner"? Did any of those thoughts do you any good? I doubt it. In fact, I bet they did more harm than good. If you are going to have some thoughts while driving (and there is no doubt you will, at least I hope so!), make them nonjudgmental thoughts.

Is there any danger in thinking about the past? There certainly can be. The nanosecond that you focus any amount of your attention on what has happened in the past, that is some attention that is not being spent on what is happening right now in the present.

Can you do anything about what has happened in the past? Absolutely not. If you make a mistake going through Turn 2, does thinking about that while heading into Turn 3 help or hinder? Hinder for sure. Does getting upset about how a competitor shut the door on you entering Turn 5 help? No. Like I said, the second you make an error, forget it. What you did, or what another driver did, is not important now.

Can you do anything about what is going to happen in the future? Yes. How? By what you are doing right now. When you focus your attention on the present,

ILLUSTRATION 23-5 What you believe about yourself is the single biggest limitation to your performance. Until you can honestly, deep down inside, believe you can take the pole, it's unlikely you ever will. Fortunately, you can build up your beliefs about yourself through the use of mental imagery.

you increase your chances of performing at the level that will result in the goals you have set.

Having no thoughts at all is far more desirable than having a mind full of thoughts. In the practice of Zen, you are encouraged to have an empty or beginner's mind. A mind full of thoughts is one that will not react instantaneously and naturally.

In his book *Zen Mind, Beginner's Mind,* Suzuki Roshi writes, "If your mind is empty, it is always ready for anything; it is open to everything. In the beginner's mind there are many possibilities; in the expert's mind there are few."

The same thing applies to the use of mental imagery or visualization. In your mental imagery, you want to actualize yourself as being open and ready for anything. Many drivers have asked me how they can visualize the start of a race and predict every possible scenario that could happen. "You can't," is my response. The same with trying to mentally prepare, using visualization, to drive a car you have never driven. How can you visualize something you have no idea of what it could possibly look like?

Instead, you need to use a kind of open-ended mental imagery, one that sees you as ready for anything. For example, mentally see yourself at the start of a race; if you can't go to the inside, you go outside and make a pass. If you don't get a jump on the field, you make up for it in the second half of the first lap. It's not that you imagine every possible scenario but that you are ready and make the right moves, no matter what happens.

Michael Jordan, one of the greatest athletes of all times, would recall images of past successes in high-pressure situations. As Jordan's Chicago Bulls coach Phil Jackson says in his book *Sacred Hoops*, "Jordan doesn't believe in trying to visualize the shot in specific detail. 'I know what I want the outcome to be,' he says, 'but I don't try to see myself doing it beforehand. In 1982, I knew I wanted to make that shot [the last-second shot he used to take his University of North Carolina team to the NCAA Championship]. I didn't know where I was going shoot it or what kind of shot I was going to take. I just believed I could do it, and I did.'"

That's open-ended imagery. That's having an empty mind, a beginner's mind, one without expectations.

24 STATE OF MIND

How important is your state of mind to your level of performance? Critical, right? Unfortunately for most drivers, their state of mind is something that just happens, and they have little to no control over it. In other words, they either get into a great state of mind or not, and it is almost totally by accident. It is rare for a driver to have a defined process or ritual for triggering the ideal performance state of mind.

Your state of mind covers many areas: your level of anxiety, happiness, anger, nervousness, fear, passion, enthusiasm, empathy, and so on. As I'm sure you already know, these states of mind play a huge role in your performance.

Where does our state of mind typically come from? Out there. Things that happen to us, things that people say to us, external happenings, whether positive or negative. And what do we typically focus on when we're in a poor state of mind? The rotten state of mind we're in. We say things like, "I'm just in a bad mood today; I got up on the wrong side of the bed this morning." And of course, that just leads to a further decline of our state of mind. In other words, it spirals out of control.

You can hope that you show up at each race in the right state of mind, but hope is not an effective strategy. Or you can learn to trigger a performance state of mind.

Your emotional or mental state of mind must be controlled if you want to be successful. If you are excited, nervous, depressed, stressed, distracted, angry, or whatever, you may not be mentally effective. Your decision making will be slowed; your mind will not be focused.

You don't need to be psyched-up. You need to be calm, relaxed, and focused.

ILLUSTRATION 24-1
Recalling the images of past successes can help lead to future successes.

Psyching-up usually makes you overly excited, and therefore, less effective. You want to drive with a "clean mind," not one cluttered with useless thoughts.

Once you get into the car, it doesn't matter what is happening outside the car. All that matters is you, the car, the track, and other competitors. Forget everything else. I think this is why many drivers find racing so relaxing. They can forget absolutely everything else that is happening in their lives.

PERFORMANCE STATE OF MIND

How do you induce a performance state of mind? The best technique is to simply ask yourself to recall, and replay in your mind, a great past performance. The interesting thing is that this great performance does not have to be while racing. It could be anything, from playing another sport to a positive business experience, or while participating in a hobby or a great personal relationship experience.

Anything that results in being extremely positive, happy, energized, and calm will do the trick. I use this technique often when coaching drivers, particularly just prior to a qualifying session. I like to find out beforehand about a great performance in the driver's past and then ask him to tell me about it just before qualifying. I've had drivers relate stories about a past hockey or soccer game, a previous qualifying session or race, or a positive business experience. In each case, I could see it on their faces that they were in a positive, performance state of mind after telling me their story.

SPEED SECRET

Replay a past success to trigger a performance state of mind.

Another technique that I like to use with drivers is to have them walk to their cars like they imagine a champion would. In other words, act like a Schumacher, a Johnson, a Franchitti. I'm sure you have noticed this yourself, how some drivers just look like they are there to win, and others do not. Most of that is in the way they present themselves, the way they walk.

If you model your walk after Schumacher, for example, your state of mind cannot help but be closer to ideal. Used in conjunction with relating a past experience, it is an extremely effective tool.

25 DECISION MAKING

Your decisions in the car cannot be made at the conscious level. If they are, you will make a lot of bad decisions. The reason, of course, is there is not enough time to make them at the conscious level. They must just happen, being made at the subconscious level.

Remember the information I presented earlier about how quickly your brain processes information—2,000 bits of information per second at the conscious level; 4 billion bits of information per second at the subconscious level. Any wonder why decisions, and just about everything else done in a race car should be accomplished at the subconscious level?

If you consider what it takes to make a good investment decision, one of the keys is having as much quality information as possible. The more, and the higher the quality, the better. Sound familiar? Many bad decisions made by race drivers are a result of a lack of quality information. Where does that information come from? Visually, kinesthetically, and auditorily—sensory input.

If you can increase the amount and improve the quality of the sensory input heading into his brain, you will make better decisions on the race track. Fortunately, that can be done. How? Sensory Input Sessions.

Sensory Input Sessions can do more to improve your on-track decision making than anything else. The more reference points you have, the sooner you can make adjustments; the smaller the adjustments need to be. The more sensory information, the more clear the picture of what's going on is, and therefore, the better your reactions.

26 FOCUS

As you read this sentence, do not think about a pink elephant. I said, do not think about a pink elephant! So what are you thinking about? A pink elephant, right? In fact, it is impossible *not* to think about something. The only way your driver cannot think or focus on something he or she doesn't want is to think or focus on something he or she does want.

For example, I have had the unenviable experience of coaching a driver who had a crew chief who would often say something along the lines of "Don't crash the car this time" or "Don't worry, if you crash I can fix the car." He would say this to him just before driving out of pit lane! Now, I know that this seems like an extreme example, but this happens more than you can imagine. In this case, what do you suppose the driver's brain was focused on? Crashing, right?

ILLUSTRATION 26-1 You need a strategy, or program, to not think about "elephants" while driving. Develop and program a preplanned thought, something that you can quickly trigger and focus on at any time while driving.

Obviously, it would be ideal if everyone around your driver could be aware that what they say can affect the driver's performance, but that is not always practical. Your driver must have a plan or strategy to manage whatever anyone says or does.

In the case of not *not* thinking about something someone says, the strategy is fairly simple. To demonstrate, imagine a blue elephant. Whenever anyone says "pink elephant," imagine a blue elephant. What you have done is developed a preplanned thought. Now, when I say, "Don't think about a pink elephant," what do you think about? I hope you thought of a blue elephant. If you didn't, you need to practice this some more.

The point is you must have a preplanned thought ready and willing to take on any unwanted thoughts thrown at you, either intentionally or not. You need to develop a Preplanned Thought (PPT), and practice using it. Perhaps your PPT could be similar to mine. I use "car dancing." Whenever anyone says anything that could distract my focus, or get me focusing on something I don't want, I simply say "car dancing" to myself. When I say that phrase, I immediately conjure up an image of driving a car at the limit on a wet track (which I just happen to love doing). Through years of practice, this image is a strong and vivid one for me, one that will take the place of practically anything anyone says or does.

SPEED SECRET

Develop and use a Preplanned Thought (PPT).

By the way, the more meaning the PPT has for you, the more effective it will be. That is why "car dancing" works so well for me but may not for you. To me, "car dancing" provides me with the image of me and the car smoothly and precisely flowing through the turns on a racetrack, dancing with the car.

This same theory can be applied to where you look when driving. Particularly if you're driving on an oval or street course track, thinking to yourself not to look or think about the walls lining the track will not do any good. When you say, "Don't think about the walls" to yourself, your mind only really registers the "walls" part of the message. And the amazing thing about the mind is that if you put an image or thought into it, it will find some way of making it happen. Even if that means driving into the wall.

So instead of thinking, "Don't look at the wall," you should think, "Look at the line I want the car to follow." The only way of not thinking or looking at what you don't want to think or look at is to think or look at what you do want. I know that is a mouthful, but it is completely true. Go back and read it again.

CONCENTRATION

Concentration is the key to consistency. When you lose concentration, your lap times begin to vary. In my early years especially, I would always check my lap times after a race to see how much they varied. If I could run an entire Formula Ford race with each lap within half a second, I was happy with my concentration level.

When you physically tire, your concentration level suffers. If you notice your lap times have slowed and become erratic near the end of a race, it may be that you became physically tired and began to lose concentration. Many drivers blame the car at this point, claiming the tires "went off," when in fact it was their concentration level that went off.

Running alone, just trying to make it to the finish, is a time in a race when many drivers lose concentration. That is when it is most important to concentrate. Often, at that point in a race, I find it best to talk myself around the track. What I'm actually doing is reprogramming my mind again. Usually after a couple of laps of talking myself around, I'm back to driving subconsciously.

Using my PPT really helps my concentration too.

There is a limit to how much a driver can concentrate on, though. You can easily spend too much concentration on one particular area when you really need to spread it over two or three areas. But when trying to go faster, work on one concentration area at a time. Don't go out on the track and try to "go faster everywhere." Your brain cannot handle everything at once. Instead, decide on two or three areas at most—two or three of the most important things that will make you faster—and work on them.

It takes more concentration to keep something from happening than it does to make something happen. Don't be concerned with making a error. You should be willing to make errors (not *wanting* to, though). The more you concentrate on resisting them (such as keeping the car away from a wall or the edge of the track at the exit of a corner), the more likely it is you will make them. Relax!

Don't let a mistake take your concentration away. Everyone makes mistakes. Learn from them, then forget them. It's important when you make an error on the track to quickly understand why it happened, so you can ensure it doesn't happen again and then concentrate on what's happening next.

In fact, sometimes, just go and drive without thinking about going faster or worrying about making mistakes. Relax and just "let it flow."

27 BEHAVIORAL TRAITS

They say you should treat everyone as equals. I disagree. The reason I disagree is that everyone is not the same. Now, I know what is meant by the saying, and actually I do agree that everyone has the same potential. However, everyone is different and therefore should be communicated with and managed differently. The reason is that everyone has different personality traits.

There are a number of personality trait profiling methods; the most popular ones being PDP (Professional DynaMetric Programs), Performax, Birkman, and Meyers-Briggs. Each of these have been used and fine-tuned by millions of people, making them extremely accurate in providing a personality profile of the individual. Although each method may use a slightly different set of traits, all of them provide a similar result.

When it comes to race drivers, one of the most useful personality trait profiling methods breaks the person's personality into four categories:

- **Dominance:** This is a measure of how dominant a person is or isn't, on a scale from not very dominant to very dominant.
- **Extroversion:** The measure of how outgoing or how much of a "people person" someone is, from being very introverted to very extroverted.
- **Pace/Patience:** The measure of what pace a person prefers to work at, or how patient they are, from very impatient to very patient.
- **Conformity:** The measure of how one conforms to doing things "by the book." In the case of race drivers, this relates more to whether the person tends to follow the "rules" or how detail-oriented they are, from not very detail-oriented to very detail-oriented.

As an exercise, ask yourself where on each scale you would like to be. Do you want to be very dominant, or not so? Do you want to be extroverted or introverted? Patient or impatient? Do you want to be detail-oriented or not?

What are the advantages and disadvantages of each? If you're very dominant, especially if you're also low on the patience scale, you may crash a lot since you will tend to be very aggressive and try to force your position. On the other hand, if you're not dominant enough, you may not win many races since you will tend to let other drivers push you around. You may want to be extroverted to make

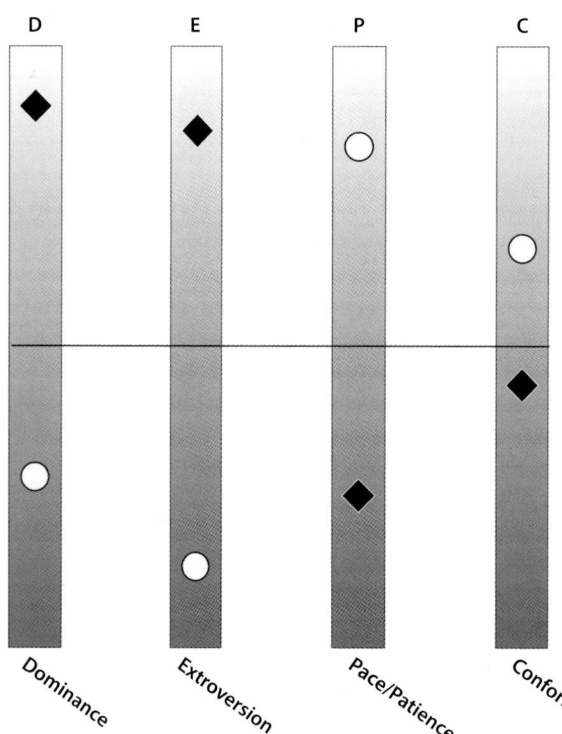

ILLUSTRATION 27-1
The higher on the personality trait scale, the more intensity of the trait. Here are two examples: The "black diamond" driver is a dominant, extroverted person with little patience; this driver's conformity level or attention to detail is middle of the road. The second, "white circle" driver is much less dominant, is introverted, patient, and is more likely to pay attention to detail.

yourself loved by the media and provide the maximum exposure for your sponsors. But one of the traits of a very extroverted person is that they love to be loved and hate to be hated. If you're too extroverted you may be a little too nice to your competitors on the track, in fear that they will not like you.

Obviously, if you're too patient you will not win often, and if you're not patient enough you will crash too often. And what about your level of attention to detail? If you're not detail-oriented at all you will not make a good test driver, nor will you be consistent. If you pay too much attention to the details, however, you will most likely be slow. Some drivers are so concerned about making sure every line they drive through a corner is perfect within a fraction of an inch, even if that means scrubbing off speed to make sure it is.

Actually, the perfect race driver is one who can adapt, who can be dominant one second and less dominant the next, one who knows when the time is right to close the door on the competition and show who's boss, or to back off and wait for a better opportunity. The perfect race driver can be extroverted when it is time for the media and sponsors, and introverted and self-centered when behind the wheel. The perfect race driver is patient when the situation suggests that would be best and has a sense of urgency when that is needed. The perfect race driver also pays attention to details to a certain point and knows when to let go of them and get on with the job at hand.

You see, the perfect race driver is like a chameleon. He or she adapts to the situation. If you look at the real great champions, you see that they did that. Was Rick Mears dominant or not? Was he extroverted or not? Was he patient or not? Was he detail-oriented or not? How about Ayrton Senna? Al Unser? Dale Earnhardt? Michael Schumacher? Dario Franchitti? Jimmie Johnson?

In fact, Earnhardt Sr. may have learned to adapt better than anyone. At one time, early in his career, he was simply dominant, introverted, and impatient. With experience he learned to adapt. He learned to dial down his level of dominance, turn up his extroversion (when it was time for a media interview or sponsor function), and dial up his patience. That is one of the reasons he was the great driver that he was. He could adapt his personality to suit the situation.

You must learn to do the same. I recommend using mental imagery to change your mental programming, which is where your behavioral traits lie. They are part of your programming. However, your programming can be changed. In fact, your programming can be that your programming can change. In other words, part of your programming can be that you can adapt your personality traits. If you imagine having four knobs on your chest, one marked D (for Dominance), one E (Extroversion), one P (Pace/Patience), and one C (Conformity), and then when the situation in your mind requires being more dominant, you just reach up and crank up your D. Or, if the situation dictates being more extroverted, you dial up the E, and so on.

For many drivers, learning to adapt their behavioral traits is the key to being a champion race driver. This only occurs as a result of reprogramming by way of mental imagery.

28 BELIEF SYSTEM

The single biggest limitation to your performance is what you believe about yourself. What a driver believes about himself or herself—the deep-seated, inner confidence—will have more of a bearing on performance than just about anything else. The bottom line is, if you do not believe you're quick, you will not be quick.

SPEED SECRET

What you believe is what you get.

Our world, especially the world of sports, has countless examples of the power of the belief system. At one time, scientists, researchers, and therefore athletes thought that if someone were to run the mile in less than four minutes, they would practically drop dead. Then along came Roger Bannister in 1954, who didn't believe these "laws of physics," and who ran the first sub-four-minute mile. Within the following 12 months, four other runners ran a sub-four-minute mile, a limit that for decades had seemed impossible. Once armed with the knowledge, and the belief, that it could be done, it was relatively easy.

Watching any world champion, you can't help but be impressed with more than their skills. It is the strength of their belief system, on a consistent basis, that is most impressive. That is what makes them as successful as they are.

If a driver honestly believes, deep down inside, that he or she has a special knack for avoiding trouble, odds are the driver will. The reverse is true as well. It is almost a self-fulfilling prophecy. Maybe even by fluke, a driver gets caught up in another car's crash. Then, a while later, it happens again. Now the driver begins to think, "Why does this always happen to me?" The driver begins to believe that whenever and wherever there is a problem, he or she will find a way of getting involved in it.

Another driver may have a similar situation occur and be almost lucky to avoid it. That driver begins to think that maybe he or she is good at avoiding trouble. Then another car crashes in front of the driver and the driver avoids it. Now the driver knows he or she is good at missing other cars' problems. And because of that, the driver is.

A driver's belief system, what the believes about himself or herself, plays a critical role in racing.

When I witness the amazing power of a person's belief system, I sometimes wonder whether a driver's inner force can actually overpower the real limits of physics as it applies to a race car. Can a driver believe so strongly in his abilities that he can make his race car perform beyond what physics tells us is possible?

A perfect example of this was Ayrton Senna at the 1993 European GP at Donnington. If you were unfortunate enough to have missed it, here's what happened. This was one of those rare times when the McLaren he was driving was rather uncompetitive, and he had qualified fifth. It was raining hard for the start of the race, but Senna literally drove around Michael Schumacher, Karl Wendlinger, Damon Hill, and Alain Prost on the outside of corners on the first lap. He went on to a victory that is part of Grand Prix folklore. Physics told me, and everyone else watching, that he couldn't do that, that he was driving beyond the limit. It was obvious that no one bothered to tell Senna about these laws. It was as if he just willed the car to do things that were otherwise impossible. He believed it could be done, so it could.

Can a driver's belief system overpower the laws of physics? I don't think so; however, do we know what all the real limits of physics are? When you start to look at what quantum physicists are studying today, you begin to understand just how little we really know. Already some things that Einstein "proved" are being shot down; just as some of the laws of physics Sir Isaac Newton proved hundreds of years earlier were.

Sometimes, being a little "ignorant" can be a good thing. Knowledge can sometimes lead to limitations. If you believe the limit of the car is at some specific point, or certain lap time, what are the chances of you ever exceeding that? Slim to none.

Maybe a driver's belief system cannot overpower the laws of physics, but it can overpower what we think are the laws of physics today. The one thing I do know for sure is this: The number one thing that limits a driver's performance is his beliefs about himself.

Nigel Roebuck, in his Fifth Column in *Autosport* (June 8, 2000) made some observations about David Coulthard, in particular his fresh approach to driving during the 2000 season.

> *In January, speaking of the season to come, he said he had concluded it was time to shed the "nice guy" image, to put on a low light anything other than his focus on the world championship. Undoubtedly there was a feeling that he needed to deliver more consistently if he were to keep his McLaren drive beyond 2000, but I wondered at the wisdom of trying to reinvent himself.*
>
> *In fact, I don't think he has. If anything, he seems to me more relaxed, more at peace with himself than at any time before, and this in spite of the awful events at Lyons airport just before the Spanish GP (walking away from a plane crash that killed the pilot and co-pilot).*

To come through an experience of that kind has to change your perspective on life, to define very sharply what matters, and what does not. I could be wrong, but my impression is that DC has actually been less intense about his motor racing of late and that his driving, his whole approach to the job, has benefited. There is a formidably insouciant quality about him now, which was never there before. Schumacher's words he should flick from his sleeve.

I believe what Roebuck, and others, observed in Coulthard was a change in his belief system. This was particularly noticeable throughout the first half of the 2000 season, when his teammate, Mika Hakkinen, was slightly off form. A driver's beliefs in himself are developed through a variety of ways. Sometimes, as I see was the case with Coulthard, it was a combination of some deliberate work (mental programming) and a "happening" (Hakkinen's slightly reduced performance level). Often times, the happening is strictly a random, lucky occurrence. Other times it is a happening made to happen. In fact, a happening made to happen is very common. In other words, did Coulthard's improvement contribute to Hakkinen's problems?

The other thing that Roebuck noticed and wrote about Coulthard was his relaxed demeanor. It is not coincidence that most athletes have that appearance when performing at their best. Rarely does anyone, let alone a race driver, perform at his or her best when "trying." Trying is a conscious act, not one that leads to maximum performance.

Tony Dodgins, in his Prix Conceptions column in *On Track* (June 8, 2000), quotes Frank Williams and Patrick Head talking about Ralf Schumacher:

The Williams team likes tough drivers—men like Alan Jones and Nigel Mansell. Schumacher fits the bill perfectly. Ask Williams what he considers to be Ralf's main strengths, and the response is instant.

"It's that very word—strength," Williams explains. "Great physical strength. And great mental strength too. Ralf's very tough in the head. Like Jacques (Villeneuve). He fears no one. Without being conceited, he doesn't believe he's second to anyone. He automatically thinks he's the best in the world, but he doesn't carry himself like that. He's intelligent and usefully experienced as well now. And blindingly quick."

The strategies—the triggers, actions, centering, and integration exercises—presented in this book have a double-whammy effect: They work in two ways. First, they physiologically and psychologically "switch you on." And second, if you believe they will help you perform better, they will. That is the power of the belief system.

Your mind—or should I say the use of your mind in a deliberate, productive manner—has the ability to shape reality. In other words, what you believe, and what you mentally "see" will come true—become reality—if you focus on it.

SPEED SECRET

**If you believe you can't, you can't.
If you believe you can, you can.**

Where do a driver's beliefs come from? They are programmed into his subconscious in three different ways:

- **Physically/Experiential:** When you experience being quick, your belief system is programmed to believe you are quick. If you are not quick, you will form a belief that you're not quick. This is how most of your belief system is developed.
- **Mentally:** You can affect your beliefs about being quick by preplaying being quick in your mind, using mental imagery.
- **Externally/Internally:** You, and other people, can have a great effect on your belief system. If someone were to continually tell you that you're quick, over a period of time you will begin to believe that you're quick. Of course, the opposite is also true. And your self-talk can also affect your beliefs. If you keep telling yourself that you're quick, it will have some effect.

If most of the programming of your beliefs came about from past experiences, how can you begin to believe you're quick before you go quick? It's a chicken-and-egg situation. Which comes first, the belief of being quick, or actually being quick which results in the belief of being quick? Of course, there is no definitive answer to that question.

Often, a driver's beliefs about his or her abilities and quickness comes before he or she ever sets foot in a race car. If the driver has been successful in other areas of his or her life, particularly other sports, the driver begins to believe that he or she is good at everything. The more success the driver has had in other areas, the stronger the driver's beliefs will be.

The good news is that you can change your beliefs about yourself.

The first step in you changing your belief system is getting an awareness of just what your beliefs are. You should make a list of both your positive and negative beliefs, being totally honest with yourself. The list should include what you believe about yourself from both the physical technique aspect and the mental side. You do not need to share the list with anyone else. This is simply for you to become more aware of yourself. After all, if you don't know what to change, how are you going to change it?

Once you're honestly aware of what your beliefs are, you can choose to reprogram any negative ones. That will rarely happen overnight. It usually requires doing a couple of mental imagery sessions per day, backing that up with some physical signs of improvement, more mental programming, more physical evidence and programming, and so on. Without the mental programming, it is unlikely your beliefs will ever change, unless, by some fluke, you happen to go quickly, in which case you will have experienced it.

Beliefs	
Positive	**Negative**
I'm great at race starts	I'm not a good qualifier
I'm a good, smart racer	I'm too nice a guy
I'm fast	I'm not confident enough
I'm assertive	I crash too often
I make good passes	I'm too tense in the car
I motivate my team	. . .

ILLUSTRATION 28-1
Your list of beliefs may look something like this. With this list you can deprogram, reprogram, and program the beliefs you want.

I'm sure you've witnessed this yourself many times. A driver who seems to have what it takes to win keeps finding ways to lose; either he or she makes a mistake near the end of the race or the car breaks. Then, practically out of the blue—by fluke—everyone in front of the driver breaks, the driver wins, and then look out! That driver begins to win everything in sight. What was the difference? Did the driver gain a bunch of driving talent all of a sudden? No, just one thing changed: the driver's belief system.

THE IMPACT OF PAST SUCCESSES

There is nothing that will build or improve your confidence like a little success. Fortunately, these successes do not necessarily have to be behind the wheel of a race car. Any success in anything will trigger your confidence level. So focus on past successes. See them vividly in your mind's eye. Recall the emotions, the feelings, the performance you experienced with these successes.

Recall, using mental imagery, past successes you've had in other sports, in school, business, relationships, or hobbies. Replay every detail about those performances from a technique point of view, and how you felt, your emotions, and your state of mind before, during, and after.

As I said earlier when talking about learning, confidence builds success and success builds confidence. It's a loop. The more confidence you have, the more likely you will have success. The more success you have, the more confidence you will have. Unfortunately, the opposite is also true!

Setting short- and long-term goals—and achieving them—are important to your confidence. To do that, they must be realistic. Do not try to move up the racing career ladder too quickly. If you get behind the wheel of a race car that you are not mentally or physically ready for, it's easy for you to lose confidence.

The same goes for something as simple as increasing your speed through a corner. If your immediate goal is to increase your speed through a turn by 2 miles per hour, that's achievable and helps build confidence so that you can improve even more. If your immediate goal is to increase your speed by 10 miles per hour, that may be too large a jump. If you don't succeed in making that jump, you may lose confidence and not make any further improvement.

Feed off your successes, no matter where and when they are in your life.

During the 1997 World Sports Car season, the average number of positions I improved during the first two laps was six. The first race was the 12-hour at Sebring, where I qualified 10th. In the drivers meeting prior to the race, the officials said they had to start the race on time due to television coverage. During the pace lap a car had a mechanical problem, which meant they had to run a second pace lap. Because of the TV time constraints, I knew right then that the green flag was going to fly at the end of that second pace lap, no matter what. Let's just say I was in second place at the end of the first lap. Prior to the next race, I kept replaying that Sebring start over and over in his mind. I got another great start. In fact, for the rest of the season, I kept playing those race starts in my head—and kept getting great starts. I knew, prior to the start of every race, that I was going to pass at least four or five cars on the first lap. Replaying those successes in my head led to many more.

SPEED SECRET

Replay your successes.

Right now, for a few minutes, imagine being in your car sitting on the grid for your next race. Imagine as many of the details as possible: the other cars, the track, weather, sounds, everything. Now see yourself just before the start, on the pace or formation lap. Stop reading and do this.

Have you ever noticed how you can "see" a person's confidence level? It's interesting how you can often notice how strong a belief a person has in himself or herself, especially in sport. You can watch drivers walk to their cars and often get a feeling which one is going to win. I've never seen a driver casually stroll to his car, looking like he doesn't care, looking like he doesn't mean business, looking like he doesn't believe deep down inside that he's going to win, then go out and win. Drivers who consistently win look like they're going to win. You can "see" the belief in their eyes and in their walk

Now, go back to your image of you being on the grid and driving the pace or formation lap. What grid position are you in?

If you imagine yourself in any position other than the pole, the likelihood of you being on pole is small. Until you can imagine yourself on the pole, you can't believe you can be on the pole, and for you to be there would almost be a fluke. In other words, until you believe you can be on the pole, it's unlikely that you will be.

The power of your belief system is amazing. This fact has been proven time and time again by world champions and amateurs alike. Do you think that Michael Schumacher consistently had a stronger belief in himself than his competitors had in themselves over his championship years? There's no doubt about it. You could hear it in the words of his competitors: They talked about having to beat him; whereas, Schumacher mostly talked about himself and his team. What about Fernando Alonso's belief in himself? You know that he knows he can beat Schumacher. He knows that he is every bit as good as, or better than, Schumacher. You can see it in his walk, you can see it in his eyes, you can hear it in his words, and you can see it on the track.

How does one develop a set of beliefs that are as strong as Schumacher's or Alonso's? Where do your beliefs come from? Yes, they certainly do come from past experience. But if you've never done something in the past, such as taking the pole, how can you begin to believe that you can? If you've never experienced winning, how can you believe that you can win? And if you can't believe you can win, it's proven that it's less likely that you will win.

The answer to these questions, and the solution to the problem of developing a belief in being able to do something before you've actually experienced it, is all in your mind. Yes, you can develop your belief in being able to take the pole by using mental imagery. In fact, by preplaying something in your mind you develop a belief in being able to do it.

But, let's assume for a minute that you've never qualified better than 10th. If you begin doing mental imagery of great qualifying performances and see yourself now on the pole, will that change your beliefs enough to enable you to take the pole next time out? Maybe, maybe not. The problem is that your mind, your current belief system, may not accept that much of an improvement. You know, you may be able to fool other people, but you can't fool yourself. Your belief system is a bit like an elastic band. You can stretch it, but if you stretch it too much it will snap. Seeing yourself moving from 10th to 1st may be too much of a stretch, and your belief system will then snap. You won't accept it.

One way to stretch your belief system is to have a reason for stretching it. If you've not made any changes in your skills or your approach, it's difficult to see why you should change your beliefs. But if you've learned something new, you have a new chassis setup, or you have a new team member, then there is a reason for your belief to stretch. It's one reason a driving coach can help so much. While a good coach will help improve your technique and mental preparation, often a big reason for improvement is that your belief system can accept there is a reason for improvement. And if you believe you can or should improve, you will improve. By simply having a coach—or a new engineer, a new shock setup, or whatever—there's a reason for your belief system to accept a stretch from tenth to pole.

SPEED SECRET

Stretch your belief system, bit by bit, through self-talk and mental programming.

29 THE INNER GAME OF RACING

Driving a race car is a series of compromises. The ideal line for a particular corner may vary slightly from lap to lap, due to rubber build up or oil on the track, the position of competitors around you, or how your car's handling changes as the fuel load is reduced. You constantly have to monitor and adjust your driving to best suit the condition of your tires. You have to consider and reconsider practically every lap what your race strategy should be. There are hundreds, perhaps thousands, more compromises and decisions to be made every lap.

The driver that chooses the best compromises is most often the winner. A driver whose mind is best prepared is more likely to make the ultimate compromises.

PERFORMANCE VERSUS COMPETITION

You're going to be more successful if you concentrate on your own performance rather than on the competition.

SPEED SECRET

Focus on your own performance rather than on the competition.

Many drivers become too focused on what their competitors are doing. They're constantly looking at what the competition is doing to their cars, at how they are driving a particular corner, and watching the mirrors to keep them behind.

Instead, if they would put that much focus and concentration on their own car and driving; they would probably be so far ahead they would never have to worry about the competition.

Concentrate and work on getting 100 percent out of yourself and your car. Don't worry about the competition. If you're getting 100 percent out of yourself, there is not much else you can do about the competition anyway. If you don't win, there is not much you can do other than improve your car's performance level or work at raising your own 100 percent. After all, your 100 percent today may be only 90 percent six months from now, because your technique has improved. And you can always improve.

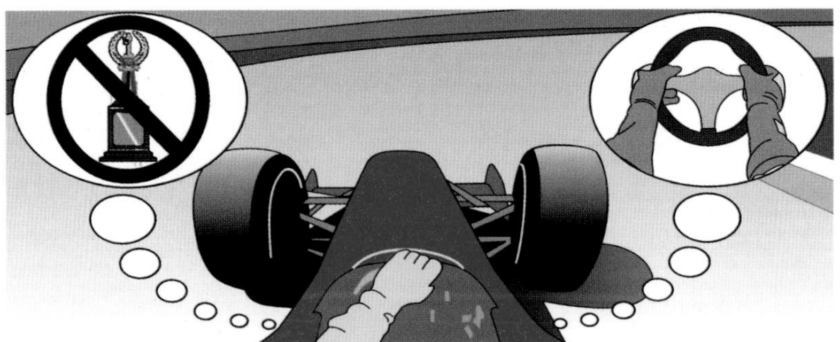

ILLUSTRATION 29-1 Once behind the wheel, there is only one thing you can control, your performance. One of the keys to consistent performance is focusing on your performance, the act of driving, and not on the result. If at any time you find yourself focusing on the result, think about one specific driving technique, such as how quickly you are turning the steering wheel entering the next turn.

By the very definition of the word "competitor," we compete against others. If your focus is on competing, however, you lessen your chances of performing well. When you focus on your performance, you increase your chances of performing well, and therefore, of winning. Ironic, isn't it? Perhaps then, instead of "competitors," we should consider ourselves "performers."

Focus on your performance, your execution, rather than the result. Paradoxically, your best results will come when you are least concerned with them, when you focus on your performance. This may be one of the most difficult "inner" concepts to accept. After all, racing cars is all about competition, beating the competition. And yet, when you detach yourself from the results, you will reduce your stress level, become more relaxed, your brain integrated—you will be "in the flow"—and the results will take care of themselves.

If you think about it, you really can't control what your competition does, anyway. You have little direct influence on them. All you can do is control your own performance. So focus on what you can control, not what you can't.

SPEED SECRET

Focus on your performance, and the results will look after themselves.

Research has actually shown that athletes focused on their own performance—their technique—have sharper vision and quicker reflexes than athletes focused on their results.

Don't worry about what other people say. Don't compare yourself to others. Compare yourself with your past performance, and strive to improve, no matter how you compare to the competition. Of all the drivers I've coached through the years, it's the ones who are constantly looking at and comparing themselves to their competition that struggle the most. The drivers who focus on themselves and don't worry about anyone else are the ones who win most often.

Only judge or evaluate yourself based upon what you've done, your performance, not on what other people say or think. Do what you think is right for you to achieve the goals you've set for yourself. Only you know what is right for you.

Winners focus on themselves today, in the present. They spend little time, if any at all, looking at or talking about what they did or achieved in the past or what they will do in the future. They look at the past only to learn from and improve. And yes, they have short- and long-term goals, but they know it is today's performance that will enable them to achieve these goals.

It is when you are totally focused on the task at hand *in the present* and not on what has or will happen, that you most effectively activate your subconscious performance programs.

Expectations, thinking about a particular lap time or a qualifying or race finish position, can really limit your performance. Often, with expectations, you are so focused on the outcome, the result, that it distracts you from the moment, from your technique, and ultimately, from your performance.

When you have no expectations, you have no limits, no preconceived ideas or thoughts to unfocus your mind. With expectations, you have pressure, stress, and anxiety that will negatively affect your performance. Plus, you will rarely ever exceed your expectations.

With many of the race drivers I coach, one of the first things I do is take the stopwatch right out of the equation. I have the drivers just go out and drive, without thinking about or worrying about their lap times or expectations. After all, why do you really care what your lap times are? If you turn a certain lap time, are you going to stop working at going even faster? I hope not! One of your objectives should be to *always* go faster.

Now, I know what you may be thinking: Your lap times are a measurement, a comparison with your competition. The point is, though, you may be focusing too much on measuring and comparing yourself with the competition, the result. If you put that much focus and attention on your own performance, you may be so far ahead of the competition that there is no need to compare.

COMFORT ZONE

When you first start racing, in whatever type of car you choose, it feels very fast. In fact, it's almost as if you can't keep up. But with experience, you become more comfortable and accustomed to the speed and feel of the car. I call this your Comfort Zone.

When you progress up to a faster car, you're once again having to push the limits of your Comfort Zone. But again, with experience, your Comfort Zone expands, and you feel confident racing at the new speeds.

Some drivers adapt more quickly to faster cars than others. This doesn't necessarily mean they are better drivers, just that they can expand the limits of their Comfort Zone quicker.

When I first drove a ground-effects car, I had to work on expanding my Comfort Zone, building my confidence. With a ground-effects car, the faster you go, the more aerodynamic downforce you have. This gives you more grip, which means you can go even faster. That takes confidence, but it doesn't happen immediately.

When I first went to Indy, it took a little time to get used to the speed. I had never run at more than 200 miles per hour, and I had to gradually work that speed into my Comfort Zone.

Often, if you feel as though things are happening too fast, as though you're being rushed, it may just mean you're not looking far enough ahead. Pick up your vision, and your Comfort Zone will expand.

To drive fast and win races you have to feel totally confident in driving at the car's limits. That means your limits—your Comfort Zone—must be at least equal to the car's limits. In fact, your comfort level must be equal to the car's performance level, otherwise it is next to impossible to drive at 100 percent. Again, this takes experience and constantly pushing the limits of your Comfort Zone.

Drivers usually operate within their comfort zones, going outside every once in a while to "push the envelope." To go faster, to improve as a race driver, you have to have a comfort zone that extends beyond the limits of your car. In other words, if you are not comfortable driving as fast as the car can be driven, you will not be able to maximize your performance. If you're not comfortable driving slightly beyond the limits of your car every now and then, you will never be able to consistently drive at the limit.

To improve your performance level, to stretch the limits of your comfort zone, you have to progressively push the limits in small increments. Some drivers never go beyond their comfort zone and never improve. Others go too far too soon. If you take too big a leap, at best you won't improve. At worst, you have a big crash.

You must feel completely comfortable and confident with the sense of speed just slightly beyond the limit. One of the best ways of doing that, although not always practical, is to drive a car that is faster than the one you race. You then become so accustomed to a higher speed that when you return to the speed of your race car, it feels slow. The objective, really, is to help slow down the feeling of speed.

Short of spending time behind the wheel of a race car much faster than yours, mental imagery is the key to developing your comfort zone. As I said earlier, this can be done by driving a track in your mind, and then accelerating the speed in your mind to "fast motion."

CONSISTENCY

The mark of a great racer is consistency. If you can consistently lap a track at the limit, with the lap times varying no more than a half second, then you have a chance to be a winner. If your lap times vary more than that, no matter how fast some of them are, you won't win often.

When you first start racing, concentrate on being consistent. Don't be too concerned with your speed. Work on being smooth and consistent with your technique lap after lap.

To do that, when driving the limit, you must remember what you did and keep doing it lap after lap. That is not as easy as it sounds. But it's not until you drive consistently that you can begin to work on shaving that last few tenths or hundredths of a second off your lap time.

If you want to change something—either to the car's setup or your driving technique—how are you going to know if it took a few tenths of a second off your lap time if you're not consistently lapping to begin with?

EFFORT

One of the most common mental errors a race driver makes is "trying." Trying is a conscious act. Not only is trying not an automatic, programmed act, but the second you try, your body tenses and therefore is unable to perform smoothly. Trying is a primary cause of errors, particularly under pressure. You must learn to relax, and let your body and the car "flow." Drive naturally, subconsciously.

As I said, one of your objectives should be to always go faster. Unfortunately, many drivers at that stage *try* to go faster. The result is rarely what the driver wants. Remember what I said about trying? Trying rarely works. Instead, don't force it. Relax and just let it happen, focusing on your performance.

As Yoda said, "Do or do not. There is no try." Either do something, or don't do something. There is no point in trying to do something. By the very definition of the word, trying gives you a way out, an excuse. Trying means "to attempt." To us, that doesn't sound positive. It doesn't to your brain either. Remember the muscle check. The second you try, you become tense. The second you become tense, your performance suffers.

As you know, driving a race car well—performing at your own 100 percent—comes from driving subconsciously. It comes from your "program" in your brain. *Trying* to drive fast is just like *trying* to make a computer with no software do something. It just isn't going to happen. Trying is driving consciously. Instead, focus on giving your bio-computer more input. Focus on what you can see, feel, and hear; become aware; and visualize the act of driving.

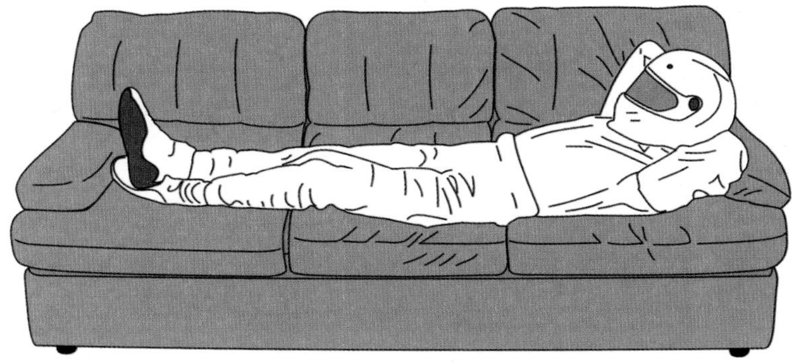

ILLUSTRATION 29-2 It's easy to get too "comfortable" behind the wheel of the race car, and never push beyond your comfort zone. Of course, that will never lead to a great performance.

Have you ever noticed how practically every great athletic performance looks almost easy and effortless? Great performances, and therefore the best results, are always achieved when the right amount of effort is used in the right places. This right amount of effort is usually less than you think necessary. Like what I said about psychomotor skills, the less *unnecessary* effort you spend, the more successful you will be. The key is to use *appropriate effort* or economy of movement.

Doing the wrong thing with more effort rarely results in a good performance. Great race drivers use less effort to produce great performances and great results. The more intense the competition, the more they relax and just let it happen.

SPEED SECRET

Relax, use less effort, and just let it happen.

Think back to some of your great performances in your life, whether in a sport or anything else. Were you tense and aggressively trying, forcing yourself to perform well? Or were you relaxed, calm, focused, assertive, and simply doing what seemed to come naturally? I bet you were in the latter mode, not even aware that you were trusting your subconscious programs to perform.

Again, focus on what you can control. Focus on what you want. Focus on where you want to go. Focus on the moment—your execution, your form, your technique—rather than on how much more there is to go, how much faster you need to go, or what position you are in.

PRESSURE

One of the most frustrating things to observe is the pressure that is often placed on young athletes by the media. It's almost like they go around with a "can of pressure" and spray it all over the athletes. This is particularly seen in and around the Olympics. The media seems to love to remind Olympic athletes of past failures and mistakes and ask them if they will be able to "put that out of their minds." If we really cared about the performances of our Olympic athletes, we would keep the media away from them. Of course, that's not going to happen.

The same is true of many race teams and sometimes with the race media. If a team manager believes that he is enhancing the performance of the team and the driver by going around "spraying" little cans of pressure on everyone, especially the driver, perhaps the team manager needs to become more aware of the strategies that really will enhance performance.

For the race driver, having an understanding of and a strategy for controlling pressure is critical.

Many drivers' performance levels are limited by their fear of failing, their fear of losing. So much of their focus is upon "not losing" (again, an outcome), and what losing will mean (at least, what they think losing will mean), that they almost guarantee losing. It's unfortunate—not to mention, destructive—the amount of pressure, external and internal, some drivers put on themselves or that others have put on them.

Internal pressure is not necessarily a bad thing, as long as it is focused in the right direction. In fact, this sometimes helps to drive or motivate a person. But most of the time, pressure to deliver an outcome only results in increased tension, stress, and anxiety, and decreasing performance. Make sure any internal pressure you place on yourself is focused on your performance, not the result.

External pressure rarely, if ever, increases a driver's performance level and most often leads to not winning. External pressure is the pressure a driver puts on himself or herself to win in front of others, or to live up to other's expectations. It's also the pressure others, whether it is family, friends, sponsors, or crewmembers, put on the driver to win.

ILLUSTRATION 29-3 Learning how to handle the pressure many team owners, sponsors, friends, and family members "spray" on you is a critical part of your job.

People who place this external pressure on you must realize how much they negatively affect you with their expectations. Understand that there is a difference between having and showing confidence in you and having high expectations. Having confidence is a performance-related thing. High expectations are outcome related. Of course, you know which is best.

The same thing applies to you. If you think about what people expect of you, what they will say about you, what their expectations are, you decrease your chances of winning. If you focus on your performance and forget about whether you may lose and the consequences of that, you increase your chances of winning.

POSITIVE TALK

Turn everything you can into a positive. For example, just saying you love racing in the rain over and over again will make you a better driver in the rain. If you take every situation that other drivers consider a problem or unpleasant (rain, boring track, uncompetitive car, too much traffic, and so on) and turn them into positive challenges, you will perform better. Turn them into "watch this" situations, chances to tackle a challenge head on and show what you can really do. It's simply a matter of turning negative thoughts and questions into positive talk.

A study showed that the average person has approximately 66,000 thoughts every day, with 70 to 80 percent of those being negative. I doubt that this study included any champion race drivers! From my observations, great race drivers seem to be able to turn almost everything into a positive. I would suggest that at least 70 to 80 percent of their thoughts are positive (although some people might joke that race drivers are not capable of 66, all of them about driving, let alone 66,000 thoughts per day!).

The more times you repeat a phrase, the more it will become a part of your belief system. If you tell yourself over and over again that you are a great qualifier, eventually you will truly believe you are a great qualifier, and your chances of being just that are greater. It really is a self-fulfilling prophecy.

Early in my career I spent a lot of time racing in the rain. Hey, it's part of life in the Northwest! As I began racing against other drivers who hadn't had as much experience in the rain, I realized I had a slight advantage. But more important, the more I told myself I had an advantage in the rain, the more I really did. Of course, that resulted in some great performances in the rain, and therefore I enjoyed it even more. I still love to tell other drivers how much I love racing in the rain, and it still results in an advantage. I do the same with everything that other drivers look at as problems. I love to make a point of letting other drivers know how much I enjoy and look forward to things they don't like. And when I am driving a car that is obviously uncompetitive, I look at that as an opportunity to really show off, to do more than anyone expected of me.

INTENSITY

Every time you get behind the wheel of a race car, you must perform at the same intensity level you want during a race. There is no point practicing in a casual, "I don't really care how I do this session; this session is not very important" attitude and then expect to perform any differently in qualifying or the race. Remember, practice is programming. If you program driving with low intensity, that's how you'll perform in the heat of the battle.

Particularly in some of the "stepping-stone" series, such as Formula 2000, Formula Mazda, Indy Lights, Midgets, and the NASCAR regional tours, the competition is intense, fierce, in fact. One technique I recommend to raise your intensity level is this: Every time you leave the pit lane, whether on a private test day, practice session, or qualifying, drive like you mean business. Accelerate hard out of the pits and get up to speed as quickly as possible. Push as hard as possible right away. Be intense (but not tense!).

If you slowly roll down pit lane and then gradually build up to speed, you may lose valuable time. Plus, and more important, it may take too long for you to mentally get up to speed, to dial up your intensity level. Quickly accelerating out of pit lane, being the first car out, driving as hard as possible right away, sets a tone for you and sends a message to your competition: you're here to do business. It is a trigger for your mental intensity.

Being intense often requires energizing yourself. Not being energized or intense is not a problem for many people in a sport like racing, but still, it is not uncommon for drivers to be too calm, relaxed, or even fatigued. If that is you prior to practice, qualifying, or a race, you need a program for energizing. See yourself alive and energized. Then use some physical warm-up exercises (cross-crawls), clench your fists, flex your muscles, yell or scream, use powerful words when talking to people, get your heart rate pumping, take some deep rapid breaths, or listen to some loud rock music.

The same thing applies to your level of being "psyched." Some drivers need to psych themselves up, while others must psych or calm themselves down. You need to determine what level of "psyching" results in your best performance. Each driver has his or her own optimum level of being psyched, the optimum level of emotions, tension, anxiety, nervousness, and energy. The key is to be aware of it at all times, and when it results in a superior or peak performance, to use that to program it so that you can recall it over and over again.

ASSERTIVENESS VERSUS AGGRESSIVENESS

Traditional wisdom says that in any sport you must aggressively dominate your competition. Observe the great athletes: Roger Federer, Michael Schumacher, Michael Jordan, Alex Rodriguez. They are not aggressive; they are assertive (well, most of the time). It may seem to be a subtle difference in language, so subtle that many people use the word "aggressive" when they really mean "assertive" and vice versa; but there is a significant difference between being aggressive and being assertive.

The dictionary defines assertion as "a behavior that emphasizes self-confidence and persistent determination to express oneself in a positive way." Aggressive, meanwhile "implies a bold, energetic pursuit of one's ends, connoting, in a derogatory usage, a ruthless desire to dominate, and in a favorable sense, enterprise, initiative, etc.; and rarely suggests the furthering of one's own ends."

Aggressive behavior is usually the result of a driver trying to hide something, a weakness. Your competitors will recognize that and most likely take advantage of it.

An aggressive start is wild, not controlled, and often results in disaster. Being assertive means placing your car where you belong. It always appears in control, because it is.

SPEED SECRET

Be assertive, not aggressive.

Did you ever see Michael Schumacher or Michael Jordan look out of control? Perhaps Schumacher was out of control when trying to hold Jacques Villeneuve behind him at the last race of the 1997 Formula One season with the world championship on the line. Was that an assertive move? Or was it a desperate, aggressive move? I'd say he resorted to aggressiveness. Even the best can make mistakes. (In case you missed it, Schumacher bounced off the track while trying to defend an *assertive* passing attack from Villeneuve, who went on to clinch the world championship.)

I'm sure you have heard the saying, "Nice guys finish last." Some people relate being "nice" to not being assertive, but that's not the case. You can be nice and still be assertive. But I doubt you can be unassertive and finish first. Perhaps the saying should be, "Unassertive guys finish last."

RISK AND FEAR

No doubt about it, racing is a risky endeavor. If you really want to succeed, you will have to take risks, not only on the racetrack but with career decisions as well.

Whether it is on the track or off, taking a calculated, planned risk and failing is better than not risking at all. Of course, the goal is not to fail. The point is, you had better plan to deal with the risks. As the saying goes, "If you fail to plan, you plan to fail."

Of course, calculating and planning risk is the key. It would be foolish to risk taking a corner at 80 miles per hour that you normally take at 70, just as it would to accept an offer to race for the Williams Formula One team in the next Grand Prix having just graduated from your first racing school. In either case, it is best to work your way up in calculated increments.

If anyone ever tells you they never have any fear in a race car, they are either lying or are driving nowhere near the limit. There's not a successful driver in the world who doesn't scare himself or herself every now and then. Fear, or at least self-preservation, is the only thing that stops you from crashing every corner. If it's the kind of fear that makes you panic and "freeze up," then that's not good. But if it's the kind that makes the adrenaline flow, your senses sharpen, and makes you realize if you go another tenth of a mile per hour faster you'll crash, then that's good.

Really, it's more of a sense of self-preservation. Usually, you are going much too fast to be scared at the moment. However, there are times when I realized after a corner just how close I was to crashing, and there's a little fear there, knowing I came oh-so-close to losing it. That probably means I was at or a little beyond the limit.

Fear comes in many forms, good and bad, or more accurately, useful and useless. It can be the fear of physical injury resulting from a crash. That fear usually limits your speed. I prefer to look at that as self-preservation, though, which is a good thing. In fact, the only thing that deters you from going over the limit and crashing at every corner is this so-called "fear."

Fear and desire are usually the opposite sides of the same coin. Some drivers want to achieve something but are too afraid of it not working out. They focus on the fear of failure, which is another form of fear—a result—rather than the desire of making it work. When faced with an ultra-fast sweeping turn, or a difficult career decision, they think about what may happen if they made a mistake.

If you concentrate instead on the solution or the goal and your performance, rather than the problem or the result, fear of failure disappears. If you keep a clear mental picture of what you want to achieve, your mind will find a way of making it happen.

The fear of failure produces tension, which disintegrates your brain, slows reflexes, and generally hurts your performance level. Of course, that usually produces the result you feared most.

Keep in mind how valuable feedback and awareness is to learning and improving your performance. There is no such thing as failure, only the results of doing something. And those results are simply feedback, or corrections guiding you toward your objective. Failure is just a result you didn't want, one that you can learn from and help you improve your performance.

MOTIVATION

If you want to be fast, if you want to win, you must be motivated. No one, no matter how much talent they have, will ever be a consistent winner if they lack motivation.

If you want to win races, you have to be "hungry." You have to want it more than anything else.

It's important to identify for yourself why you want to race. And then, do you want to win? What is it about the sport that you enjoy? Be honest. It doesn't matter what it is. What does matter is, once you've identified it, then *focus* on it. To be motivated, you must love what you are doing. Remember and relive what you love about racing. If that doesn't motivate you, nothing will.

Understand, if you want to win, you will have to take some risks. You almost have to decide how much risk you're willing to accept. If you're not motivated, I'll bet you're not willing to accept much risk.

If you are not 100 percent motivated, it is doubtful you will perform consistently at 100 percent. Focus on what you truly enjoy or love about the sport. Motivation mostly comes from the love of what you're doing. As part of your regular mental imagery sessions, see yourself enjoying the art of driving, experiencing the thrill of racing, loving every second of it.

Something most racers do not have to worry too much about because of the high costs of testing and racing, can actually help increase your motivation level: moderation. Taking a break from the sport, to the point where you miss it; that may be just what the motivation doctor ordered.

Racing can be such an all-encompassing passion that many drivers spend practically 24 hours a day, 7 days a week eating, breathing, and living the sport. If that is you, when you finally get behind the wheel of your race car some of the passion and burning desire to drive may be gone.

Keep your racing in perspective and a balance in your life. Remember why you race. Do not take yourself, your career, your racing too seriously. Have fun. After all, that is why you started racing, wasn't it? You may have to remind yourself of that every now and then!

SPEED SECRET

Think about what you love about racing.

Having said that, it is your level of commitment and desire—your burning desire—that will determine, more than anything else, how often you win and how far you go career-wise in professional racing (if that is your goal).

Much of your motivation comes from your expectations as to how you will do. If you believe you will not do well in an event, most likely your motivation to do what is necessary to maximize your performance will not be there. Of course, this leads to a self-fulfilling prophecy. You don't expect to do well, so you don't prepare, which leads to a poor performance, which leads to a poor result, which meets your expectations.

Again, this is why it is so important to focus on your performance and not the result. There is never any reason you cannot perform at your maximum, so there should never be any reason to become unmotivated.

Having goals or objectives prior to each session or race event can certainly affect your motivation. Positive, achievable, but challenging, performance-related goals give you something to strive for, something to go after. Conversely, unrealistic or easily achieved goals will most likely discourage and de-motivate you.

PERSEVERANCE, COMMITMENT, AND DEDICATION

Did you know that Michael Jordan was originally told he wasn't good enough to play for his high school varsity basketball team? Of course, we all know he didn't take that evaluation and walk away from the sport. Instead, he practiced every day until he made the team, and the rest is history. The point is, he persevered. He never gave up.

To make it to the top in motor racing takes a tremendous amount of work, sacrifice, commitment, perseverance, and dedication. Don't ever fool yourself: No matter how much talent you have, you will never be a winner in the top levels of professional racing (F1, Indy Car, NASCAR, Sports Car, and so on) without those elements.

If I had to pick just one thing that a person requires to make a professional career in auto racing, it would be perseverance. Bobby Rahal was once quoted as saying that it takes 10 percent talent and 90 percent perseverance to make it in racing. I agree.

During a race, never give up. No matter how far behind you are, no matter how hopeless it seems, if you keep pushing there is a chance your competition will have problems. Focus your mind on that possibility. If you haven't pushed hard, if you've given up, you may not be close enough to take advantage of others' problems.

Commitment and perseverance alone will not guarantee success, but without them you can guarantee you won't perform to your maximum. Sure, there have been many drivers who have made huge commitments, who have persevered, who have made the sacrifices, and who have not made it to the top. But I also know of no driver who has made it who hasn't made the commitment, who hasn't sacrificed and persevered.

PREPARATION

Mental preparation for racing, as in any sport, is a key element. All the skills and techniques in the world are not going to make you a winner if you are not properly prepared mentally. My experience tells me that the most successful drivers, no matter what the level, are the ones that prepare more and better than the others.

Your mental approach to driving may just have the single biggest effect on your success. What you do to mentally prepare before a practice session or race is somewhat individual. It's difficult for me to tell you what will work for you. You have to experiment to find out for yourself what works, and what doesn't. For some drivers, sitting alone, not talking with anyone is the trick. For others, that results in more nervousness, and possibly talking with friends or your crew will take your mind off the pressure of the next practice or qualifying session or the race.

People often talk about the "natural talent" of athletes like Michael Schumacher, Michael Jordan, Wayne Gretzky, and Tiger Woods. If there is one thing all these great athletes have in common, it is how hard they have worked, how much they have practiced, and the amount of time and effort they put into preparing for their sport.

There is a true story about Ayrton Senna that is a great example of how true natural talent mostly comes from hard work. A couple of hours after winning his first Formula One Grand Prix in Portugal in 1985, Senna was seen driving around the track in a street car. Remember now, this was after he had just totally dominated the entire race with one of his magical performances in the rain. And what was he doing driving the track in a street car? Trying to figure out how he could have performed even better. That's commitment to being the best. That's preparation. That is what is often confused for natural talent.

Michael Jordan would often show up early prior to a game, well before his teammates arrived, and practice his three-point shot. If someone of Jordan's abilities knows the value of practice and preparation, shouldn't you?

Winners go way out of their way to ensure they have prepared in every way possible. That includes your diet, physical exercise program, mental training program, even planning travel to suit yourself, ensuring your clothing is appropriate for sponsor functions, proper public relations, and so on.

SPEED SECRET

Preparation is not just one thing; it's everything.

Race driving is all about control and discipline. Most, if not all of the all-time great drivers (Senna, Schumacher, Petty, Earnhardt, Mears, Andretti) controlled their lives and everything around them. Their attention to detail was paramount. Their commitment to looking after their driving equipment was a good example. I doubt you'll often find a world champion who doesn't like things to be organized, controlled, disciplined, and prepared.

I strongly suggest giving yourself a few minutes immediately before each session to visually drive the track (more on this in a minute), seeing the changes and adjustments to the technique you've planned. In fact, prior to every session, plan out what you're going to change. Laps around a racetrack are valuable. Make them count. Make a plan and then work on that plan.

Now, the obvious: As a race driver, your goal is to constantly strive to go faster, faster than all your competitors. That's all it takes to win!

However, once you've decided you need to go faster (and who doesn't) and how you're going to go about it, consider everything that could then happen. The car may not turn in when entering the corner 1 mile per hour faster; it may begin to oversteer during the transition phase because of unbalance and too much speed; and so on. This enables you to be mentally prepared for the consequences. This also helps your confidence level because you have it under control. It doesn't take you by surprise, but don't dwell on it.

In fact, focusing on negative thoughts or ideas will most likely slow you down. Thoughts like, "If I go this much faster, I'm going to crash," takes some concentration and attention away from the ideal, positive thought, such as, "I can enter Turn 4 a half mile per hour faster."

To go faster, you should have an open mind about learning more, about how to improve your driving, about new techniques, about how to make the car go faster, to constantly strive to go quicker and quicker. It's definitely one of the most enjoyable challenges in the world.

Getting advice from more experienced drivers or other knowledgeable individuals is a good practice. Many drivers will be flattered that you chose them to talk to, and respect you for making the effort to improve.

Talk to and watch successful drivers. Reading biographies of the best drivers in the world can help. Analyze what they are doing and saying. Obviously, you can't believe everything they say, but listen. Many times they are not intentionally trying to lead you astray with wrong advice, but they may not actually know what it is that makes them successful. That's why it's important to watch for yourself, and really *think* about all the aspects that come into play. Watch how other drivers take a particular corner that may be a problem for you. They may have found the "secret" you haven't. But be careful; they may be worse than you! Check their times and talk to some of the more experienced drivers.

When watching other drivers, notice the line they take and the "attitude" or balance of the car. Ask yourself why the car or driver is doing what it is doing. Understand the strategy and technique being used.

A word of warning: Listen to the advice, but you be the judge. Just because it works for someone else, doesn't mean it will for you or your car.

FLOW

With experience—"seat time"—comes flow. This is when you are driving subconsciously, naturally, without trying. Often, after being passed or passing someone, it may be difficult to regain your flow. It's important to concentrate on getting back into the flow, regaining your rhythm. A couple of laps of talking yourself around the track may help. Or better yet, using a trigger word for a performance state of mind.

You know when you're in the flow, and when you're not. It feels great when you are. Often, when you're not, it's because you're trying. You can't try to get in the flow. It comes naturally. Just let yourself feel like you're part of the car; become one with the car. Everything you do becomes automatic. The shifting, braking, and turning all becomes subconscious.

I think everyone has experienced being "in the flow" at some time in their life. It may have been while doing a job, playing a sport or musical instrument, or just going about a normal day. It's that time when everything just seems to go right, everything you do works perfectly, almost without thinking about it. Unfortunately, everyone has probably experienced the opposite, when no matter how hard you try, it just doesn't seem to work. And often that's the problem, trying too hard.

I bet that every time you've ever been in the flow or zone in the past there were two factors at play: You felt challenged, and you had confidence in your ability to handle the challenge. In fact, a combination of challenge and belief in yourself will do more for triggering flow than just about anything else.

SPEED SECRET

Challenge + belief = flow

If you're not feeling challenged, you'll almost feel bored. You're unlikely to have the intensity it takes to get into the flow and perform at your peak. Of course, if the challenge seems to daunting, it's also unlikely you'll perform at your best. You'll feel overwhelmed, and likely not have the inner belief that you can handle the challenge. But if you feel challenged by what you're doing, and you feel confident that you can overcome the challenge, you're more likely to perform in the flow.

Framing racing as a tough challenge, but one you can handle, will help you perform better. Sometimes it's a matter of doing just that, framing it: looking at it as not easy, but something not beyond your capabilities, something that you can handle. With that mindset, you're more likely to get into the flow, the zone, that magical state where you perform almost effortlessly, totally focused, where time seems to slow down, and where you just enjoy being in the moment.

WINNING

I talked a lot in this chapter about removing the emphasis or focus from winning and placing it strictly on the act of driving, on your performance, and how, by doing this, you will increase your chances of winning. So, does this mean that winning is not important? Of course it is! It matters very much. That is what racing is all about.

The objective of racing is to win, but the purpose of racing is to race. Winning races is the objective of the sport, but it should not be the focus. Winning is the ultimate result of a great performance. A losing but great performance has a more deep and long-lasting personal satisfaction level than a winning but crummy performance. Use the strategies presented throughout this book to ensure you have great performances, rather than crummy ones. Do that and winning will look after itself.

Of all the drivers I have worked with through the years, it is the ones who have the absolute burning desire—a need—to win, who seem to have the knack of performing at their best consistently. They are the ones who would do whatever it took to win. They spent the time preparing, physically and mentally. And they did win, more often than anyone else. But their focus always seemed to be on improving their own performance; the winning just looked after itself.

Some of the drivers I have worked with did not take losing very well, to put it mildly. Unfortunately, this attitude often led to more poor performances and more losses. When discussing how upset they became after a loss, they would all claim it was because they were so competitive; they hated to lose. I don't know of anyone in racing who enjoys losing. But the drivers who look at a loss as something to learn from—and no, not something to enjoy or be satisfied with—are the ones who most often came back to win next time out.

Competitive people—people who want or need to win—are the ones who most need to learn from their losses. Every race you compete in will have more losers than winners; it's the nature of the sport. If you become overly upset and focused on a loss and never learn from it, you are bound to lose again and again.

Sometimes a driver has to learn how to win. Often, it takes a win, whether it be a total fluke or a deserved win, for a driver (and team) to learn that they can win, for them to really believe they can win. Once that happens, a driver often gets on a roll, and the victories just seem to happen.

I don't know how many times I've seen it happen where a driver and team has everything it takes to win but just can't seem to do it. Then, after almost fluking out a win, watch out. All of sudden, you can't stop them. They start winning everything in sight.

That's why I feel it's important for a driver to race where he can win. If you're racing in a competitive series where you can't seem to pull off a win, don't be afraid to go back and do a race or two in a lesser or easier series just to practice winning. Then take that winning attitude back to your main focus.

One of my favorite quotes is by Henry Ford. He said, "If you think you can, or think you can't, you're probably right." You must have total confidence in yourself and the people on your team to be a winner.

STRATEGIES

In all your mental imagery, see yourself in and out of the race car relaxed and calm. See yourself focused on performing at your maximum, not particularly concerned with your competition. See yourself completely comfortable in your surroundings out of the car and with the speed in the car. See yourself confident in your ability to perform and that you belong at the front of the field. No matter what, you are "going to the front." See your ideal level of "upness," not too psyched, intense, or energized, but not too laid back either. See yourself as assertive and making "smart" racing decisions. See yourself racing for the pure love of it, fully motivated to do whatever it takes to perform well. See yourself fully prepared; you've eaten well, you've physically and mentally trained; you're ready. See yourself facing some adversity, but overcoming it by persevering, demonstrating your commitment to yourself and others. See yourself dealing with pressure placed on you by others by focusing on your performance and letting that take care of the results.

Program all these feelings, these attitudes, states of mind, beliefs relaxed but intense. Calm but energized. Psyched up but in control. Focused but aware.

See, hear, and feel yourself performing better than you ever have. And notice the result—winning—something you want more than anything else in the world but knowing it was your performance that produced the result.

Write down on a piece of paper what success in racing means to you. What do you want to achieve? How do you want to feel? For some drivers, becoming world champion is the only objective. For others, it's to get paid to race cars, no matter what type or level. Others still only want to race for the pure enjoyment of it, and whether they race at the amateur club level or make it in professional racing does not matter.

Then write down why you want to achieve that level of success. Is it to make lots of money, have lots of fame, feel good about yourself, for the sense of accomplishment, to fulfill the dreams of a parent, for the thrill of controlling a car at speed, to beat other drivers, or because you haven't found anything else you are really good at?

The point is, the reason doesn't matter. One reason is not any better than another. The key is to know why for your own personal motivation. The more honest you are with yourself, the more effective this information will be to your motivation level. When you need that little pick-me-up, focus on your ideal level of success and the reasons you want them to come to fruition.

As I mentioned, success and feelings of success lead to further success. Take some time to recall and write down at least three of the best performances of your life. These do not have to have anything to do with racing or have resulted in a victory or high grade. They can be how you performed in school, in another sport, something you accomplished in a job, or about a relationship. Make note of how you felt before, during and after these performances. Recall every detail you can about them. Relive them and write them down. Then, go back and read them every now and then, or update them with new experiences.

30 MANAGING ERRORS

Every race driver makes errors. Being able to recognize and then analyze your errors are important. Until you can do that, you cannot even begin to correct them and improve. I'm not suggesting that you dwell on them, but here are a few of the most common ones so that you might be able to recognize some of your errors a little earlier.

I know this part really well. I've made enough errors myself. In fact, I think one of the things that separates a good driver from a not-so-good driver is that the good driver has made more errors and learned from them. I know that I can consistently push closer to the limit than some less-experienced drivers, simply because I've gone beyond the limit enough to know how to survive. I know how to recover from a mistake. That only comes with experience.

Probably the most common error for race drivers of all levels is turning into the corner too early, before reaching the ideal turn-in point (see Illustration 30-1). Ultimately, this will result in an early apex and running out of track on the exit. This means to avoid running off the track, you will have to ease off the throttle to tighten up the corner and regain the ideal line. Obviously, this is going to hurt your straightaway speed. The trick to correcting this error is to use an easy-to-identify turn-in point, know exactly where the apex and exit points are, and be able to see them in your head before getting there.

Often, turning in too soon is caused by braking too early. A driver brakes too early, slows the car down to the speed he thinks he should take the corner at 10 feet before the turn-in point, and then turns. Obviously, the easiest way to cure this problem is simply brake a little later.

Another common error, with the same result, is turning into the corner too quickly or sharply. What happens is you turn in at the correct point, but turn so sharply that it results in an early apex again. The correction is in knowing in your head where the apex and exit points are before you begin to turn in. That, and learning to turn the steering wheel slower.

As I mentioned earlier, using all of the road on the exit is important; however, it can be misleading if you drive to the edge without having the speed to force the car out there. When you drive to the edge of the track surface for the sake of driving the ideal line, you fool yourself into believing you are going as fast as you can because you don't have any more room. Instead, sometimes, hold the car as tight as possible (without scrubbing speed or "pinching" it) coming out of the corner so you have an accurate feel of where that speed takes the car. Then, once you feel you are using all the speed and track, you can work on letting the car run free out to the exit point again.

Many small errors can result in a spin, an off-track excursion, or a crash. Most are caused by a lapse in concentration, leading to an error with the controls (usually upsetting the balance, and therefore traction, of the car) or a misjudgment in speed or positioning. The result of the error is usually determined by how calm you stay and your experience. Learn from your errors.

If the car should start to spin (severe oversteer), once you correct the first slide, be ready for one in the opposite direction caused by over-correcting. If it happens, gently correct for it by looking and steering where you want to go and smoothly try to ease the speed down until you get the car under control again.

As you know, weight transfer has a great influence on how your car behaves in a skid or slide. Smoothly controlling that weight transfer is the real key to controlling a spin. As this is an oversteer situation, just look and steer where you want the car to go.

If the car begins to spin and you can't control it, you are going to spin out completely. Nothing wrong with that, if you stay relaxed, watch where you are going, depress the clutch and lock up the brakes, and don't hit anything. That is about all you can do, besides avoiding the spin in the first place.

In fact, many believe this is the best way to *really* find out if you're driving the limit. So, if you do spin, learn from it.

If you spin, you should immediately hit the brakes, locking them up. This will cause the car to continue in the general direction it was heading before locking the brakes, while scrubbing off speed. At the same time, try to depress the clutch and keep the engine running by blipping the throttle. It is hoped you'll be able to drive away after the spin. Remember the saying, "Spin, both feet in," on the clutch and brake pedals.

ILLUSTRATION 30-1 Here's an example of what happens when you turn in too early for a corner. You end up with an early apex and then run out of track at the exit. Of course, if you realize you turned in and apexed early, you can try to slow down gently (remember what happens if you lift off the throttle suddenly while turning) and tighten your radius to get back on line.

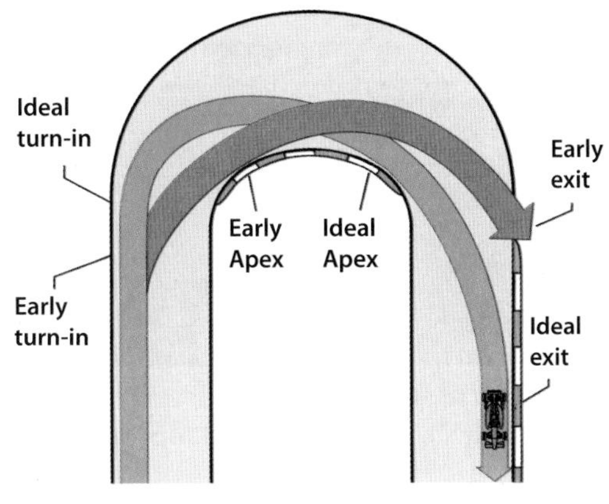

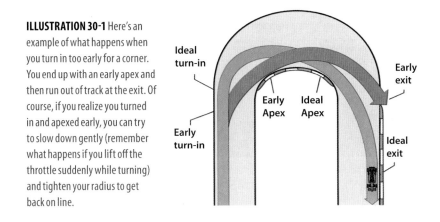

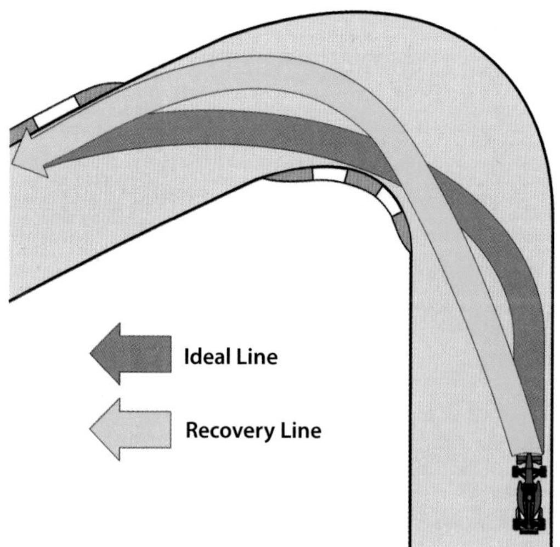

ILLUSTRATION 30-2 If you brake too late or if you are running out of brakes and you find yourself entering a corner too fast, aim the car for a very early apex. This effectively lengthens the distance you have for slowing the car.

And no matter how bad it seems, always look where you want to go. Never give up trying to regain control.

In all the excitement of a spin, drivers will often stall the engine when trying to get going again. Take your time, look around to avoid being hit (and watch for signals from the turn marshals), and get going again, using lots of engine revs. Remember your tires may have stones and pebbles stuck to them, severely reducing their grip. So take your time until they clean off, otherwise you'll find yourself spinning again.

On an oval, once you've reached the point in a slide where there is no way you're going to be able to correct it, you have to let it go. If you continue trying to correct it, you will probably spin or drive straight into the wall. You're better off admitting you're going to spin, and let it. It will most likely spin down to the bottom of the track.

What if you enter a turn slightly *too* fast to the point where it is impossible to make the car turn in properly? Most drivers' reaction is to continue braking. But you'll actually have a much better chance of making the corner if you ease off the brakes slightly. Why? For two reasons:

- The car is better balanced (not too much forward weight transfer), which allows all four tires to work on getting the car around the corner, instead of having the fronts overloaded.
- Your concentration and attention is on controlling the car at the traction limit as opposed to "getting the car slowed down" or "surviving."

Believe it or not, knowing and using this plan will do more for making you go faster than many other "tricks." You may discover the car will actually go around the corner faster than you thought.

As well, if you are entering a turn too fast (probably because you left your braking way too late), aim for an early apex. This allows you more straight-line braking time to slow the car.

If you run out of track on the exit of a corner (probably due to an early turn-in and apex), you may drop a couple of wheels into the dirt off the edge of the track. If you do, the first thing to do is straighten the front wheels and drive straight ahead, even if it means driving toward a wall for a few seconds. What this does is allows you time to get the car slowed down and back under control.

If you try to steer the car back on the track immediately, you will most likely end up with the two front wheels back on the pavement and only one rear on. This usually results in a quick spin back across the track, quite often into another car. Or, if the wheels catch the edge of the pavement at the wrong angle, it may actually "trip" the car, causing it to roll over. Again, keep the front wheels pointing straight until you get the car back in control. Don't panic and "jerk" the car back on the track. It won't work.

MINIMIZING ERRORS

As I talked about in Chapter 22, experienced drivers don't typically make any fewer mistakes than inexperienced drivers; they simply recognize them sooner due to taking in more sensory input, and therefore, reference points, and react to them sooner and in a more subtle way. In other words, they minimize errors.

So the more reference points you have around a track, the fewer errors you will appear to make. In fact, you may not even notice the errors yourself; you seem to correct them before they even occur.

This also applies to decision making. If you are making a financial investment, are you more likely to make the right decision if you have more information? Absolutely. The same thing applies to racing wheel to wheel. If you are heading into a turn in a pack of cars, the more information you have, the better your passing decisions are going to be. There are top-level professional drivers who have reputations for making bad decisions, all because they are lacking some little piece of information. For some reason, they are missing some sensory input.

Just because a driver has more experience doesn't automatically mean that he has more reference points, and therefore is better at minimizing the effects of his errors, or that inexperienced drivers have fewer reference points. This is more just a trend, for as a driver gains experience, the driver usually gets better at absorbing information to input into his or her brain/computer. But, believe me, it is not always the case. I have seen some novices who are better at taking in information than other drivers who have been at it for 20 years.

SPEED SECRET

The more reference points you have, the less errors you will appear to make.

ILLUSTRATION 30-3 The more reference points—the more sensory input—you have, the earlier and more subtle your error corrections will be. When faced with the same section of track, two drivers may not perceive or pick up the same amount of references. Notice how much more information the driver in the upper scene is taking in than the driver in the bottom scene.

Let's try to tie all this altogether. Where and how does a driver acquire more reference points? Is it only through more experience, more seat time? Well, that usually helps. But a driver can speed up the process. How? Simply by focusing on it. By practicing being more sensitive. By increasing the quality and quantity of sensory input.

Of course, I'm talking once again about doing Sensory Input Sessions. Most drivers have never done these. The ones that have often end up with a reputation for being very fast, not making mistakes, and being great test drivers because of their sensitive and accurate feedback on what the car is doing. The ones that don't do these, well, you know what their reputation is.

Let's go back to the most common—"the classic"—error made by race drivers: apexing a corner too early. This is caused by either turning in too early or too abruptly. In either case, the ultimate result is a line that ends up with your car running out of track at the exit point and either dropping wheels off the outside of the track, spinning back across the track (when the driver makes a last second correction by turning the steering wheel more), or hitting the wall. That is, unless you make a correction in the middle of corner. Which shouldn't be too difficult if you recognize the problem early enough.

The problem is that some drivers, way more than I think should ever do this, either do not recognize the error soon enough, or don't know how to correct it. Which is surprising, considering how easy it is to do both, recognize and correct it.

If your car is against the inside of the corner before the apex, you have made an error and have early apexed. The cure is simple: an adjustment in speed or steering is all that is required. If you are normally at full throttle and unwinding the steering wheel at the apex, then you are going to have to ease up on the throttle slightly (a big lift is probably going to make the car spin) and hold or tighten the radius until you get back on your normal, ideal line.

Of course, this is assuming that you know exactly where the apex of the corner is. If not, then you had better get a well-defined and recognizable apex reference point. If you don't already have one for every corner on the track, that may be the root cause of an early turn-in. In this case, it is a matter of not knowing exactly where you are going. The old saying, "You're never going to get somewhere if you don't know where it is you're going," certainly applies to this situation.

Again, this is why having as many easily recognizable reference points as possible is so critical, and that comes from absorbing them through your senses during practice.

SPEED SECRET

Minimize errors through maximizing sensory input.

One of the most frustrating things I see race drivers do is assume that if something doesn't work once, that it will never work. It's a common "error."

Let's say you think that you can carry a bit more speed—1 or 2 miles per hour more—into Turn 6 and still get great acceleration out of the corner. You know this will make a significant improvement in your lap times. You head out onto the track, and after a couple of laps you carry a bit more speed into the turn, the car understeers a little wide, and you can't get the car down to the apex. Your conclusion is that the car cannot handle that little bit more speed, and you go back to your original corner-entry speed.

Sound familiar? I bet it does. And no, this conclusion is not made at the conscious level but at the subconscious level. You may not even be aware it is happening, but

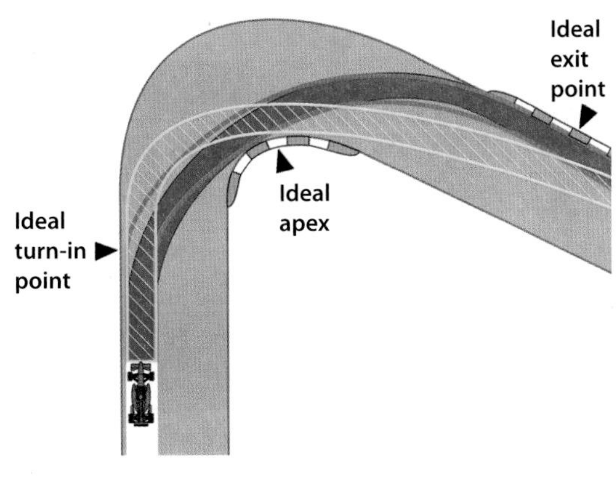

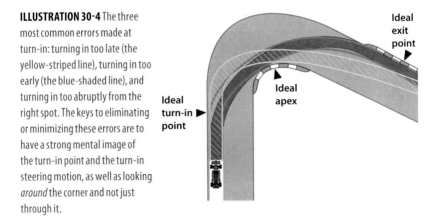

ILLUSTRATION 30-4 The three most common errors made at turn-in: turning in too late (the yellow-striped line), turning in too early (the blue-shaded line), and turning in too abruptly from the right spot. The keys to eliminating or minimizing these errors are to have a strong mental image of the turn-in point and the turn-in steering motion, as well as looking *around* the corner and not just through it.

ILLUSTRATION 30-5 A common error that many drivers make is "crabbing" into the corner, easing the car away from the edge of the track prior to turning in. That's cheating, and it's going to cost you. Be aware if your car is right against the track edge at your turn-in point.

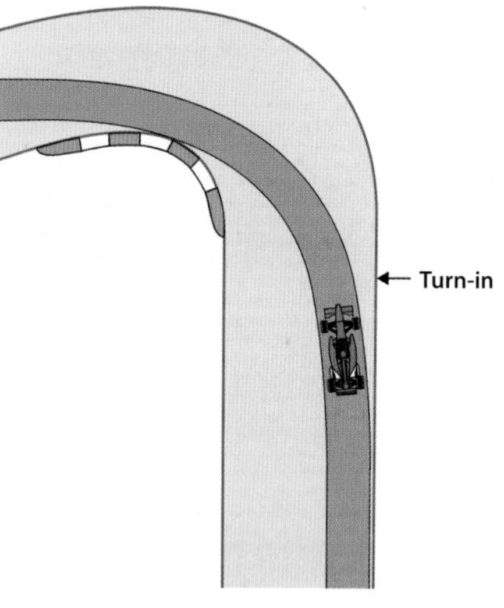

your sense of self-preservation will automatically adjust your speed back to the original "I won't crash at this speed" corner entry.

In many cases, the problem is not the added speed but rather a lack of technique adjustment to go along with the speed. In this case, perhaps if you had trail-braked a little more as you carried more speed into the turn, the car may have rotated toward the apex nicely, and you would have been much quicker. In fact, sometimes, when drivers work to carry more speed into a corner, they simply ease off the brakes earlier. They trail brake less. Yes, that will result in more corner-entry speed, but without the ideal amount of trail braking, the car may not handle it.

It may be a case of blaming the wrong "cause."

So, just because you can't make a small increase in speed seem to work, don't discount it entirely. Rethink how you approach that section of the track. Maybe, by altering your technique slightly, you can make the car stick with the added speed.

Remember the four stages I mentioned earlier: the line, corner exit, corner entry, and midcorner. As your corner-entry speed increases, you may need to alter your line, for example.

SPEED SECRET

Just because a change doesn't work the first time, rethink and retry it again. Alter your technique to make the increase in speed work.

You see, it is not a matter of making less errors, it is simply managing them. Once you realize this, you will undoubtedly make fewer of them. Why? Because, in most cases, you will stop trying to avoid making them. Trying not to make an error is a sure-fire way of ensuring you make them. When you buy into the idea that your job is simply to minimize the effects of them, life gets much easier.

31 COMFORTABLE BEING UNCOMFORTABLE

Great race drivers know that driving at the limit requires driving over the limit at times. As I mentioned earlier in the book, it's a matter of driving a little over, under, over, under, over, until you average out being right at the very knife-edge of the limit. Sure you need to be smooth and tidy, but there are times when you can take that approach a little too far. If you ever feel like the car is great—it's well balanced, not excessively understeering or oversteering—then maybe you just need to push the limit a bit more. Make the car show its weakness. If you drive right at the limit, or just that tiniest fraction below the limit, the car will often feel well balanced. You may not know what its weakness is. That's the signal for you to drive the car a little more assertively.

If you think the elite-level drivers never get crossed up, never make mistakes, and never over-drive the car, you're wrong. Champions do all of these things and more, but are able to make the most of the situation. In fact, what separates them from others is their ability to over-drive, make mistakes, get way crossed up and have big moments and still make it look easy and tidy.

If you've ever spent much time watching an in-car video of Formula One, Indy Car, NASCAR, sports cars, or any other driver at the limit, you know that the drivers are constantly on the verge of crashing. It's like driving at the limit is a series of near misses. Even the very best, in any discipline of the sport, constantly is making mistake after mistake. They are constantly catching one minor error after another. Some of these errors are visible, and many are not, no matter how close you watch. If a driver never makes an error, I'll guarantee he or she will be slow.

SPEED SECRET

Make the car show its weakness.

I've talked a lot throughout this book about how driving at the limit is meant to be a bit ragged, over the top, on the edge. The mental image I want you to have is of being slightly uncomfortable. But, and this is important, you need to be comfortable being uncomfortable. Some drivers understand that they need to be uncomfortable, but they are not comfortable with that. It's like they know they need to be on that ragged edge, but they don't feel at home that way.

For other drivers, they are so uncomfortable with being uncomfortable, they don't drive the limit. They have a mental image of what driving at the limit is, and it's not the right one. Their mental image is one of driving comfortably, and it's not fast.

The mental image you need in order to be fast is one of being comfortable being uncomfortable.

SPEED SECRET

Be comfortable being uncomfortable.

Use mental imagery of other situations where you've not felt comfortable in the past. Anything that felt uncomfortable in the past, replay it, and feel your comfort level matching it. Then see your comfort "envelope" stretching, with everything you did feeling more and more comfortable. But, and this is very important, see yourself craving a little bit of discomfort. Feel your body tingle with that uncomfortable feeling, but still breathe normally and getting a big kick out of the feeling.

Part of being comfortable is being prepared and ready, and that applies to any condition or situation. For many drivers, the sight of rain clouds sends them into a fit of nervousness and undermines their confidence. They "know" they are not good in the rain, and therefore they *are* not good in the rain. And yet other drivers get a big grin at even the thought of it beginning to rain. It's related to being comfortable being uncomfortable, but part of it is just feeling ready, which is another thing that you can use mental imagery for.

See, feel, and hear yourself being fast in all the tough stuff that other drivers don't like as much. Build a program for feeling comfortable when you're uncomfortable, and know that when you're uncomfortable, other drivers are also uncomfortable. The difference is they are not ready for it. They haven't prepared for it, and therefore they are not comfortable being uncomfortable.

The best drivers love any condition or situation. It's as if they get better when the conditions are worse. The tougher and more demanding they are, the more at home they are. You want to be comfortable being uncomfortable, with the car dancing on that ragged edge.

32 USING YOUR STRENGTHS

It's almost funny the way some drivers are really strong for the first few laps of a race, and then fade, while others only really get started once they get to a certain point in the race. Obviously, the perfect race driver is fast at the beginning of a race and stays that way throughout. From my observations, more drivers struggle with making a fast start and then get going, rather than the other way around. But there are many drivers who do fade as a race unfolds.

Earlier in the book, I talked about a season where I was extremely successful over the first lap or two, gaining at least three or four positions on the first lap of every race. As that season wore on I developed a strong belief that I was the fastest starter, but then I began to doubt my abilities to continue to be fast throughout the race. It's odd how we humans think: If someone is exceptional at one thing, they can't be exceptional at something else. Sure, there are exceptions to that rule in the form of superstars who seem to be great at everything, but they are rare. For most people, they think that if they are good at something, such as the start of a race, they might not be good at something else, like the end of a race.

This kind of thinking goes on in every sport. Have you ever noticed how bicycle racers are either climbers or sprinters (unless you're Lance Armstrong), basketball players are either great at offense or defense but not both, and some business people are great in marketing and not great in finance, while others are the opposite? Well, I started to go there myself. I started to wonder how I could be good throughout a race if I was that good at the start.

The good thing about racing is that we can look at data, either from a data-acquisition system or just from a lap timer. During my season in question, I looked at my lap times from the first lap to the last for every race and felt a lot better. My first laps were within a half second of my qualifying lap, and they stayed at that speed throughout the race. Many of my competitors that year took a few laps to get that close to their qualifying laps. So it wasn't me slowing down; it was my competitors taking a few laps to get up to full speed. Therefore, I was able to manage my belief or confidence in my ability to maintain consistent laps throughout a race.

I once coached a driver who was confident and quick over the first lap or two of a race, but with each lap he lost confidence. Unfortunately, his lap times matched his confidence. As his confidence or belief in himself faded, so did his lap times. When I first coached him, I felt the reason for this was simple. I thought it was just that he was so busy racing over the first lap or two that he didn't have time to lose confidence. But as the racing began to settle down and he had time to think, he realized what was going on and figured he didn't deserve to be as fast as he was. While I was right about his mind being so busy that he didn't have time to realize that he shouldn't be confident, the cure was not just telling him to stop thinking that way.

Telling a race driver who is not confident to be more confident is like telling a depressed person to be happy. That just doesn't work. Until a person changes his or her mental programming, the kind of change this driver needed wasn't going to happen.

Ultimately, helping him was actually fairly simple, because the cure to his problem was right in front of him. Because he knew he was good over the first lap, I suggested he just drive a series of "first laps." As simple as that sounds, it had never occurred to him that that was all that any race is: a race-length number of first laps. Just saying that to a driver who is 100 percent confident in his or her abilities at the beginning of a race, you can see a sense of relief on that driver's face.

This driver's mental programming consisted of driving hundreds of "first laps" over and over again. He saw, felt, and heard every detail of driving the first lap of a race. He felt the emotions of driving the first lap of a race. He could feel his state of mind when driving the first lap of a race. He drove the first lap of a race like it was the last first lap he would ever drive, and he loved it. He loved the feeling he got from dragging every last ounce out of the car right from the start of the first lap, to the end of the first lap. And he did it over and over again in his mind. With each first lap he drove in his mind, the better he could see, feel, and hear himself driving hundreds, or even thousands, of first laps at the ragged edge, at the limit. He felt more than confident. He owned the first lap, no matter how many first laps he drove.

Over the next three weeks, this driver would spend a minimum of 20 minutes a day driving "first laps" in his mind. And every single time he passed the start or finish line in his mind he said, "First lap." He built a trigger to launch his first-lap program, and then he just let it go once he got on the track.

An almost amazing thing happened when this driver drove his next race. Not only did he drive every single lap of the race within a couple of tenths of each other, his real first lap was more than a half second faster than he had qualified. And since every lap was within a couple of tenths of that first lap, every lap of his race was faster than he had qualified at. So, guess what he did in his next mental imaging session? You're right, he began programming his qualifying laps as if they were "first laps," and at the next race he took his first pole position.

SPEED SECRET

Use your strengths to help you program over a weakness.

33 LEARNING

If you had to choose just one thing that separated the real superstars of any sport from the rest, what would it be? Superior eye-hand coordination? Desire to win? Work ethic? Natural talent?

While all of these traits, and many more, are factors, the one thing that truly separates the greats from not-so-greats is the ability to learn. In my opinion, the superstars of any sport are not necessarily born with any more natural talent than anyone else. It is what they do with that natural talent that really makes them great.

Rather than just assuming that you have so much talent in your mind or body, why not learn more; why not develop your talents? This chapter is all about learning how drivers learn.

If asked what the main objective is when heading onto a racetrack, most drivers will respond with, "to go faster," "to win," "to develop the car," or something similar. Obviously these are all appropriate objectives. However, I believe there is a more important objective, one single one that will ultimately lead to achieving all other objectives. That objective is to learn. If you continue to learn, becoming a faster, better race driver comes naturally.

When you learn, you will continue to improve your performance. By improving your performance, you increase the chances of developing the car, going faster, and winning.

THE LEARNING FORMULA

Perhaps the most important thing that I have learned personally is what I call the Learning Formula. It may just be the single most valuable piece of information in this book, if you use it.

The Learning Formula is this: $MI + A = G$.

MI represents Mental Image, *A* represents Awareness, and *G* is your Goal (what you are trying to learn). If you use the Learning Formula every time you are trying to improve upon something (which should be at least every time you head onto the track, if not every moment of every day), you will be amazed at your ability to learn and improve.

A few years ago I was coaching a young driver during his first experience on an oval track. One of the big challenges every driver goes through when driving an oval for at least the first time is allowing the car to run close to the walls at the exit of the corners. If you don't unwind the steering and let the car run close to the walls, at best you will be scrubbing speed, and at worst you will spin the car. So, over the course of a couple of hours of on-track practice, I continually reminded and told him over the radio to get closer to the wall exiting the turns.

None of my telling and reminding did any good. He never got within 4 feet of the wall.

Then it dawned on me. I asked him where he needed the car to be exiting Turn 4. He said, "About 1 foot from the wall." I asked him to get a clear mental image of what that would look like from the cockpit. Because it was a private test day, we took the opportunity to walk out onto the track and physically get a picture of what that would look like. He then spent about 10 minutes relaxing, closing his eyes, and mentally "seeing" his car exiting Turn 4 just 1 foot from the wall, over and over again. He developed a "mental image."

For years I have used and taught other drivers to use visualization to learn and improve a variety of techniques. It works. Using visualization greatly increases the ability and the speed at which learning took place. That is why athletes in every sport rely on the technique. I knew that this mental programming can take some time. I had begun to get impatient. I wanted him to drive the car close to the wall, now.

Around this time I had begun to truly understand the value of awareness in the learning process. So I decided to add this to the mix. I asked my driver to go back onto the track after developing his mental image, and simply become aware, without trying to do anything else, including trying to drive near the wall. Each time he exited Turn 4, I asked him to radio to me with the distance his car was from the wall. Basically, I was forcing him to become aware, to add his A to the equation.

On his first lap, he came on the radio and said, "4 feet." On the second lap he once again said, "4 feet." Then, on the third lap he said, "2 feet." And on the fourth lap he came on the radio and said, "1 foot." In four laps he had made the change we had been working on for 50 or 60 laps. In a matter of minutes he cured the problem that I had been telling him to fix for more than two hours. All by adding *MI* to *A*, he had reached his goal, his *G*. He never again had to work at allowing the car to run close to the wall.

As you can imagine, I jumped on this technique and have used it extensively in coaching ever since and with myself. Whenever I want to make a change to my driving technique, I use visualization to develop a clear *MI*, and then I go onto the track and simply become aware. I build my *A*, mostly by asking myself questions.

These awareness-building questions are things like, "Can I carry more speed into the turn?" or "How assertive am I?" I might ask myself, "How far from the end of the curbing am I turning in?" or "How far into the turn am I at full throttle?" When used along with the appropriate *MI*, these *A*-building questions help me achieve my goal, quickly, efficiently, safely, and enjoyably.

I don't know of a faster way of learning anything than with the Learning Formula. Practice using it both on and off the track.

Here's another example of how it works: Let's say you're consistently over-slowing the car when entering a particular corner. You could just tell yourself over and over to "carry more speed into the turn." Will that work? Not likely. You could do some mental imagery—visualizing or mental programming—of braking a little lighter, not taking as much speed off, and entering the Turn 2 or 3 miles per hour quicker. Will that work? Most likely, but it will take a while.

The reason it could take some time to take effect is that you may *only* have the mental image of what you want to do. You may not have any awareness of what speed you are currently carrying into the turn. So rate your corner-entry speed on a scale of 1 to 10, with 10 being the ragged edge, the limit, 10 tenths. Then, go back onto the track and ask yourself how close to a 10 you are each lap, how close to the mental image of his goal you are. By combining the mental image with awareness, you will begin to carry more speed into the corner before you know it.

Again, if a driver has a clear mental image, and an awareness of what he or she is doing right now, the driver's mind will bring the two together. You may be surprised at how rare it is for a driver to have both of these components. Some may have a clear mental image of what they want to accomplish, but have no awareness of what they are currently doing. Others do not have a mental image of what they want, mostly because they are overly aware of what they are currently doing. They are so focused on what they are doing wrong, they can't get the mental image of what they want.

SELF-COACHING

I wish that every time you went onto a racetrack that you had a qualified coach to work with you to continually improve your performance. I hope you wish for that as well. However, it is doubtful that will be the case, either due to financial restraints or the fact that there are probably not enough qualified coaches in existence to cover every driver. Therefore, you need to learn to coach yourself.

Self-coaching is the technique of guiding yourself toward maximizing your performance and improvement in everything you do.

Part of self-coaching is debriefing with yourself. The primary objective of debriefing yourself is to increase your awareness, for without awareness, you will have a difficult time knowing what you should work to improve. Nor will you make any improvement, or be aware of any progress.

One of the best ways I know of to increase your awareness level is by asking yourself questions, and rating your performance and abilities in a variety of areas. For example, I like to use a 1-to-10 scale to rate my overall performance, smoothness, level of intensity, how close to the limit I am, etc. To help facilitate that I use a debrief form like the one in Illustration 33-2.

The idea of the debrief form is to rate how close you have the tires to their very limit. On a scale of 1 to 10, with 10 being the very ragged edge of the traction limit and 1 being a long ways from that, rate each section of the track. Do this for the braking zone, the entry one-third of the corner, the middle third, and the exit third of the corner.

Date:	Event:
Driver:	Car:
Session:	Best lap:
Misc.:	

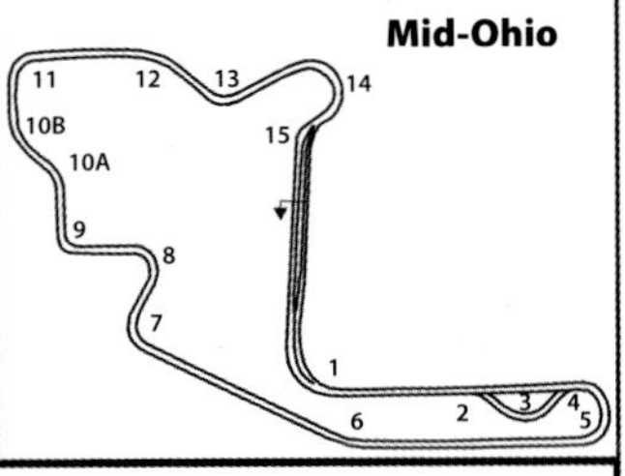

Mid-Ohio

11 12 13 14 15 10B 10A 9 8 7 1 6 2 3 4 5

Driver Notes					
Turn	Gear	Braking	Entry	Midcorner	Exit
1					
2/3					
4/5					
7					
8					
9					
10A/B					
11					
13					
14					
15					

Comments/Areas to work on:

ILLUSTRATION 33-2 Use this debrief form to coach yourself by raising your awareness level. After each on-track session, sit down and rate how close you drove the car to its limit in each section of the track.

SPEED SECRET

Increase your awareness by debriefing, even with yourself.

Obviously you can debrief yourself without a form like this, but I've definitely experienced that a driver who actually writes this information down gains more from this strategy. The physical act of writing something seems to increase your awareness level, as well as being more accurate. You will be more honest with yourself when you write the rating numbers down on paper.

In addition to using the debrief form, I recommend asking yourself a number of questions after each session. I've listed a number of example questions in Appendix C, ones you should ask after each session and ones to ask yourself before you even get to the track. If you answer them, even if you go through the process of digging deep in the attempt to answer them, you will coach yourself to a much higher level of performance behind the wheel of a race car. The overall objective of these questions is to help you become more aware of exactly what you are doing. If you are aware of what you are doing, and you know what you want to be doing, you will quickly and naturally make the necessary improvements.

LEARNING STYLES

Every person on this Earth who learns anything (and that includes race drivers!), learns in a different way. We all have our own preferred or dominant learning style. Some drivers learn better when things are presented or approached visually, others learn best auditorially, and others still learn best when they experience things, a kinesthetic style.

The main point I want you to understand is if your driver's preferred learning style is kinesthetic, and you proceed to tell him how to do something, don't blame him if he doesn't get it. If his preferred learning style is auditory, and you use a picture to illustrate something to him, you are the one to blame for him not understanding it.

If you want someone to learn something, present it to him or her using that driver's preferred learning style. If not, the driver may have a difficult time learning it.

So how do you know what your driver's preferred learning style is? Often you can simply observe when the driver has really learned something quickly and effectively, and relate that to how it was presented to him or her. But the easiest way is to ask the driver. And if your driver doesn't know? Ask him or her to think back to when he or she learned something in life quickly and easily. How was it presented? Was it told to the driver (auditory)? Was it shown to him or her (visual)? Or was it not until he or she actually did it or experienced it that the driver learned it (kinesthetic)? Armed with that information, you can begin to experiment with the different styles to see which is most effective.

Having said that, the most effective way of learning is to actually combine all three learning styles. Use the driver's preferred style, and back it up with the other two. To have your driver truly learn where the apex of Turn 2 is, tell the driver where other drivers are apexing, draw a picture of where it is, and physically drive or walk the driver through the corner. But, make sure you use, and probably start with, the driver's preferred learning style.

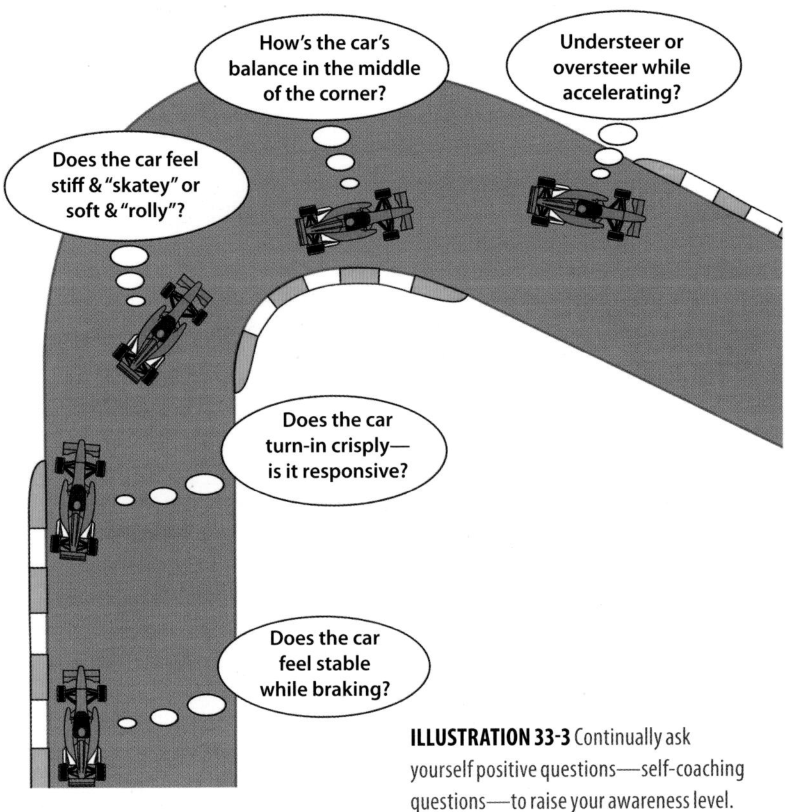

ILLUSTRATION 33-3 Continually ask yourself positive questions—self-coaching questions—to raise your awareness level.

LEARNING STAGES

Whether it is learning to walk, throw a ball, or drive a race car, all human beings go through four stages:

- Unconscious incompetence
- Conscious incompetence
- Conscious competence
- Unconscious competence

It may be easiest to relate these four stages to a baby learning to walk. In the beginning, a baby is at the *unconscious incompetence* stage. The baby hasn't yet discovered that people can walk. In other words, the baby doesn't know what he or she doesn't know how to do.

At the *conscious incompetence* stage, the baby has now seen his or her parents walking, and wants to, but can't. The baby knows what the baby doesn't know how to do.

The next stage, the *conscious competence*, is where the child who is first learning to walk has to think about each step. The child knows what he or she knows how to do but is having to do it at the conscious level.

Finally, at the *unconscious competence* stage, the toddler no longer has to think about walking; it now happens automatically. The child just does it. He or she doesn't think about what he or she knows. The child knows and does and doesn't have to think about it.

Every driver goes through each of these four stages with every technique the driver learns. As an example, blipping the throttle while downshifting. At one point, your driver didn't know the technique existed, knew nothing about it, and knew nothing about why a driver does it. Stage 1: *unconscious incompetence*.

Then, the driver became aware of the technique, but didn't know how to do it. Stage 2: *conscious incompetence*.

As the driver began to practice it, he or she had to think through each detail of it. Stage 3: *conscious competence*.

Finally, after practicing it over and over, it became automatic, a habit, and the driver no longer had to think about, he or she just did it. Stage 4: *unconscious competence*. Obviously, to drive a race car quickly, your driver must reach this stage. This is driving at the subconscious level.

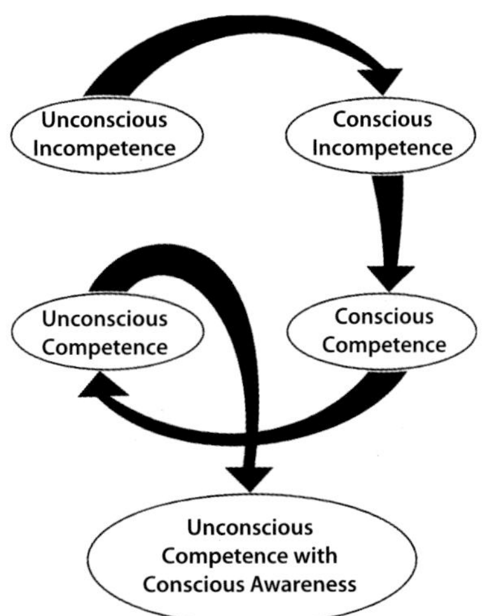

ILLUSTRATION 33-4 With every skill and technique a driver learns, he or she goes through these learning stages. The conscious competence stage is where the driver is having to consciously think about everything; the unconscious competence stage is where the driver is doing it without thinking, driving at the autopilot or subconscious level. But without the fifth stage, without conscious awareness, the likelihood of the driver improving is low. Driving at the subconscious level while being aware at the conscious level (unconscious competence with conscious awareness) is practically the definition of being "in the zone."

These stages of learning can be found in most textbooks on learning strategies. I would, however, like to add a fifth stage that is not included in the textbooks, but that I have witnessed and experienced at the racetrack. At least when it comes to driving race cars, and I believe with most everything else, there should be an *unconscious competence with conscious awareness* stage.

The *unconscious competence* stage is much like the experience of driving somewhere, only to get there and not remember actually driving there. I'm sure you have experienced that at least once in your life. You are driving completely on autopilot; your mental programming is handling the chore, while your conscious mind is off in another world.

Yes, at this level you are operating about as efficiently as possible, but you are *unaware*. Think of it this way: You've been driving the same route to work for years. You drive to work but don't remember doing it. You are certainly competent at it,

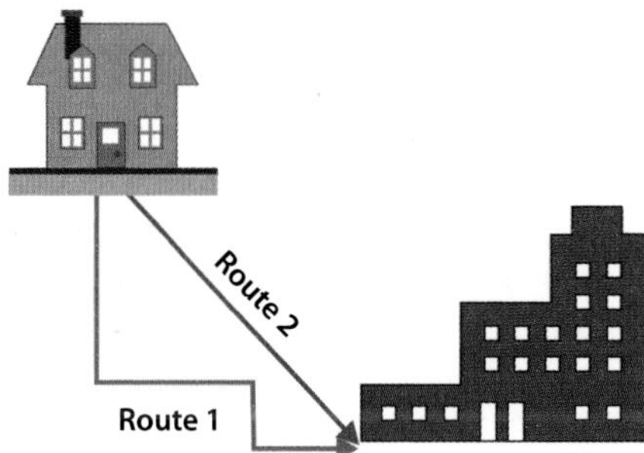

ILLUSTRATION 33-5 Have you ever driven from your home to work, and upon arriving, realized you don't remember driving there? Most people have. That's illustrated by Route 1 above, and it's what happens when you operate only at the unconscious competence stage. You're doing it well, but the chances of doing it better are slim. Without conscious awareness you are likely to be unaware of a new super-highway, Route 2, having been built. It's when you're at the unconscious competence with conscious awareness stage that you perform at your peak and continue to improve.

so much so that you didn't even have to consciously think about it. However, while your conscious mind was off in another world, the highway construction crew built a new road, a shortcut, that would cut your commute time in half. Because your conscious mind was unaware, though, you never noticed.

Unless your conscious mind is aware, you will never make any improvements. Yes, you will continue to drive well at the subconscious, programmed level, but you will never upgrade the performance of that programming. Many times, this is the reason for a plateau on the learning curve, a complete unawareness of what could or should be improved.

The ultimate goal for any race driver is to drive at the subconscious level, relying and trusting his or her programming, while using his or her conscious mind to observe and be aware of any ways to improve the programming. It should be as if the conscious mind is looking over the driver's shoulder, much like an in-car camera would see it, watching for opportunities to upgrade the software.

LEARNING FROM THE INSIDE OUT

Drivers, and everyone else in the world, learn in one of two ways: from the inside out or from the outside in.

Learning from the outside in is what most people typically think of as learning. This is how we were taught in school. Learning from the outside in is what happens when you "teach" your driver something. It is the definition of you telling the driver what to do or of giving the driver information. It is the information or knowledge coming from the outside (from you) and getting in the driver.

Actually, the getting in is not much of a problem; it is the staying in that is the challenge. And, without staying in, the driver really hasn't learned it.

On the other hand, learning from the inside out is what coaching and driver engineering is all about. Learning from the inside out is when drivers discover or learn something for themselves—often through your guidance or stimulation—rather than just being told what to do.

As an example, in the past I have spent countless hours teaching drivers about the line through various corners. I have talked to them about the turn-in point, the apex, the exit, and why it all matters. I've discussed the reasoning behind the line until I'm blue in the face. And yet, many times the drivers I'm talking to do not drive the line consistently (if at all).

Of course, what I've been doing is attempting to have them learn from the outside in. It has only been in the past few years that I've learned myself just how effective and efficient learning from the inside out really is. Recently, when trying to get a driver to understand why one specific line through the corner is more effective than another (or no line at all!), I took an entirely different and unique approach. Instead of telling him where he needed to drive, I used a tool (I'll share that with you soon) that helped him become aware of what the car needed, and let him discover (learn) for himself where the line is. His comment after using this tool was, "Oh, that's why you wanted me to drive that line!"

This type of learning is so much faster, efficient, and long-lasting. In fact, once drivers learn from the inside out where the line is, they will know it forever. They will also be able to apply it in other corners, at other tracks, and with other techniques.

If you have ever wondered why some drivers seem to have a knack for knowing "instinctively" where the line is for a corner, while others struggle with it, or need to be told by someone, this is the reason. Once drivers truly understand *why* the car needs to be driven on a certain line through a corner, they will have learned the feel for it and will be able to apply that to each corner they face.

Of course, this approach works equally well for all aspects of driving, not just learning the cornering line. You can tell your driver to carry more speed through a corner, be smoother with his or her downshifts, or turn the steering wheel more progressively, but as long as you are telling the driver how or what to do, he or she will not fully learn it.

TRIAL AND ERROR

Is there anything wrong with making an error? I know, I know, it depends on how much the error costs and who's paying for it. But if you want to improve, or even just to perform at your very best right now, you must be willing to make a few errors. Why? Because, looked at and used in the right manner, errors are a valuable learning experience.

If you think about it, when we were young, trial and error was our most common and effective learning technique. Take the act of learning to walk, for example. Imagine if, after falling down the first few times you tried to walk, your parents said, "We don't want you trying that anymore; you always seem to fall down." Or you, yourself, thinking, "I can't seem to get this right, so I think I won't bother trying anymore."

Pretty absurd, right? And yet, we do this as race drivers all the time. The second we make a mistake, we tell ourselves (sometimes subtle, sometimes not so subtle) to never make that error again. How often does that help? Not very often!

Of course, crashing a race car is much more expensive than the damage caused when you fell down while learning to walk. The point is, though, the more you resist errors, the more likely it is you'll make more.

One of the biggest differences between great drivers and not-so-great drivers is *not* that the former makes less errors. In fact, they both make about the same amount of mistakes. The difference is that the great drivers recover from, learn from, and know how to minimize the consequences of most errors. That only occurs when there is an atmosphere that allows errors and learning from them.

Errors are simply a form of feedback that helps you home in on the desired goal. They are error signals that help you continue to improve.

LEARNING THROUGH OSMOSIS

Why do you think second-generation drivers like Michael Andretti and Al Unser Jr. were so good (and their third-generation sons)? Is it because of their genes, what they inherited from their fathers? While I won't say their fathers had nothing to do with their abilities behind the wheel, I don't believe it had much, if anything, to do with their DNA. But I do think they acquired most of their "natural talent" from their fathers prior to ever getting in a race car.

Michael Andretti and Al Unser Jr. acquired much of their talent by keenly observing their fathers, by absorbing everything they were exposed to as children. And, in both of these cases, that was a lot to absorb. They learned through osmosis.

All race drivers learn through osmosis. The more they are exposed to, the more they learn.

Tennis coaches in England have noticed for years a direct correlation between their students' abilities and the television coverage of Wimbledon. For a couple of weeks immediately following the tournament, tennis players' performances improve significantly. Did they practice more, change their swing, or buy a new tennis racket? No. They simply learned by watching.

You can learn a lot by watching other drivers. Of course, it makes sense that you should observe and learn from the best you possibly can. You will not learn as much from watching drivers who are not as good as you, although it is still possible to get something from that experience.

LEARNING PROGRESSION

Every driver, no matter how much natural talent you think the driver has or doesn't have, will continue to improve throughout their career. Even drivers reaching the ends of their professional careers are still improving in some areas (which, unfortunately, are often overcome by other factors, such as lack of motivation or desire or a deterioration of physiological functions). How quickly, and how consistently, you'll improve all depends on you and the environment you're operating in.

One thing is clear, though. No two drivers have ever—or ever will—progress at the same rate. Some drivers' learning curve is mostly a steady upward incline, while others are full of steps of all shapes and sizes.

What is common is the plateau. Often, you and the people working with you become frustrated with the lack of progress, and the plateau lasts longer. In my experience, most of the time plateaus are followed by a sharp incline in progress, if you control the frustration and focus on improving the your awareness.

Many times the plateaus appear to even be steps backward. It is like one step back, two or three forward, one step back, two or three forward, and so on. I like to compare them to the calm before the storm. In this case the calm is the apparent lack of progress, and the storm is the whirlwind of learning.

If you think back to the learning stages, you'll understand why. Most often, for you to progress, you must go back to the conscious competence stage, where you're thinking through each step in a mechanical manner. This results in too much conscious thought and an apparent step backward. If you don't become frustrated by this, and with a little patience, this new technique, skill, or mental approach becomes part of your programming. You then progress to the unconscious competence stage, where it becomes something that you seem to do naturally. At this point, there is a significant step up in the learning process.

LEARNING CURVE

Watching children learn just about anything is an educational experience in itself. One thing I've learned from close observation of my daughter is the steps a child takes in the learning process. Just when there seems to be no progress whatsoever, bam, she masters it. It certainly is not a steady progression. No, the learning curve is more like learning steps.

Excuse me for a moment while I use my daughter as an example. When she was four years old, I decided that it was time she learned to ride her bicycle without training wheels. Notice I said, "I decided." So I went ahead and took off the training wheels, and then spent the next few hours trying to get her to learn how to keep her balance. It was certainly good exercise for me. The bottom line was, she was neither willing nor ready to take this next step. The training wheels went back on.

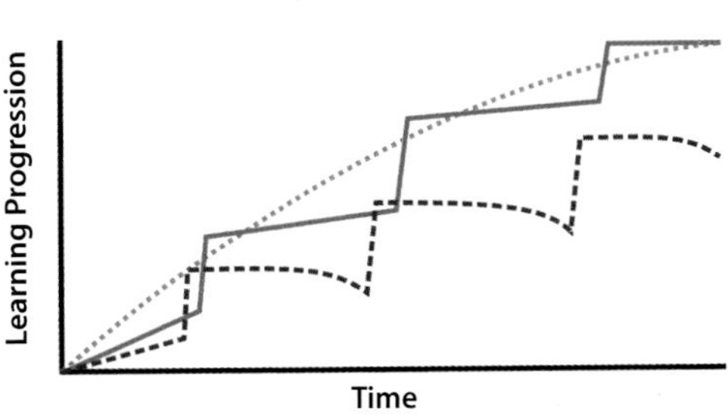

ILLUSTRATION 33-6 Rarely, if ever, does someone's learning truly follow a curve. Instead, it happens in steps, what I call learning steps (the solid green line) as opposed to the commonly thought learning curve. There's an all-too-common variation: the frustration steps (dashed blue line). This is where a driver gets frustrated by the lack of improvement, a plateau, and begins to try real hard and actually gets worse. It's only when the driver "gives up" and relaxes that the driver begins to improve again.

A couple of months later she came to me and asked to have the training wheels taken off again. *She* had decided it was time to learn to ride a two-wheeler. Within minutes she had practically mastered it. Within 30 minutes she was showing me how she could ride up and down steep hills while holding on to the handlebars with one hand.

There was no learning curve here, at least not when observed from the outside or probably even consciously on her part. It appeared as though her learning curve was absolutely flat, then took a perfectly vertical step. In reality, even though neither of us were aware of what was going on, she was progressively learning.

And you know what is most interesting? All the race drivers I've observed or worked with closely follow the same pattern in their development.

The only drivers that do not seem to follow this pattern are the ones that get frustrated when they are on a plateau, the flat part of the curve. They get to that point where they don't feel they are getting any better, get upset or frustrated, and stay at that level or even get worse. The one piece of advice I can give about the learning process is that if you seem to be stuck at one level for some time, be patient. If you are using the strategies suggested here, you are about to make a big step up to the next level. You are about ready to take the training wheels off.

SPEED SECRET

If it seems you are not improving, you are about to.

Over the past few years a lot of people have talked about the importance of karting to the success of today's top drivers. Look at the grid, at the front of the grid, anyway, of almost any of the top forms of road racing today and you will see drivers who have grown up racing karts. And, check out what they do between races, and you will most likely find them at a kart track of some type.

One question comes to mind to most casual observers: How does driving 30- or 40-horsepower karts relate to driving a 500- or 800-horsepower race car? The usual response is to point out the kart's power to weight ratio, the cornering grip, and just how fast things happen on a kart. All these things are suppose to help a driver keep in shape, mentally and physically. And they do. However, there is another area where driving karts of any type and speed can help a race car driver: *learning how to learn* how to go fast.

Every time you drive on a track, you are constantly trying to figure out how to drive quicker. At least you should be. If not, you are not a real race driver.

As you drive through a corner, whether in a 700-horsepower car or a 5-horsepower rental kart, your mind should be sensing how the car or kart is reacting. You should be traction sensing, becoming aware (subconsciously) of whether the car or kart has any traction in reserve at any point throughout the corner, allowing you to go faster. You should be analyzing whether turning in earlier or later, taking a shallower or larger radius through the turn will enable you to carry more speed. You should be experimenting and discovering whether a change in when you brake, how much you brake, when you begin accelerating, how hard you accelerate means a quicker lap time.

In other words, you are constantly trying to learn how to go faster. And it doesn't really matter what you are driving to learn this. I drive a shifter kart to stay in shape and tune up. I also drive rental karts on indoor tracks. I sometimes wonder which I learn the most from. Yes, the shifter kart relates more to my race car in terms of speed, but learning how to make a 6-horsepower kart get around a slippery indoor track is just as challenging from a learning perspective. If learning how to find that last little "trick" to break the track record at the local rental kart track is your objective, you will become a better race car driver.

LEARNING AS AN OBJECTIVE

Given a choice, which would you rather have, a race win or a great learning experience? That may be an unfair question, but I want to get you thinking. Most racers would do or give just about anything for a win, but would you give up learning for one?

No matter how talented you are, or how successful you are, the more you improve, the better your chances of winning in the future. I've known and seen far too many drivers who thought they were so talented and knew so much about driving that they would be successful wherever they raced. Every single one of them eventually reached a point where they no longer had an advantage, and they were no longer successful. If they had focused on learning and improving, they would have built on the advantage they had and continued to be successful.

SPEED SECRET

The more you learn, the better you get;
the better you get, the more you win.
Focus on learning, and you'll win more often.

Let me tell you about a driver, who will remain nameless, who I knew. He had always been a competitive person, in business, in sports, in anything. When he and his friends did anything, from buying homes to cooking the best barbecued burger, he had to be the best at it. He hated to lose.

The competitive nature had been a good and bad thing for him. It drove him to practice as often as possible and be aggressive in traffic, but it's also inspired him to do things that he regretted later, such as trying to force another driver off the track. And because of that inner drive and competitive nature, from the beginning he won more than his share of races. And that's what led to the problem.

The more he won, or even came close to winning, the more he focused on winning. While that's not a bad thing in itself, it is if that's the only focus; as others in his class improved, he began to win less often. The less often he won, the harder he tried to win. The harder he tried, with all of his focus simply on winning, the worse his driving performance became and the less he won. Whenever anyone tried to talk with him about his approach, he would snap back with, "I've won in the past; I'll win again if I just try harder."

So here's a question for you: Do you typically turn your fastest laps when you're trying really hard or when you're relaxed? I'm not suggesting that relaxed is not caring, being unfocused, or not performing at your maximum. But there is a huge difference between trying hard and being relaxed and focused. And that's where my acquaintance made his mistake.

SPEED SECRET

Focus on your performance; the result will take care of itself.

ESTABLISHING STRATEGIES AND OBJECTIVES

A car engineer would never be successful at developing the car without some type of plan. The same thing applies to learning and developing your driving. Without a plan, one of two things will happen. Either no change—and no improvement—will take place, or the wrong changes will take place. That is why establishing objectives prior to every on-track session is critical, even if the objective is to not make any change so that you can make note of some subtle change to the car.

It is a complete waste of time if you head onto the track without two or three specific objectives. These objectives may relate just to the car, in terms of feedback on a particular setup change, or they may all have to do with a change in driving technique. The point is, without making a change, it is doubtful at best that the car or your driving will improve.

One of the best ways of doing this is by figuring out what questions you plan to ask yourself after the session. If, for example, you're going to ask where you're beginning to brake for Turn 1, what you do with the steering wheel just after turn-in for Turn 4, and whether the car understeers or oversteers at the exit of Turn 8, you have helped establish three specific objectives for the session. You have helped yourself focus.

34 ADAPTABILITY

The driving styles of the greatest racers in the world have always had one thing in common. Whether it's Jackie Stewart, Alain Prost, Ayrton Senna, and Michael Schumacher in Formula One; Mario Andretti, Rick Mears, Helio Castroneves, and Dario Franchitti in Indy-car racing; or Richard Petty, Darrell Waltrip, Dale Earnhardt, and Jimmie Johnson in NASCAR, the key to their success has always been smoothness and finesse (even if it doesn't always look like it in the rough-and-tumble world of NASCAR).

With experience you will develop your own driving style, one that suits your personality and your car. Everyone, in fact, has their own driving style. I hope yours will be one of smoothness and finesse, as well.

Driving styles are why a car set up for one driver may not suit you. If, for example, the car you're driving is understeering slightly in slow corners and you want it to oversteer, think about how you can alter your driving style to help the situation. What normally happens is you get a little frustrated with the understeer and try to force the car to go faster. About all that does is make the understeer even worse, slowing you even more. Usually, you're better off being patient with an understeering car. Slow down a little more on the entrance to the turn, working the weight transfer to your advantage, and concentrate on getting good acceleration out of the corner onto the straight.

What I'm saying is, when the car is not handling the way you would like, think it through. Think about whether there is a way you can modify your driving style to suit the car. It may be easier and less expensive than trying to modify and adjust the car.

Your driving style or technique may actually be the cause of what you consider to be a handling problem. So whenever you are having a handling problem with your car, don't just think about how to adjust or modify the car's suspension and aerodynamics. Consider your driving style, or perhaps your driving errors. The first thing to determine when dealing with a handling problem is whether you are causing the problem. Take a real good look at your driving style and be honest.

You influence the weight transfer and tire traction at each corner of the car and at each and every turn on the track in a variety of ways. If you are too hard on the throttle in the middle of a turn (probably because your corner-entry speed was too low, and now you're trying to make up for it by accelerating too hard), you may cause the car to either understeer or oversteer. How and when you use any of the controls can often cause, or cure, a perceived handling problem.

For example, when entering a turn, if you turn the steering into the corner too quickly (not giving the front tires a chance to gradually build up their traction forces), you may experience an initial turn-in understeer. This is particularly true if you do not trail brake enough.

Is this initial understeer a handling problem, something for which you should modify the chassis setup? Or maybe you're turning too gradually, never getting the car to take a set in the corner until you're half way through it.

Sure, you should work on adjusting the suspension to help cure any problem. But by doing that you may cause another problem elsewhere (such as an oversteer during the midcorner or exit). Instead, it may be better to adjust or improve your driving style or technique. The key is analyzing and recognizing the problem.

Now don't get me wrong. I don't suggest trying to overcome every handling problem by altering your driving. Always consider how you can improve the car, but don't fool yourself. Look at your driving technique as well.

ADAPTABILITY

One of the key areas that separate good race drivers from great race drivers is their ability to adapt their driving to suit a car's handling or from one type of car to another.

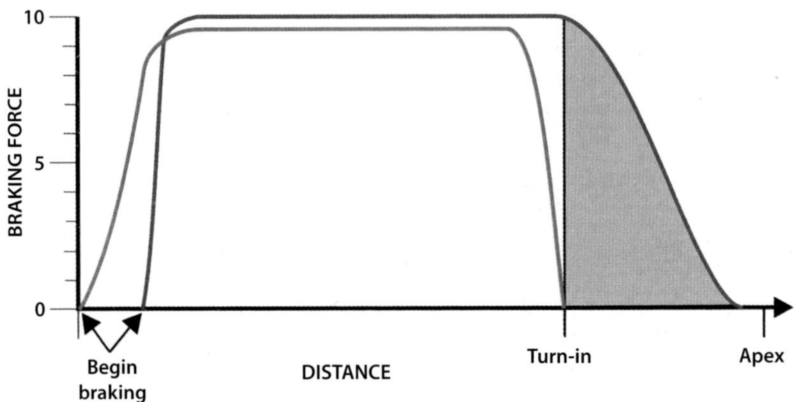

ILLUSTRATION 34-1 If you graphed the perfect, theoretical brake force application, it would look something like the solid green line in the graph above. The shaded area is the trail-braking phase. The red line shows three common braking errors. First, the initial application of the brakes is too soon and too slow. Second, the driver is not using all of the brake force available. It's just under 10. And last, and probably the worst thing, the driver finishes braking too soon. By not keeping the front of the car loaded by trail braking, the car would probably understeer beginning right at the turn-in point. Is that a chassis setup problem or a driving style problem?

Some drivers, despite how the car is handling, will only drive it one way—their style. And guess what? A driver's style will never suit every handling characteristic. If you cannot adapt your style to suit the car's handling, a change in track conditions, a mechanical problem, or a different type of car, I doubt you will ever be a real champion race driver.

In 1994, Michael Schumacher finished second in the Spanish Grand Prix, despite the fact his Benetton was stuck in fifth gear only. What was really impressive was that other than in about two laps when he first encountered the problem (as he figured out how to adapt his driving to the situation), no one but his team even realized he had a problem. His lap times barely changed. That is one of the reasons he was the champion he was.

Although it would be impossible to list every problem scenario you may someday face when racing, I am going to attempt to identify the most common ones and give you some suggestions as to what you may be able to do to adapt your driving to help the situation.

The overall objective with the following suggestions is to give you some knowledge as to what you may be able to do to reduce the effect of the problem. In other words, what can be done so the problem has the least effect on your lap times and your ability to race your competitors? Of course, if at all possible, you would adjust the car: the anti-roll bars, brake bias, weight jacker, and so on. But if you don't have any more adjustment, or any adjustment to begin with, it is all up to your adaptability once you are in the race.

As I said, being able to jump from one type of car to another is also an important element to being a great race driver. For example, having the skill and knowledge to be able to drive a rear-wheel-drive purpose-built race car running on slicks and a front-wheel-drive production-based car running street tires will greatly increase your chances of being a hired gun. Therefore, I will also try to cover the basics of the differences in driving style required for different types of cars.

SPEED SECRET

Improve your adaptability through knowledge and practice.

I'm giving you this information or knowledge at the conscious level. For you to truly use it you will need to make it a part of your subconscious by programming it using mental imagery. Until it becomes a subconscious program, the information will be next to useless; you will not be able to access it efficiently while at speed under racing conditions.

CORNER ENTRY UNDERSTEER

The best place to start when trying to figure out what you can do with your driving technique to help any handling problem is to think about the weight balance of the car. If your car is understeering in the entry portion of a corner, consider what you can do to induce some forward weight transfer and what you can do to lessen the weight transfer to the rear.

To increase the forward weight transfer, you can increase or lengthen the amount of time spent trail braking into the corner. That means not trailing, or easing off the brake pedal so quickly, keeping a little more pressure on the pedal for a little longer. And, if it is a corner that requires little to no trail braking, then it may be a matter of waiting a little longer—being more patient—before beginning to accelerate, or squeezing on the throttle a bit more gently.

One of the challenges is when you are chasing another car. As you approach and enter the corners, the distance between your car and the competitor's car is reduced. Visually, it seems you are catching the other driver (even if the time gap has not changed, it is just that you are traveling at a slower speed). So your natural instinct in trying to catch your competitor is to actually ease off the brakes and get back to throttle a little sooner. That, of course, exaggerates the understeer, slowing you a little more. Then, you "try" a little harder, carrying more speed into the corner, causing more understeer, overheating the front tires more, causing more understeer, you "try" even more, and so on. As you can see, the problem just gets worse and worse.

The key, then, is to be patient. You will probably end up entering the corner a mile-per-hour or so less. If you focus on increasing the forward weight transfer and decreasing the rear weight transfer, you will be able to get the car rotated (turned) earlier in the corner and get back to throttle solidly, without having to come back off it to control the understeer. That will improve your acceleration out of the corner and down the straight, giving you a better chance at passing the opposition.

Avoid the temptation to turn in a little earlier. If anything, you want to turn in a little later, opening up the exit line so that you can focus on the acceleration phase of the corner.

One of the benefits to trail braking more is that you may actually be able to begin your braking slightly later (since you are doing more of the slowing down in the entry phase). However, that may also be part of the cause of the understeer problem: overloading the front tires. If you are carrying a lot of speed into the corner, still have a fair amount of braking going on, and trying to get the front tires to change the direction of the car, you may be asking too much from them. In this case, the cure is to begin braking a little sooner and trail brake a little less. Be patient.

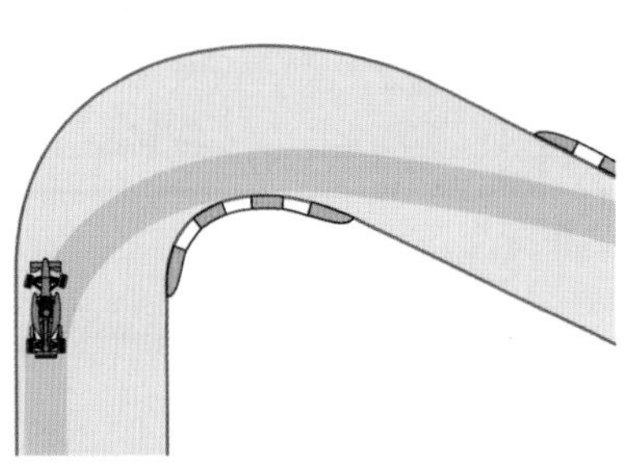

ILLUSTRATION 34-2 Corner-entry understeer.

If the weight balance of the car is not the cause or the cure of the understeer, then you have to consider one other thing. It doesn't matter where in the corner the understeer is; think about what you are doing with the steering wheel. Often, corner-entry understeer is caused by the driver cranking in too much steering input or cranking it in too abruptly. Try turning the steering wheel a little less and a little more gently. I know it may not feel right; the car is not turning enough (understeering), so you turn the steering less? Exactly. Keep the front tires at an angle they can work at. If you steer the front tires too much, they can't help but give up their grip and begin to slide.

Again, be aware of how much steering you have input and try taking some out. Or turn the steering wheel a little slower, a little more gently as initiate your turn-in. Give the tires a chance to build up their cornering grip.

CORNER-ENTRY OVERSTEER

Oversteer is often a result of too much weight on the front tires and not enough on the rears. If that is the case during the entry phase of a corner, that probably means you are braking too hard into the corner; you are trail braking too much.

The cure then is pretty simple. Just begin braking slightly earlier, and trail off the brakes a little sooner as you enter the corner. Perhaps, especially if it is a corner that requires little to no braking, it is a matter of beginning to accelerate sooner (but very gently), transferring more weight onto the rear tires.

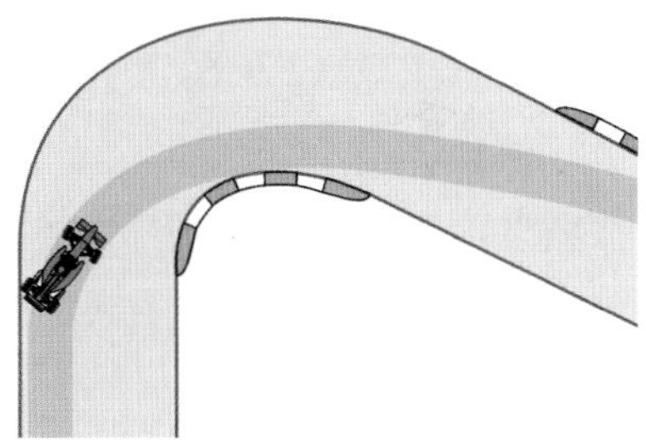

ILLUSTRATION 34-3 Corner-entry oversteer.

Again, when you are chasing another competitor, it is easy to fall prey to the "brake late and I catch him" impression. Always keep in mind that you will gain more, both on your competitors and in reducing your lap times, by early acceleration than you will by late braking.

One other thing that may help reduce corner-entry oversteer is to turn the steering wheel less abruptly. Ask the car to change directions, from straight forward to a curved path, a little more progressively.

MIDCORNER UNDERSTEER

Usually the best way to handle midcorner understeer is simply by smoothly modulating the throttle to change the weight balance of the car. In other words, just breathe. Gently lift off the throttle to cause some forward weight transfer, giving the front tires more grip.

Often the understeer is not related to the car's setup, but you have gotten back on the throttle to begin accelerating just a little too abruptly or early. Again, just breathe the throttle to transfer some weight forward.

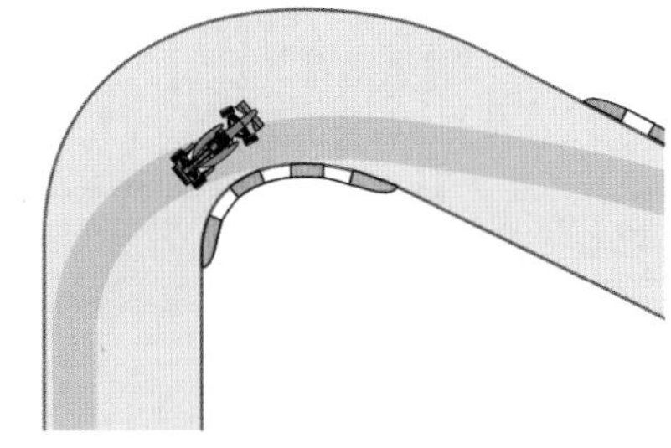

ILLUSTRATION 34-4 Midcorner understeer.

Also, just like with the entry understeer, be aware of the amount of steering input you have dialed in. Perhaps the cure to your car's midcorner understeer problem is just unwinding the steering a little bit to allow the front tires to get some grip.

MIDCORNER OVERSTEER

Dealing with a midcorner oversteer is almost always done by changing the car's weight balance. In this case, that means squeezing more throttle on. However, one of the reasons the car has begun to oversteer is that the speed you are carrying is slightly more than the rear tires can handle. So the last thing in the world you need right then is a bunch more speed. That is why it is critical to squeeze on just a little more throttle.

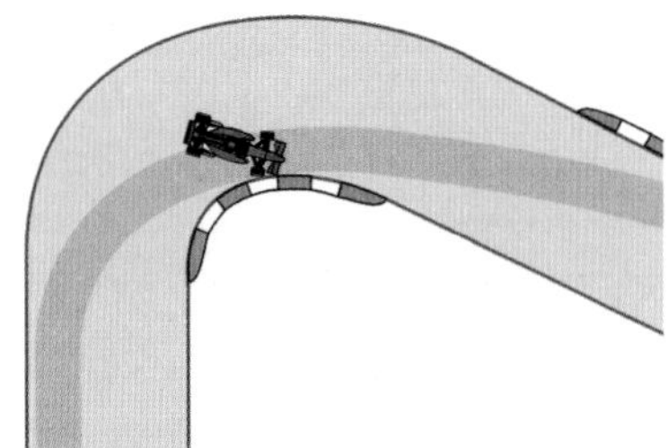

ILLUSTRATION 34-5 Midcorner oversteer.

The midcorner oversteer could also be caused by wheelspin (in a rear-wheel-drive car, of course), which should be dealt with by being a little easier on the throttle.

If your car's setup is the cause of the wheelspin, about all you can do is be as gentle as possible with the acceleration and possibly alter your line slightly. If possible, try driving the car a little deeper into the corner before turning in, make the initial turn radius a bit sharper, aim for a later apex, and then let the steering unwind as early as possible. This makes for a straighter acceleration line, meaning that there will be less cornering force to combine with the acceleration force that you are asking from the rear tires.

EXIT UNDERSTEER

If your car has an exit understeer problem, the best thing you can do without reducing your acceleration is to alter your line. Your prime objective is to lessen the amount of time you are turning the car while accelerating. So if you turn in a little later and sharper (even if this means slowing the car down a little), and aim for a later apex, it will allow you to unwind the steering a bit earlier. That means you will be accelerating in a straighter line, reducing the harmful effects of the understeer.

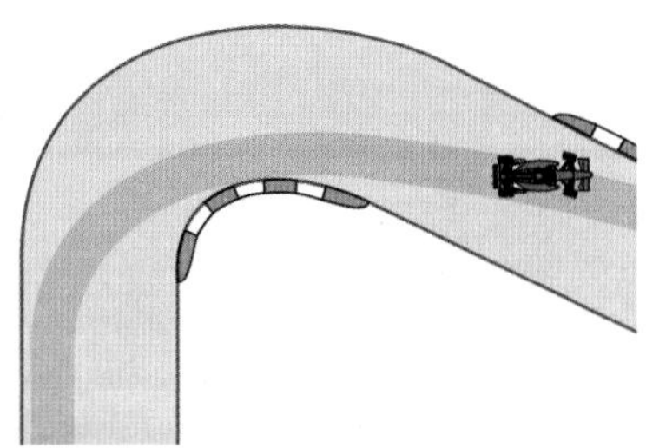

ILLUSTRATION 34-6 Exit understeer.

And one more thing: The more gentle you are with the acceleration, the less understeer you will have. If you jump on the throttle, the understeer is going to be exaggerated. So squeeze the throttle.

EXIT OVERSTEER

Exit oversteer can be related to one of two things: Either it is power-oversteer, caused more by the car's inability to put its acceleration traction to the ground; or it is due to the weight balance.

Usually, the way to deal with either type of exit oversteer is much the same as with exit understeer. The goal is to open up the exit of the corner, increasing the radius of the corner as soon as possible, by using a later turn-in and exit.

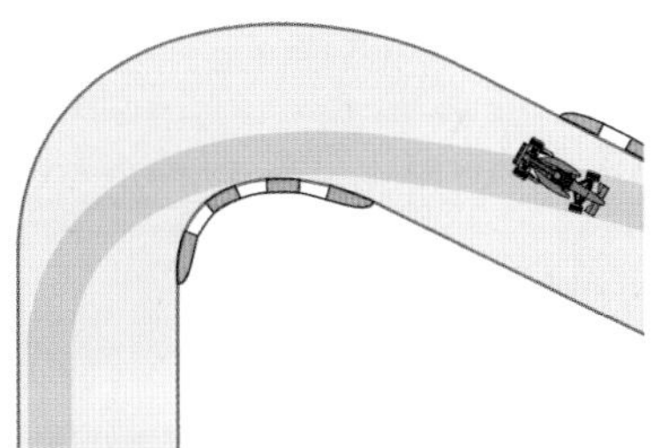

ILLUSTRATION 34-7 Exit oversteer.

One of the other things you have to keep in mind with exit oversteer is to be gentle with the throttle under acceleration. If you stand on the throttle, even if you have altered your line, you are going put a big load on the rear tires. In time, this will overheat them, making the oversteer problem worse, even causing it to oversteer in other parts of the corner.

There is one other approach to dealing with a car that oversteers at the exit of a corner, especially one that has an extreme oversteer problem, and that is to almost give up, or sacrifice that part of the corner. Instead of slowing the car down and using a later turn-in and apex, you pretty much do the opposite. As you approach the corner, you brake later and carry much more speed into the corner, taking an earlier apex, and then get the car straightened out and pointed down the straight well after the apex. The idea here is that since the car will not accelerate out of the corner very well, you might as well try to take advantage of where the car is working—the corner entry.

Before using this technique, I would make sure that every other technique didn't work, as you will not be setting any track records using this approach. It is a bit extreme! Once in a race, however, it may help you hold off a competitor behind you, at least for a few laps. Perhaps the biggest challenge in using this technique is that it is unlikely you have a mental program for it. Therefore, you may just want to try it on a test day or during a practice session sometime so that you are prepared for it.

BRAKE FADE

Brake fade is one of the scariest things a race driver ever experiences, but unless you want to just give up and pull into the pits every time it happens, you are going to have to live with it at some point in your career.

Typically, there are two reasons for the brakes to fade, both having to do with overheating. The first and most common is when the brakes get so hot from the repeated use that the brake fluid in the system begins to boil. As it boils, air bubbles are created. Unfortunately, air is much easier to compress than brake fluid, so that your brake pedal becomes soft and spongy, sometimes to the point that the pedal travels all the way to the floor without applying much pressure to the brake pads.

The second reason for brake fade has to do with the overheating of the pads themselves. In this case, the temperature of the pads has risen to a point beyond their designed operating range. When that happens, a gas is actually boiled out of the pad material, but it doesn't just float away. It forms a layer between the pad and the brake rotor surface, acting almost like a lubricant. With this situation, the brake pedal stays nice and firm no matter how hard you push on it, but the car just doesn't slow down very well.

In either case, the problem you have to deal with is the overheating of the brake system. The only thing you can do is allow the brakes to cool down, which is not an easy thing to do when trying to drive at the limit. In reality, there is no way you can drive at the limit, at maximum speed, while cooling the brake system. There are, however, some ways that you can allow the brakes to cool somewhat without it affecting your speed too much.

As I've talked about previously, braking later for corners does not gain you much. There is more to be gained when accelerating than there is when braking. Therefore, braking a little bit earlier will not hurt your lap times that much, as long as you brake lightly so that your corner-entry speed is just as high as it was before. Braking lightly means less heat going into the brake system.

The overall goal, actually, is to put as little heat into the system, while letting it cool as much as possible by allowing air to flow through them.

If there are any places on the racetrack where you now come off the throttle and brake for a short period of time, this is an opportunity to help the brakes cool. Instead of touching the brakes, just come off the throttle a little more, or longer. Even if you had been using the brakes for a fraction of a second, by not using them here you are allowing the air to cool them without adding any more heat to them.

AILING GEARBOX

The first thing to consider is why the gearbox is beginning to fail. Is it because you are not blipping the throttle enough on your downshifts (probably beating up on the dog rings)? Are you lifting on your upshifts? Have you missed one shift, damaging the dog ring, and now it pops out of that gear?

If you are driving a car with a sequential shift, there is not much you can do about how you place the gearbox in gear. It is just a matter of pulling backward or pushing forward all the way, being firm and positive with it. But you can control the use of the throttle. On upshifts, make sure that you exaggerate the throttle lift, which takes the load off the dog rings prior to moving to the next gear. And on downshifts, make sure you are blipping the throttle enough.

If you do not have a sequential shift gearbox, there are a couple of other things you can do besides what I just mentioned about the use of the throttle (which apply here as well). The first, and most obvious, is to make sure you do not miss any gear changes, even if that means shifting slower and more deliberately. But just as I said with a sequential shifter, be firm and positive with your shifts. Just as many gearbox failures have occurred by "babying it" too much, as from being too rough.

With a troublesome nonsequential gearbox, you can work at "placing" it in gear a little more. Be precise, but firm and positive. If it begins to pop out of gear, that means the gearbox dog rings have worn. About all you can do is make sure you do not miss any more shifts and try holding it in the gear that is having the problem.

Whether you use the clutch or not to shift is an important factor. If you do not use the clutch, and the gearbox begins to get difficult to shift, won't go into a gear, or pops out of gear, you may want to try using it. If you have been left-foot braking and now need to change to right-foot braking to allow your left foot to work the clutch, that will be quite the change. If you do not have the mental program to drive that way, it may be a bit too much of a change to make in the middle of a race. It may be something you want to try in a practice or test session someday.

If you are driving a car with a synchromesh transmission, like a production-based car, you may want to try double-clutching if it begins to get difficult to get it into gear.

DIFFERENT CARS

Some would say that driving a production-based car, such as a front-wheel-drive sedan, is all about adapting, and that "real race cars"—purpose-built open-wheel cars—don't require adapting. Their point is that the purpose-built race car should do what you want it to do, at least if you've done your job of setting it up correctly. While I would agree that production-based cars often require a little more adaptation, I disagree that a purpose-built car does not require you to adapt your style.

I can tell you from personal experience that production-based race cars can be some of the best all-round balanced and setup cars in the world, while some purpose-built formula cars are far from it. And yes, setting up the car perfectly is ultimately your responsibility. However, it's rare that you will drive a car that is perfect. In fact, by definition you cannot have a perfect car in every corner. If you get a car, purpose-built or production-based, to work perfectly in one corner, it's next to impossible to be perfect in every other corner. The laws of physics will back this up.

So how do you adapt your driving to suit the car and track? Here's a list of just some of the things you can do:

- Alter the timing of when you turn into a corner.
- Alter how quickly you turn the steering wheel into a corner.
- Alter how long you keep maximum steering input in the car through the midsection of the corner.
- Alter the timing and how quickly you unwind the steering out of a corner.
- Alter the timing of when you release the brakes when entering a corner.
- Alter the rate at which you release the brakes when entering a corner.
- Alter the time you spend between fully releasing the brakes and the initial application of the throttle.
- Alter the rate at which you apply the throttle exiting a corner.
- Alter how much and how abruptly you modulate the throttle.

Each one of these approaches can be altered in isolation, as well as in combination with any other(s). So, how well can you adapt and alter each one of these techniques? If there is even one of these that you are not adept at adapting to, you are not as complete a driver as you could be. Learn to alter each one and to combine them in the perfect way to make the best of every corner. Some drivers are great in certain types of corners and not in others. The reason is that they can adapt in some ways, but not in others. They are not adaptable drivers.

As a strategy, take parts of your test sessions and just practice these different approaches. Try turning in a little earlier and slower or later and crisper. Try gradually feeding in steering all the way to the apex and then unwinding it, as well as feeding in all the steering early in the corner, holding it there for most of the turn, and then unwinding it at the end of the corner. Try trailing off the brakes more slowly and then coming off them relatively abruptly. Try being patient from the time you come off the brakes and before you begin applying the throttle, and then making those two almost overlap. Try playing with the rate at which you feed in the throttle and modulate it. Make note of what all of these approaches do. If you know what one of these do, when you need the car to do something, you will automatically and naturally give the car what it needs.

SPEED SECRET

Give the car what it needs so it will give you what you need.

UNDERSTEER-OVERSTEER PROBLEM

How many times have you complained about the car understeering early in a corner and oversteering toward the exit? I don't know of a driver who has not driven a race car at some point in his or her career who does not do this. If drivers haven't yet, it is only a matter of time before they do.

Engineers hate this problem. After all, the cure for half of the problem often exaggerates the other half. The biggest problem, the real problem, is that the problem is not always the fault of the car. The problem often lies with the driver.

The solution? Ask yourself (or have someone else ask) some awareness-building questions. A common cause of this handling problem is that as you experience understeer early in the corner; you turn the steering wheel even more. Think about it. You're entering a corner at 100 miles per hour, you turn the steering wheel, and the car pushes toward the outside edge of the track. What would you do? Probably the same as many drivers do: crank in more steering angle, trying to get it to turn. It is human instinct. It is survival instinct.

If you ask yourself exactly what you're doing when the understeer occurs, you begin the awareness-building process. Ask, "What position do I have the steering wheel in when the understeer occurs?" Give yourself time to think about it. Close your eyes and visualize what was happening.

At first you may answer no. You know that is the "right" answer. But keeping thinking about it and ask yourself questions. Don't be in a hurry for the answer. Think it through, preferably visualizing what you had been doing.

Often, you will become aware of the fact that you actually turn the steering wheel more when faced with the early-in-the-turn understeer. You will then realize that when you do that, it causes the oversteer later in the turn. What is happening, of course, is that as you turn the steering wheel more and more, the front tires begin to scrub off some speed, then suddenly regain traction, causing the car to snap to oversteer.

By simply becoming aware of what you're doing, as a result of your questions, you have discovered the solution to the problem, not cranking in more steering input. Your next step is to develop your *MI*. You could ask yourself what you should do and how that would look, sound, and feel. Describe in as much detail as possible. Close your eyes and imagine it in as much detail as possible, and do that over and over, daily, weekly, and monthly.

The key is defining the real cause of a problem. There are many times when the car is not the problem; it is you that is causing the problem. Far too many teams have been led down the wrong path by not digging to the core of the problem. They chase the effects of the problem. Before you tweak away on the car's setup, raise your awareness to determine the real cause of a handling problem.

35 OVAL TRACK

It used to be that a book like this only talked about driving road courses. But more and more forms of racing in North America are including oval track races. I'm not going to go into great detail about strategies and so on, but let's take a look at some basic techniques and tips specific to driving on an oval.

SETUP

First, car setup. Generally, the car should be set up to understeer slightly, not oversteer. It's next to impossible to control an oversteering (loose) car on an oval, due to the consistently high speeds. You may be able to control it for a couple of laps, but eventually it will catch up with you, sending you spinning into the wall. On an oval, you want to base your car's setup around what the front end is doing, not the rear, as you would on a road course.

Driving an oval—particularly superspeedways—requires more smoothness, finesse, and precision than road racing. Concentrate on turning the steering more gently and smoothly, arcing into the turn. Getting the car to take a set in the turns is critical on an oval, so don't turn too slowly.

POOR HANDLING

Before my first Indy race on an oval, I was given some good advice: "If the car doesn't feel right on an oval, don't force it." On a road course you can overcome a bad-handling car somewhat by changing your technique slightly. This is difficult—and dangerous—on an oval. This means that the car's setup is more critical on an oval. Also, if it feels as though there is a mechanical problem with the car, come into the pits and have it checked. The result of a mechanical breakage on an oval is serious.

If the car is not handling well, do not try to force it to go fast. If you try to make a bad-handling car go fast on an oval, at best you will spin, at worst you will end up in hospital. This is especially the case with an oversteering car.

Oval tracks are not very forgiving, and a bad-handling car can get you hurt. You need to decide whether to continue or not with a bad-handling car. You won't get any more credit for driving a car with evil handling to a mediocre finishing position than you will for bringing it in to make adjustments to it. You will get a lot less credit for stuffing the car into a wall than you will for making a smart decision.

Remember, it is your car's life, and sometimes your own life on the line. It's your decision.

If your car is oversteering on the oval, and you choose to continue, you need to have smooth, gentle hands. In other words, don't turn the steering any more than you have to, and let the steering unwind as soon and as much as possible. With an oversteering car, a natural reaction is to hold the car away from the wall at the exit of the corners. That is about the worst thing you can do. The more you hold it away from the wall, the more likely it is that you will hit it.

Also, with an oversteering car, make your initial turn-in as gentle and progressive as possible. Do not make an abrupt turn-in.

With an understeering car on an oval, turn the steering even less, perhaps by driving a higher line through the turns. Let the car run free; release it from the turns.

When adapting to understeer, use the same weight balance adjustment techniques I suggested above, while becoming aware of what you are doing with the steering. As the car pushes, it will be instinctual to turn the steering a little more. As you know, this is not a good thing. Typically what happens is eventually the front tires scrub off enough speed that they begin to grip again but now you have so much steering input dialed in that the rear end breaks loose and the car spins. It is amazing how often a spin that appears to be caused by oversteer is actually the result of understeer.

THE LINE

The ideal line on an oval varies depending on the turn's banking, its shape, and the handling of your car. You need to "feel" your way through the corners more so than on a road course; you need to let the car run where it needs to go. Everything I said earlier about the cornering compromise, reference points, and control phases apply to oval tracks as well. And just like on a road course, your straightaway speed is dictated by how well you exit the corners.

In fact, momentum is everything on an oval. The smallest error or lift of the throttle will have a tremendous effect on your lap speed. Don't over-slow the car entering the turn. Try easing off the brakes just slightly sooner than you think possible and let the car run; carry its momentum.

As on a road course, on an oval you want to wait as long as possible before getting off the throttle and beginning to brake. This means you are going to brake into the turns, even more than you would on a road course. But gently. Remember the traction circle. You can't brake as hard while turning as you could in a straight line. You have to ease the brakes on.

Looking far ahead is especially important on an oval. When driving an oval, I try to look as far ahead as I can, then just think about getting there as quickly as possible. This may sound obvious, but it helps. Often, a driver's natural reaction is to look at the wall or the point you're just about to get to. That's not enough. You won't drive a smooth, flowing line if you don't look far ahead. And looking

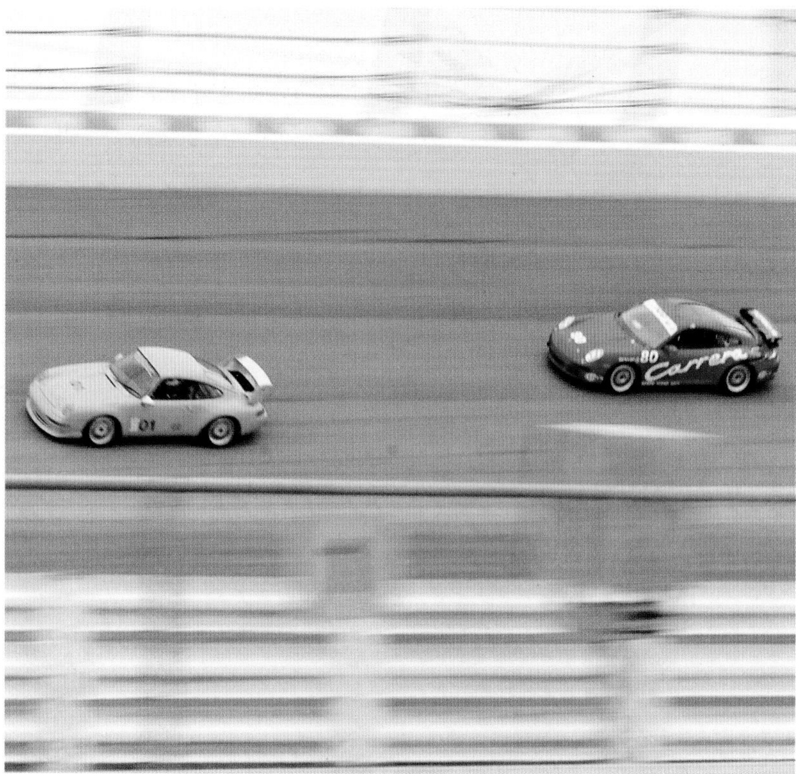

Being fast on an oval requires looking farther ahead than you could on a road course, being very precise and smooth, and trusting the car to do the work of gripping the track. *Shutterstock*

well ahead, and concentrating on getting to where I'm looking, seems to really help me.

OTHER CARS

Traffic on an oval is an entirely different experience than in road racing. Especially on the smaller ovals (1 mile or less), you are constantly dealing with others cars, either passing or being passed. Using your mirrors and peripheral vision is especially important in oval racing.

Turbulence from other cars is a tricky factor on ovals. When trying to pass another car, it may be difficult, as the closer you get the less downforce you will have (the leading car blocking the airflow to your car), slowing your speed. You have to ease off and try going into a corner a little slower, then accelerate earlier to get a good run out of the corner and slipstream past on the straight.

Particularly on superspeedways, a car closely following you can affect the handling of your car. When a car gets close to your rear wing or tail, the airflow over the rear is disturbed, decreasing downforce and causing the car to oversteer.

36 THE UNKNOWN CORNER

"How do I estimate the speed and braking through a corner I have never been through before?" There's a question every rally and autocross driver would love the "secret" answer to. Even road and oval track racers could use it, although it is not as important since they have the advantage of being able to go round and round until they figure it out, using trial and error. Let's take a look at what you're up against when faced with a corner you've never been through before and see what we can do to discover that "secret."

Approaching an "unknown corner," there are four interrelated factors that come into play:

- **Speed sensing:** Your ability to sense, determine, and establish a particular speed. Obviously, this must be done at an intuitive level, not by looking at the speedometer.
- **Traction sensing:** Your ability to feel or sense whether and how close your tires are to their limit.
- **Database:** You have a database of information from the hundreds, thousands, or millions of corners you've driven in your life. Your database is primarily made up of visual images of what corners look like, along with the resulting speed and traction-sensing information. If your speed-sensing and traction-sensing skills are poor (a lack of sensory input), the database will not be accurate or as useful as it could be. Of course, you could say your database is just experience, or seat time, and you would be right to some extent. But why, then, do some drivers with little experience seem to have a larger database? The better your speed sensing and traction sensing, or sensory input, the better (richer) your database will be. In other words, your database is made up of tens, hundreds, thousands, or millions of reference points, ones you see, ones you feel, and ones you hear. It's as if the file of information on each corner is much thicker or deeper, thus the more sensory information you have taken in.
- **Car control:** Your ability to "dance" with the car, handling the controls in such a way as to keep the car on your desired cornering line, while keeping the tires at or near their limit.

So heading into a corner, your speed-sensing skills sense the speed (obviously) you're traveling at that instant, and your traction sensing says, "I am 'this' close to the limit." All the while your database compares the visual image of the corner with others in its files, recalls one that most closely matches, and makes the best estimate of the speed required. Then, it relies on your speed-sensing skills to manage the adjusting of speed (slowing down) to "that" speed.

At that point your traction-sensing skills begin to take over, sensing "how close to the limit" you are. That is when your car-control skills come in. If the speed estimate is too high, or your speed sensing didn't do a good job of matching the estimate, then you have to control and manage the extra speed as best possible. Of course, if the estimate was too low, or your speed sensing over-slowed you, then your car-control skills will have to do what it takes to increase your speed.

Once you've driven the corner once, it is added to your database. Once it's in the database, you can begin to work with it using mental imagery. Using the information in your database, along with awareness, you can update the database without necessarily driving. Ask yourself, "How close was I to the limit?" "If I carried 1 mile per hour more into the corner, what would happen?" "Two miles per hour more?" "Three miles per hour more?" Close your eyes, relax, and picture a mental image of that speed, but do more than just visualize it. Include more than just visual information. Also imagine how it feels and what it sounds like. That is true mental imagery, or mental programming.

Then, the next time out, simply compare the two: your mental image with your awareness of how close you are to that mental image of the ideal speed. This is using $MI + A = G$, the simplest, quickest, and most effective way of learning and improving that I know of. The stronger, more vivid your mental image, and the more awareness you have, the more effective this will be, the easier it is to achieve your goal.

If I had to simplify all of what I just said into "the secret," it would be the following. Work at improving your speed sensing and traction sensing by practicing taking in more sensory input, from your vision, your kinesthetic sense (balance, feel, touch, the g-forces, vibrations, pitch and roll of the car, and so on), and your hearing.

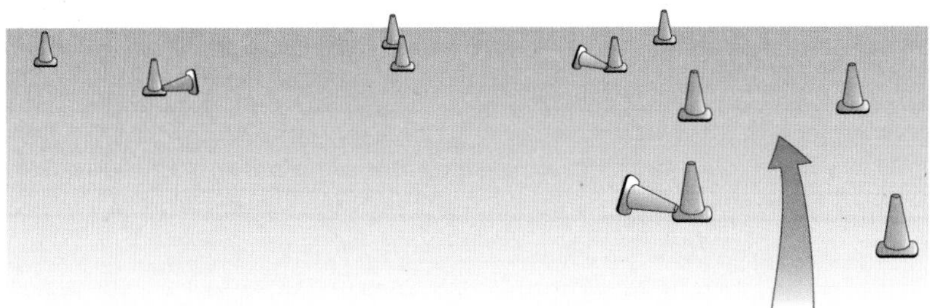

ILLUSTRATION 36-1 Approaching a turn you've never seen before, like one on an autocross or slalom course, your mind goes through an amazing process to determine what speed you should slow to. It begins by comparing a visual picture of the turn with all the ones stored in your brain's database.

Practice listening to the car. What is the engine note saying? What are the sounds coming from the tires telling you? Does the tire noise continue to get louder and louder, or does it taper off after the tires reach their limit? Are they growling, howling, squealing, screeching, or screaming? What does that tell you about their grip levels?

Practice feeling the dynamics of the car. Does the steering get heavier or lighter as the tires reach their limit? How much body roll is there before the tires begin to lose their grip? Remember, the tires are talking to you. Are you listening?

Practice seeing more. Take in more visual information. Act like a sponge, soaking up sensory information. Then, simply be aware. If you combine awareness, or sensory input, with a mental image of what you want to achieve, you will reach your goal of driving at or near the limit, even if it is the first time you've driven through the corner.

And that's the secret to knowing what speed to enter a corner you've never seen before. Use your database, your speed sensing, your traction sensing, and then your awareness as you do it so that you can add that to your mental image later. Of course, that adds to your database, and the whole cycle continues, getting better and better each time you drive. That is, really, the secret: improving every time out.

SPEED SECRET

Sensory input and awareness are the keys to driving fast, no matter what the corners look like.

For those of you who thought I was going to provide you with a secret like always slow down 2 miles per hour for each foot of turn radius, and always turn in 6.73 feet before the pylon, I apologize. It's just not that simple (as if you didn't already know that).

If you look at every run, every stage, or every lap from this perspective, that you are soaking up information to add to your database, my bet is that you will be immediately quicker. There are two reasons for this: First, when you give your brain more information to work with, it will produce a better result. And second, with this approach it is more likely that you will relax and drive more at the subconscious level, rather than "trying" to go fast.

37 CAR CONTROL

Car control is the ability to control the car at the very limit. It is the most important skill a driver can ever learn. The ability to make a car do what you want it to—brake, steer, accelerate, oversteer, understeer, neutral steer, and so on—comes from coordinating all the basic control skills together with the correct timing, precision, and application.

Great car-control skills allow you to drive at, and just beyond if you wish, the "theoretical" traction limit. This is where the car is in a slide all the way through the corner balanced on the edge of control. However, it's easy to slide the car too much, which actually slows you down by scrubbing off speed. Remember the examples in the slip angle section in Chapter 5.

A lot of this great car control will come with experience, getting comfortable with being able to "throw" the car into a turn and feeling confident with being able to then catch it at the limit. Of course, this must be done smoothly.

I believe a driver who has great car-control skills but who is not driving the ideal line through a corner will be faster than a driver with the opposite abilities. Therefore, if you want to go fast, if you want to win, develop your car-control skills through practice. Then make sure you are driving the ideal line as well; that should be easy.

FIGHTING THE CAR

Although driving the ideal line is important in terms of maximizing your cornering speed, fighting for perfection may actually slow you down. Don't fight the car, if that's what it takes to make it drive the ideal line. If you do, you will actually scrub off more speed than if you drove slightly off-line.

Too many drivers fight the car to clip past the perfect apex, only to slow the car down doing it. This shouldn't be an excuse to drive off-line. But, if you do turn in to a corner and realize you are going to be slightly off-line, don't fight it. Let the car go where it wants. Don't fight the car to stay on the ideal line if you have to force it. The car will tell you if you are driving the right line or not. And again, don't "pinch" it into the inside of the turn on the exit. Let it run free as you unwind out to the exit.

I remember the first time I drove the Milwaukee oval I learned quickly that it's better to let the car run where it wanted to, rather than fight it to run on the theoretical ideal line, which was bumpy at Milwaukee. Often, driving the ideal line is not the fast way around a turn or track.

38 THE LIMIT

Consistently driving the limit, as fast as you and the car can possibly go, is the ultimate goal. So how do you get to that limit? How do you learn where the limit is? You simply make changes that result in you driving the limit or in lower lap times. And you either make these changes by:

- Analyzing and planning prior to going on the track, or
- Trial and error experience on the track

The first may be dangerous if you analyze and plan with misleading information or without having sufficient background, while the second can *only* be done on the track (costing money).

Keep in mind when looking at a map of a track that trying to learn the ideal line by studying a diagram can sometimes fool you. Elevation, banking, and track surface changes aren't evident, not to mention the accuracy of the map may be less than perfect. This sometimes leads to a misconception of how to drive a corner. Therefore, before you can actually learn the right way on the track, you have to unlearn the preconceived ideas. This can take up a lot of valuable track time.

You must be able to observe what you are doing so you can improve on it. Errors should be looked at to see what influenced or caused them. This is not to suggest dwelling on every single mistake you ever make. But, study the decision or action that led to the error, to ensure it doesn't happen again.

Observing what you do is the key to learning from your errors. In fact, sometimes, let small errors happen; learn how a different line works, or doesn't. You should consider that in most cases, by the time you notice an error, it may be too late to correct it anyway. About all you can do then is minimize its effect. In fact, that is the key: minimizing the effect of an error and doing so as soon as possible.

Mistakes are a natural process. Don't fight them. Instead, consider what you can learn from an error, then reprogram or see yourself doing it the correct way and forge ahead.

Imitation is the ultimate learning technique. Copying is the most instinctive, simple, and natural way to learn. After all, that's exactly how we learned to do practically everything as a child.

Learn by observation, appreciation, and imitation. If you want to learn a skill, find someone who is good at it, then watch this person very carefully. As you watch, feel yourself moving in the same way, then practice by visually imitating. That doesn't mean just what the driver is doing in the car. How a driver acts outside of the car is just as important. "Acting as if" you were Michael Schumacher, Lewis Hamilton, Jimmie Johnson, or Dario Franchitti outside the car will improve your ability to drive like them.

Even if you aren't able to imitate someone perfectly, your attempts will increase your awareness of what skills, techniques, and mental approach you still need to develop. Of course, you must first be prepared to imitate someone. Don't try copying the advanced techniques of a world champion before mastering the basics.

And remember, every driver's learning curve is different. Some learn and progress quickly; others learn much slower. This is not an indication as to how much talent a driver has.

DRIVING THE LIMIT

How do you really know when you're driving right at the very limit, getting the last ounce of speed out of your car?

Ultimately, and simply speaking, your speed is limited by three things: engine output, aerodynamics, and traction. With more engine output you will be faster on the straights; with more traction you will be able to brake harder on the approach to a turn, go faster through the corners, and accelerate harder coming out of corners; and aerodynamic downforce helps traction while the drag slows you down. Once you're in the car, you can't do much about engine output or the car's aerodynamics, but you may be able to do something about traction. You may not be able to increase the amount of traction your car has, but you can drive so that you use all the traction effectively.

As I mentioned earlier, the more gradually you turn into a corner for example, the more traction the tires will have since a tire's traction limit will be higher if you progressively build up to it. As well, balancing the car will increase your useable traction.

When trying to drive "the limit," you are actually dealing with three different limits: the car, the track, and yourself (the driver). You must recognize and maximize each if you are to go faster. Although there isn't anything you can do about changing the track's limit and raising the car's limits are for mechanics and engineers—with your input of course—maximizing *your* limits is something to strive for.

Let's go back to the beginning, the obvious. Driving at the limit means having the tires at their very limit of adhesion (traction) at all times, during braking, cornering, and acceleration. Think, for a moment, of dividing up your driving into those three phases: braking, cornering, and acceleration phases. Now we know that with most cars, we are nowhere near the limit of traction during acceleration at anything above first gear (how many cars can you spin the tires consistently in second, third, fourth, or fifth gear?), which makes the acceleration phase fairly simple.

However, remember from my previous comments about the traction circle that there should be an overlap of the three phases. Overlapping the acceleration and cornering, and even more so the braking and cornering, is where the skill comes into play.

To drive the very limit, you must brake as late as possible at the traction limit all the way to the corner turn-in point. Then as you begin the cornering phase, ease off the brakes (overlapping the braking and cornering to keep the tires at their traction limit) until you are at the cornering limit. At this point, you begin squeezing on the acceleration while unwinding the steering (again overlapping cornering and acceleration to stay at the limit of traction).

Now, if all this is done properly, you will be driving the car at the very limit of adhesion. And remember, at the limit, the tires are actually slipping a certain amount, so don't worry if the car is sliding through the corner. It should be. As you drive through the corner, the car should be sliding slightly, with you making very, very small corrections to the brakes, steering, and throttle to keep the tires at their optimum slip angles or traction limits.

But *your* traction limit may not be as high as the next driver's. Why? Because you may not be balancing the car as well as the other driver. Remember that the better the car is balanced (keeping the weight of the car equally distributed over all four tires) the more overall traction the car will have. So it is possible for you to drive your car at *your* limit and still have someone else drive your car faster. Or your limit can be higher than someone else's. It all comes down to balancing the car.

For example, when Ayrton Senna and Alain Prost (both multi-time world champions) were teammates in the McLaren F1 team. In the same car, Senna was often quicker. It wasn't because Senna's car was faster, or that he was braver, or drove a better line through the corner. It's certainly not because Prost wasn't driving his own limit. It was because Senna was able to balance the car so delicately, so perfectly, that his limit of traction was slightly higher than *even* Prost's. That allowed him to enter the corner at a fraction of a mile per hour faster, or begin accelerating a fraction of a second sooner, meaning he was also faster down the straight.

At all times you are receiving information from the car. The more sensitive you are to receiving that feedback, the more able you will be to drive the car at the very limit. People always talk about the feedback a driver gets through the "seat of his pants." Well, I don't know about you, but I have many more nerve endings in my head than in my rear end. You receive more information through your vision than through any of the other senses (smell and taste have relatively little to do with race driving, hearing does play a role, and feel is certainly important, but not as important as vision).

Imagine yourself looking at the roadway just over the nose of the car. Now, if the car begins to oversteer you will be looking in a slightly different direction. But if you were looking farther ahead, almost to the horizon, you would notice a much larger change in sight direction. In other words, the farther ahead you look, the more sensitive you will be to slight changes in direction or the car sliding. Much of the feel of driving comes from your vision.

But how do you really know if you are driving the limit? The only way to really know for sure is to go beyond it every now and then. However, that can be a little hard on equipment, unless you are able to go beyond the limit and still catch it before you end up in the weeds. That's the tricky part.

In fact, before you can consistently drive at the limit, you have to be *able* to drive beyond the limit. Think back to the four hypothetical drivers in the slip angle section of Chapter 5. Remember how the second driver would drive beyond the ideal slip-angle range, the car sliding more than what was optimum for maximum traction. Yes, that was not the fastest way to drive, but you have to be able to do that before you can really know where the limit is. Once you've driven beyond the limit, and kept the car on the track somewhere near the ideal line, it is much easier to dial it back a little, back to "at the limit."

If you can't overcompensate, or overdrive, you'll never be able to "home" in on the limit. If you can't overdrive a car, you'll never be able to drive the limit consistently.

Every corner on every lap you want to strive to brake as late as possible (at the last possible moment and still be able to get the car to turn in properly, which is a mistake many drivers make, braking so late they can't get the car to turn into the corner correctly), then enter each corner at a speed slightly above what you think is the limit, and then make the necessary corrections to balance the car as it slides through the rest of the corner while beginning to accelerate as early and as hard (still gently to keep the car balanced) as possible to maximize straightaway speed. It may be easier to do than it is to explain.

Oh yes, and don't forget to drive the absolute perfect ideal line, or at least within a quarter of an inch of it. Many drivers can do this for one corner or one lap. But to do this consistently lap after lap is the goal. You can drive the limit on the wrong line, but you're not going to be a winner.

The difference between a slow driver and a fast driver is that the slow driver is not consistently driving at the limit all the way around the track. The difference, then, between a fast driver and a winner is that the winner drives consistently at the limit on the ideal line.

I have my own little mental check to see if I'm driving the limit. If I ever felt as though I could turn the steering wheel a little more—tighten the radius—at any point in the corner, without causing the car to spin or slide more, then I knew I wasn't driving at the limit. Next lap, I would try a little faster, to push it a little closer to the limit.

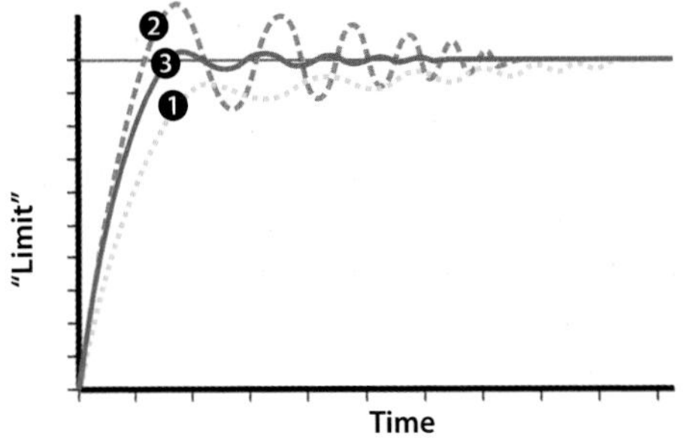

ILLUSTRATION 38-1 Three approaches to driving the limit: #1 creeps up on it slowly, taking small steps; #2 takes big steps, often going way beyond; and #3 takes medium steps, going slightly beyond the limit and bringing it back, homing in on the limit.

39 GOING FASTER

We're now entering the area for more experienced racers. Once you perfect the basics, you'll no doubt be asking yourself, "How can I go faster?" If you could only get an answer that would shave a few fractions of a second off your best time, you would be a happier driver. The following ideas might just help you figure out how to go faster.

During my first visit to Indy, during my rookie orientation, I spent time with Rick Mears. His explanation of how to go faster was interesting; it was the way I've always approached it. To go faster you should inch up on the limit, going a little bit quicker each lap until you feel you're going beyond the limit, taking little bites of speed to reach the limit rather than taking large bites. If you take big bites of speed, you may go from just below the limit to way beyond in one step.

When trying to go faster, never ask yourself, "Why can't I take that corner faster?" That's a negative question. Instead, ask yourself positive, constructive questions like "where," "when," "how much," "what," and so on. Don't use negatives such as, "I didn't brake hard enough" or "I didn't have a good line through that corner." Instead, say, "I started braking at so-and-so, and I think I can brake later if . . ."

I never like to ask myself negative questions such as, "Why can't I take Turn 4 faster?" Instead, I ask myself, "Where can I go faster?" and "How much faster can I take Turn 4?" These are all positive thoughts.

A driver needs more of a plan than "I'm going to take Turn 4 faster." You must have a plan of *how* you are going to go faster, how you're going to take Turn 4 faster. After each session, sit down and think it through. Take a map of the course and visualize yourself driving it as you just did, making notes on areas where you may be able to improve.

To do that, think about what you're doing at and during each corner's reference points and control phases: braking, turn-in, trail braking, transition, balanced throttle, apex, progressive throttle, maximum acceleration, and exit. Then ask yourself how you can change what you are doing to go faster.

Corner-entry speed is critical. If it is not correct, you spend a lot of time and concentration trying to make up for the incorrect speed. But you need as much of your concentration as possible to sense traction, balance, and the line at this point. So make sure your corner-entry speed is correct.

Following the advice, "Enter the corner slow and come out fast" may just be the problem. Now I know I have said this earlier, and it's still true. You can take it too far, however. It's possible (and likely) that you may be entering the corners too slowly. Then, as you accelerate to get up to the correct speed, you exceed the traction limit of the driving tires and get wheelspin. The overall result is you're slow, even though it feels as if you're at the limit because of the wheelspin. Plus, once you've realized you are entering the turn too slowly, it takes time to react and correct your speed.

It's important not to slow the car too much with the brakes on the approach to a corner. Remember the saying "Brakes are like lawyers; they cost you every time you use them." Every time you slow the car with the brakes, you have to work hard at regaining your speed or momentum.

Let's look at an example. If a corner could be entered at 52 miles per hour, and you slow to 50 miles per hour at the entrance, then try to accelerate back up to 52 miles per hour. You may exceed the driving tire's traction limit, resulting in power oversteer with a rear-drive car or power understeer in a front-driver. If you had entered the corner at 52 miles per hour, you wouldn't have to make up for the error in speed. The change in speed wouldn't have been so drastic.

In fact, the more you slow the car at the entry of the turn, and the longer you wait to get back on the throttle, the more likely you'll want to make up for the lack of speed by accelerating hard, but probably too hard. That results in demanding too much from the driving tires, leading to either power oversteer or power understeer. Again, the change in speed is too extreme.

Racing in the rain taught me a valuable lesson, one I use in the dry as well. I found that if I purposely made the car slide slightly from the second I entered the turn, I was automatically smoother and more relaxed, and therefore faster. This is because I had no fear of the car suddenly taking me by surprise by starting to slide. I was operating within my comfort zone. The moment I learned this, I started winning races.

You should aim to enter the turn just slightly faster than the traction limit dictates (as long as you can still make the car turn in to the corner properly), so the car slides (scrubs) while you are transitioning off the brakes over to the throttle to begin acceleration. This accomplishes two things:

- While the car is scrubbing a little speed, it allows you time to transition to the throttle without wasting speed (instead of having the car lose speed while you sense you are going too slow, and then having to react and try to correct your speed).
- It mentally prepares you for the slide, so that it doesn't take you by surprise.

Don't judge your corner-entry speed by your mistakes. Just because the car wouldn't turn in at 52 miles per hour doesn't necessarily mean that's too high a corner-entry speed. It may just be that it's too fast for the way you've balanced the car and the way you've turned the steering wheel into the corner. Try working on your corner-entry technique for a while, trying to get the car to turn in at 52 miles per hour or even faster.

Remember that most of the time the fastest straightaway speed comes on laps with the fastest midcorner speeds. To have a fast midcorner speed, you need to enter the corner as fast as possible—at the limit.

This is another one of those compromises you need to make in your driving, deciding whether you're better off entering a corner slightly slower and getting on the throttle earlier or carrying more speed into the corner. Usually, if your increase in corner-entry speed delays when in the corner you begin to accelerate, you are better off slowing down slightly to get back on the throttle early.

When trying to go faster, work on problem areas and leave strong points alone. Work on one thing at a time. Record all lap times and have someone take segment times (divide the track up into segments and time yourself and others through them. This will determine where you are gaining and where you are losing) to determine where you are fast and where you are not.

When I'm learning a new track or car, I concentrate on finding the big chunks of time first, trying to improve two or three pieces at a time. There is no point in going out on the track and trying to go faster everywhere. The mind can't handle too much information at one time. So I pick two or three places on the track where I think the largest gains can be made. And I work on them only until I've gotten them dead on, then pick two or three new places or things to work on. Any more than three and my brain tends to go into "overload." Of course, it's the final little pieces of time that are the hardest to find.

Making changes to the car is one obvious way of going faster. It's also a way of going slower. Don't fool yourself. Don't pretend to feel a chassis or aerodynamic change if you don't, just to make it look like you know what you're doing. Not every change is noticeable.

And don't make changes to the car before knowing the track and getting into a flow. Take your time. Make sure you are consistently driving at the limit before making drastic changes. That way, you'll know if it's the car or you that's making the difference.

FAST CORNER STRATEGY

Whether on a road course or an oval, perhaps the most difficult corner for any race driver is the fast one, the one that can ultimately be taken flat out. The biggest problem for most drivers is that the self-preservation program in the right foot (at least that is where it seems to be) takes control, causing it to lift off the throttle. As soon as that happens, the balance of the car is not ideal and now it feels as though it is on the limit. And it may just be. Of course, it is the lift of the throttle that causes the car to be at the limit. If the driver had kept his foot flat to the floor, the car would have been better balanced, and it would have stuck—at the limit—through the corner.

In this situation, it is as if the driver's right foot has a mind of its own. And, rather than just telling yourself to keep your foot flat to the floor, you need a better strategy.

The real problem with fast corners is that the car works best, it's balanced with good grip, when you are flat on the throttle all the way through the turn.

If you're on and off the throttle, or even progressively squeezing on the throttle through the turn, the car often feels uncomfortable. It hasn't taken a set, and therefore has less grip. However, it takes a lot of confidence to take a really fast corner flat out right away, so most drivers lift just prior to the corner. And that's what upsets the car's balance. It takes a lot of practice to be able to drive into the corner without lifting.

At Indy, I was told of another approach and have used it a lot since. At first, ease out of the throttle well before the corner on the straightaway to reduce speed enough to make you confident. Then get back on full throttle prior to turning into the corner and continue flat through the turn. This way the car is balanced and comfortable through the corner. With each lap, gradually reduce the amount of lift before the corner until you're able to take the turn without lifting at all.

This way, you will feel comfortable entering the corner at full throttle because the car will be going slow enough. And since you are now driving through the corner flat to the floor, the car will be balanced, telling you it has lots of grip, building your confidence. Therefore, the next lap you will feel comfortable lifting off the throttle a little less on the straightaway, the next lap lifting even less, and so on, until you are completely flat through the corner.

It may sound as though this strategy takes more time, but in reality it doesn't. From my experience, both using it myself and coaching other drivers with it, you will be taking the corner flat out much sooner than if you just kept telling yourself to take it flat.

The key is in building and maintaining your comfort level. Without that, you will never take the corner at full throttle.

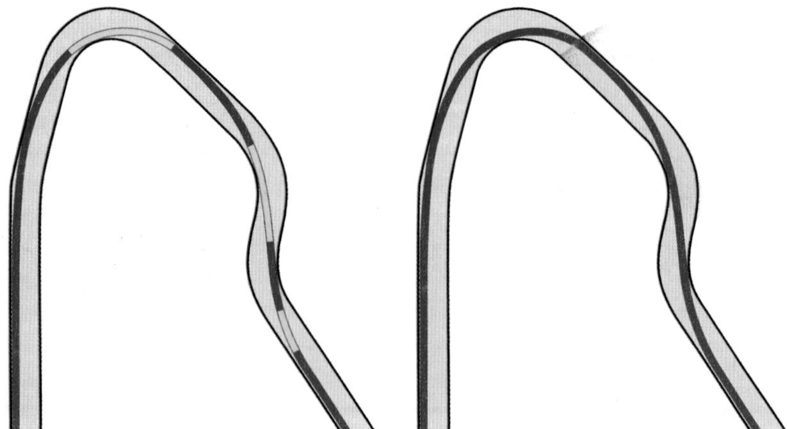

ILLUSTRATION 39-1 Putting a complete lap together means having the car at the limit on every inch of the track, not just on certain sections of track. In this illustration, two drivers drive the same section of track. The green shaded areas represent where the car is at the limit, and the yellow areas where it's not quite at the limit. The driver in the illustration on the left is not hustling the car enough; he's not using every last ounce of grip. Sure, he's driving at the limit—some of the time, but not all of the time.

HUSTLE

One area many drivers give up a lot of time in is those short sections of track where they think that part throttle is as much as the car can take, that it is "good enough," when a short burst of full throttle is possible. Many drivers "coast" for a fraction of second, thinking that 80 percent throttle is good enough. They are not "hustling" the car.

Anything you can do that increases the time spent at full throttle is a good thing. Even if it's for a fraction of a second between two turns, or instead of slowly trailing off the throttle at the end of a straightaway you come off the gas quickly (not forgetting smoothness). That's *hustling* a car.

When trying to shave that last few tenths or hundredths of a second off your lap time, you really have to look at where you are not hustling the car. This is in those short sections on the track where you think 80 percent throttle is good enough. To be a winner, good enough just won't cut it. You have to use 100 percent throttle, flat out. You also have to be aggressive with the car, smooth but aggressive. You have to attack the track.

I know this sounds obvious, but your feet should either be on the brakes, squeezing the throttle down, or flat to the floor on the throttle. Okay, there are some rare exceptions to this generalized rule, but they are the exception.

SPEED SECRET

You should either be on the brakes, squeezing on the throttle, or at full throttle.

Of course, just recognizing that you are not hustling the car is not enough, as easy as that is using data acquisition. Neither is telling yourself to hustle the car, to use full throttle in a few key parts of the track. Often that just frustrates you by pointing out your weaknesses. This really is a place to use the Learning Formula. Get a clear *MI*, and then become aware of how close are to this ideal *MI* by rating your own level of hustle on a scale of 1 to 10.

As you rate your hustle level, it will naturally progress from a 1 or 2 to at least an 8 over a short period of time, as long as you have spent the time to get a good clear *MI* in your mind. Going from 8 to 9 and then 10, the last couple of hundredths, may take a little longer, but will happen.

A "GO FASTER" PLAN

To end this chapter, let's take a look at three specific plans (there are thousands) to go faster.

- **The Late Braker:** For the average racer this is the most common and most overused technique. Most drivers think that by going a little deeper into the corner before braking, they will gain a lot by maintaining the straightaway speed longer. It's only natural to think this way. After all, when running side-by-side with another driver, whenever you brake later you end up in front.

In reality, however, by braking later most drivers brake harder than before, meaning the car enters the turn at the same speed as before. Just braking later, while not carrying more speed into the corner, will gain you little. All it does is maintain your top speed for a few feet longer on the straight. This is okay for picking up a few hundredths of a second, but not much more. Carrying more speed into the corner (as long as you can still make the car turn in and accelerate through the turn) will make a much bigger improvement.

Consider this: On an average road-racing circuit, if you can enter each corner even 1 mile per hour faster, then you will have made up to a half second improvement in your lap time. That's a huge gain.

The big problem with late braking, though, is that you end up spending too much concentration simply on braking, when some should really be spent on more important things. In fact, quite often you've focused so much on the braking that you overreact and lock up the brakes. Usually, you've left your braking so late that all you're doing is thinking about surviving and not about braking correctly and what you have to do when finished braking.

- **The Light Braker:** This is usually the first step in the right direction in trying to go faster. You brake at the same point as before but with a slightly lighter brake application. This means you will carry more speed into the corner (remember, if you can carry just an extra mile per hour, it's going to lead to a great reduction in lap time).
- **The Late, Correct Braker:** This is the goal. You brake later than previously but at the original (threshold) braking rate. So now you gain by maintaining your top speed on the straight longer (small gain), as well as carrying more speed into the corner (big gain). As well, you haven't spent all your concentration on just braking; you are thinking about corner entry. That's how to go faster. Remember, of course, there is a limit.

40 PRACTICE AND TESTING

Your mental approach to testing and practice is important. You want to simulate the competitive spirit and environment as closely as possible. You want the same intensity and aggressiveness in practice as you do in the race. If you practice at 99 percent, that's how you will perform in the race. It's difficult to get back to 100 percent.

SPEED SECRET

Practice how you plan to race, and then you'll race as you practiced.

This programs your mind so that, under actual race conditions, you instinctively respond. Treat practice and the race with the same respect and intensity. When you practice, it should be with the same mental focus and determination as if it were in a race. Then, during a race, you will be as relaxed and calm as if you were practicing.

There is no point in ever going on a racetrack if you're not going to drive at 100 percent. If you're testing or driving an endurance race where you may not want to drive right at the limit, you should still be 100 percent focused, have 100 percent concentration. There is no reason to ever think that a sloppy turn-in is "good enough." You don't want to make "good enough" a habit. The only way to ensure that doesn't happen is to always drive at 100 percent.

We often believe the more we practice a skill or technique—over and over again, many times—the better we'll get. This is not necessarily true. Experience is not always all that it's cracked up to be. In fact, every time you practice a technique incorrectly, you're increasing your chances of doing it wrong again. It's easy to become experienced at repeating the same mistakes.

SPEED SECRET

Practice doesn't make perfect; only perfect practice makes perfect.

So don't practice too much at first, or you're likely to develop incorrect patterns or movements. Instead, begin with a few laps, maintaining intense concentration and motivation. Continue practicing only while concentration and interest is strong. If you begin to repeat an error, or if your concentration or attention start to fade and you start to become casual, stop. Clear your head, get your concentration and motivation back, then go again.

A driver can practice many of the techniques required to win while driving on the street. Practice smooth, consistent braking, squeezing and easing the throttle, arcing the steering into and out of a turn, picking the ideal line through a corner, being smooth, and keeping the car balanced. You don't have to drive fast to do this. This is not just physical practice. Just like a golfer or tennis player "grooves" his swing, you are "grooving" your car-control techniques. Each time you apply the brakes or turn the steering wheel, your actions are being "programmed" into your brain. The more your technique is "programmed," the easier, smoother, and more natural they will be in the heat of the battle on the track.

A lot of race drivers practice bad habits when driving on the street. They don't hold the steering wheel properly, they rest their hands on the shifter, they don't squeeze the brake and gas pedals, and so on. How do they expect to drive any differently at speed, on the racetrack, when they've just programmed those techniques into their head? And, if you can't do something at slow speed on the street, you'll never be able to do it naturally on the racetrack. It's the same with any sport. What do you think would happen if a tennis pro practiced hitting one-handed backhands all year, and then went to Wimbledon and played using a two-handed backhand?

One of your objectives during practice is finding the right chassis setup for the race or qualifying. For the race, you want a comfortable, consistent, reliable setup. For qualifying, you may want a setup that is less "comfortable," perhaps with less aerodynamic downforce, but is fast for one or two laps. A good race setup allows you to know you can move up from where you qualified.

The first few laps of a practice session may be the time to bed in new brake pads or scrub in a new set of tires. Generally, with most brake pads, the trick is to gradually heat them up by braking heavily (but be careful as they can begin to fade at anytime, so brake hard but early), and then run a few easy laps to let them cool. As this is not always the best procedure for bedding pads, and some come pre-bedded, check with the manufacturer first.

Concentrate on the car's setup in practice and testing and what you can do to improve it. Part of your job is to become sensitive to what the car is doing.

Check the brake bias by overbraking at different locations to see if the front or rear tires lock up first. How is the handling in the slow corners? The medium speed corners? The fast corners? How's the initial turn-in? Does it understeer or oversteer? What about the middle of the corner?

Does it put the power down well on the exit of the corner, or is there too much wheelspin? Does the car bottom out going over bumps, or can you get away with lowering it? Does the car feel too soft? Does it roll and pitch too much in the turns? Is it too stiff? Does it feel like it's "skating" across the track with too little grip? Are the shocks too soft, too stiff? What are the effects of an anti-roll bar sweep?

How are the gear ratios? What is the maximum rpm on the longest straight? Are there corners where having a slightly taller or lower gear ratio would help?

Consider how each change interrelates; that is, if you change the handling to better suit one particular corner, will the gear ratio still be correct, or will it be too low now with the extra speed you're carrying? Consider the top gear ratio: What about in a draft? Will it be too low a gear when you pick up a few extra miles per hour in the draft?

Obviously, you can't do much of any of this development of the car's setup until you know the track well. If you're making improvements each lap in your driving, how are you going to know if a change you made to the car helped or not? This is where consistency comes in.

At the same time, practice is where you should try different things. Try taking a corner in a taller gear. Try braking later and carrying more speed into a turn. Or the opposite, brake earlier and work on getting on the power earlier in the turn. Which works best? Follow a quicker car, watching when it brakes and how it takes the corners.

Debrief with your engineer, mechanic, or just yourself after each session. Make notes on everything about the car and your driving.

The real question you need to ask yourself is this: "What can be done to go faster?"

DEVELOPING FEEDBACK

For most, if not all, engineers, crew chiefs, mechanics, team owners, team managers, and even drivers themselves, testing and practice is for one thing: developing the race car. I would like to suggest that is only part of its role. In addition to developing the car, testing and practice should also be used to develop the driver. In fact, you should really look at it as developing the car-driver package.

How important is it for you to be sensitive to any and all of the subtle changes made to the car? You can either hope you have the necessary sensitivity, or you can develop it. A big part of testing and practice should be used to develop your sensitivity.

As I mentioned earlier, one of the most effective tools you can use to improve your performance and your sensitivity to setup changes are Sensory Input Sessions. You may be thinking that track time, whether on a private test day or during an event practice session, is far too valuable to be wasted on training. You are right, if you have no interest in winning.

If you do have some interest in winning, then one of the reasons practice and testing is so valuable is the opportunity to improve the car-driver package. That means developing you just as much as the car.

If you need any further convincing, then think of it this way. The better you get at soaking up sensory inputs, the better your feedback will be about the car. I'm sure you will agree, without good feedback, you cannot do what it takes to maximize the race car. You need your feedback to do your job properly.

LEARNING

Rarely does a team go to a track for a race and have to be on the pace right out of the box. While that would be nice, at most events there are one or more practice sessions prior to qualifying and the race. So think about it. When do you (and your car) have to be as fast as you can possibly go? For one lap of qualifying and then in the race, right?

Use the practice sessions for learning, learning how to be as fast as possible for your qualifying lap and for the race. This strategy provides better results in the race.

If at all possible during a test day, try to end the day on a positive note. People tend to recall most vividly the last piece of information they had. This is referred to as the "recency effect," meaning that what was most recent is now most deeply programmed into the brain. In other words, you will recall and mentally replay your last session on the track more than any other. Mentally recalling and replaying creates programming. I'm sure you would rather program a technically correct, positive experience than the opposite.

This also suggests how important it is that you recognize when you start to become tired, either physically or mentally, so that you can stop before you begin programming errors. Remember, practice does not create perfect; only perfect practice creates perfect. It is far better to quit a test session early than to practice and get good at, making errors. Of course, if you have physically or mentally tired before the end of a test day, you need to create a fitness-training program that will ensure it doesn't happen again.

ADAPTABILITY

Certainly, there are times you want to drive the car in the exact same way as you have in the past. Without consistency it is difficult, if not impossible, to determine if changes to the car's setup helped or hindered. That is what being a good test driver is all about.

But there are times when it is important for you to be able to adapt to whatever the car is doing. That is what being a good racer is all about. For example, if the car develops a turn-in understeer in the middle of the race, you had better hope that you know how to adapt to it. Otherwise, you are going to have to watch yourself go backward in the pack.

How many times have you heard a driver complain about how his "car began to push in the middle of the race"? More than once, I would bet. But how often have you heard that same driver follow that statement up with, "and I didn't know what to do about it"? Never. And yet, it's often the case.

How do you develop the necessary adaptability? By educating yourself on the dynamics of the car, and giving yourself the time and opportunity to practice, to learn, during a test session.

The overall objective is to learn to be more adaptable to a car's handling problems. Most drivers try to force a bad-handling car to do what they want it to do. That won't work. A driver can't make a car do what it doesn't want to do. The only option is to adapt to it.

To improve your ability to adapt, spend a portion of a test day going through the following routine. Begin with a warm-up and set a baseline with the car fairly neutral in its handling. Then, work on adapting to *understeer.* Tune the car's setup to make the car understeer during the entry to the corners, the midcorner phase, and while exiting the turns. Then try adapting, lessening the negative affects of the understeer. Take time to play with different turn-in points and techniques, varying the amount of trail braking, and so on.

After that, work on adapting to *oversteer* by altering the line, changing the speed at which you turn in, when and where you release the brakes, and then get back to power.

The method you use to adapt depends somewhat on where in the corner the understeer or oversteer begins, whether the car is in a steady-state or transient state, and what you're doing to it. Therefore, when making the changes in the car's handling, try using the shocks, anti-roll bars, and maybe even springs and aerodynamics to vary the timing and severity of the handling problem.

In adapting, you need to compare rpm at a reference point on the straightaways as well as lap times to see which method works best. You may want to do that as well with the data-acquisition information. One method may work in one type of corner but not another. Though it may have helped in one corner, you might have lost at another part of the track, so the lap times will not tell the real story. That's why it is important to compare straightaway speeds as well as lap times.

Ultimately, which method works best does not matter. The main goal is for you to be able to use any and all of the methods, for at some time any one will be the best choice. You should become aware of how to do each method, how it may or may not help, and what to expect from each one. This is all about adding information and knowledge to the data bank in your head.

Earlier, I mentioned that there is not one style that suits every car and corner. Rather, a great driver will vary his or her style to suit the situation. One of the ways you learn how to do that is through practice, and this adaptability exercise is one of the best ways to learn to vary your style.

If you have time, as a complement to the above exercise, you might want to try another, one that shouldn't take too much time but will be valuable as well. This time, starting with a balanced car, *you* (with your driving) make the car understeer at entry, in the middle of the corner, and at the exit, and make it oversteer at entry, midcorner, and exit. The idea is if you know how to make it do these things, you may recognize (become aware of) yourself doing some little bit of this at some time. If you realize you may be causing some of the understeer or oversteer, it becomes easy to fix.

Exactly how do you induce the understeer or oversteer with your driving? That is up to you to find out. You should experiment with how you turn the steering wheel (the timing, how abrupt or smooth, the speed at which you turn it, and so on), the cornering line, controlling the weight transfer (with the brakes and throttle), and the car's speed. If someone told you exactly what to do to make it understeer or oversteer, you would not get as much out of it. It is the experimenting—the trial and error, the self-discovery—that will help you become aware.

Deliberately making the car understeer or oversteer makes you aware of whether you are causing or exaggerating any handling problem. It is like a golfer who consistently slices his tee shots. The typical fix is to turn his body or change his grip to compensate for where the ball is likely to go. The best fix, though, is to go to a driving range and deliberately slice a number of balls. By figuring out how to make yourself slice the ball, you have identified how to truly fix the cause of the problem, rather than just Band-Aiding it.

This approach uses the *MI+A=G* to the fullest extent but at a subconscious level. Using this approach, you will fix any problem without even trying. It seems ironic, perhaps, that by trying to make an error, you become aware of what causes the error, therefore allowing you to fix it.

PRACTICING Q-MODE AND R-MODES

It really is unfair to expect yourself to be able to qualify at your best if you've not been allowed to practice it. With many cars and tire combinations requiring drivers to put in their perfect qualifying lap on a specific number of laps into a session, your job becomes even more challenging.

The primary way for you to learn to qualify is through physical programming: practice. Yes, mental programming is also important, and you must do that as well, but it is difficult to visualize something you have never, or rarely done. You must give yourself the time to experience it and practice it. Then, you should go away and mentally program driving in what I call Q-mode.

The combination of physical and mental programming of Q-mode will result in you putting your car in the grid position it deserves.

In addition to programming your Q-mode, you should also work on various R-modes. What do I mean? In most forms of racing, a driver is not necessarily going to drive the entire race at what is often referred to as ten-tenths, at what I call R-1. R-2 is the mode where you're backed off just so very slightly, at a pace that you could maintain all day, whereas R-1 is more of a "flyer" lap, most likely used in qualifying and the first and last few laps of a race. R-1 is on such a ragged edge that you would probably find it difficult to maintain throughout a race. R-3 is backed off from R-2 a bit more and is perhaps used to save the tires, brakes, gearbox, or engine.

The important point here is that you cannot make the decision in the middle of a race to back off a bit, or to crank it back up again near the end of a race, without having a subconscious program to do that. I'm sure you have witnessed a driver who tried to back off slightly to protect a lead, only to crash. Or the driver who crashed, having had a big lead and being able to back off, only to have to crank it back up to the R-1 mode after a full course yellow ate up his lead. Even drivers like Ayrton Senna and Michael Schumacher have made big errors trying to change modes at the conscious level.

These different levels, or modes, must be programmed. Once again, the most effective way of programming them is through both physical and mental practice. So you need not only practice driving at the R-1, R-2, and R-3 levels, you need to learn to trigger them. After some physical and mental programming, spend a session saying to yourself "R-1," "R-2," or "R-3," and immediately drive at that level.

BUILDING PROGRAMS

Perhaps the greatest hockey player of all time, Wayne Gretzky, said, "No matter who you are, no matter how good an athlete you are, we're creatures of habit. The better your habits, the better they'll be in pressure situations." One of the roles of practice is to build better habits, or better programming.

Of course, just practicing driving around a racetrack may not be the most efficient use of time. As I've said before, only perfect practice makes perfect, so any amount of track time you get should be supervised and focused. If not, you may just get better at doing the wrong thing. And this applies no matter what level you're at.

It has always amazed me how the greatest athletes in any sport are generally credited with having superior natural talent, and yet they all seem to practice more, harder, and with more focus than their competitors. It makes me wonder if all that natural talent is really just more (and better) practice.

Michael Jordan would show up for a game before other members of his team to practice his shot. During a short period of time early in 2001 when Tiger Woods was not winning everything in sight, he claimed it was because he was working on shots he would need specifically for the Masters later that year. Some people doubted his claim, until he won the Masters again. Martina Navratilova, winner of 167 singles titles in tennis, including a record nine Wimbledons, said, "Every great shot you hit, you've hit a bunch of times in practice." And here I just thought it was all her natural talent that won those tournaments.

The stories of Michael Schumacher's commitment to practice and being the best is already part of his legend. After a day of testing at Ferrari's test track, where he had just completed the equivalent of two full Grand Prix race lengths, he would spend a couple of hours in the gym working out.

The point is that no athlete, not even you, can be expected to be the best if you don't practice, both on and off the track. That practice is all about building better programming. The more you build better programming, the more natural talent you will be credited with.

One final comment about practice: It's dumb to crash in practice. This session is really to learn the track and to find the right setup for the car so you will be quick in qualifying and the race. Don't waste it by crashing. There's not much satisfaction in having someone say, "It's a shame about that crash; you almost won that practice session!"

41 QUALIFYING

Qualifying can be an art in itself. Being able to pull off one extremely quick lap is what it's all about. Obviously, it's important to qualify well. The closer to the front of the grid you are, the fewer cars you have to pass. Plus, psychologically, it gives you an edge on everyone you outqualified.

It's often best to wait for a clear gap in traffic during qualifying. There is not much point in driving in a group of cars, only to have those cars slow you down. Sometimes you spend more concentration on "racing" the cars around you, rather than focusing on what *you* need to do.

That said, some drivers actually perform best when there is a little extra incentive, such as chasing another car. Plus, you may be able to get a good draft off the car in front. But be careful you don't get too caught up in what the competition is doing. Again, focus on your own performance.

As I said earlier, you may want to set up the car a little differently for qualifying. Sometimes, setting the car up a little looser (so that it oversteers more) or with less downforce is best for a couple of quick laps but would be difficult to control for the length of a race.

There also comes a time in qualifying when you may have to go for what former World Champion Niki Lauda called a "chaotic lap." This is where you push for that last extra tenth or hundredth of a second. This may mean leaving your braking that fraction longer, entering a turn a fraction quicker, or taking that "almost flat-out" corner absolutely flat-out. Obviously, this can be the most dangerous driving you ever do. It will likely be the most thrilling as well, and the most satisfying when it all works out.

Qualifying for an oval track race, where one car at a time makes a qualifying run, is probably the most pressure-filled moment of your life. But, like anything, the more experience you have doing it, the easier it gets.

It's important to be focused for qualifying, whether it's by yourself on an oval or in a pack on a road-racing circuit. This is where you really have to shut out everything else around you and visualize yourself doing everything perfectly on the track, pushing for that last ounce of speed. Then, once you're on the track, just let it flow. Don't "try." If you're focused and you've visualized what you want to do, it should come naturally. Let it happen.

Putting a lap together for qualifying can be a challenge for some drivers, even some who are no doubt fast. The challenge can be driving the car at the limit every inch of the track. All fast drivers drive the car at the limit, for most of the track. Super-fast drivers drive at the limit the whole lap, not just parts of it.

A few years ago I was coaching a driver competing in the Formula Atlantic series race that was supporting the Canadian Grand Prix in Montreal. As luck would have it for me, the team's shirt I was wearing that day was almost identical to the ones worn by the track security people. I first noticed this when I walked up to the front of a grandstand to watch the Formula One cars during a practice session. As I stood there with a radio attached to my belt and a hard-card pass hanging around my neck, three people showed me their ticket and asked where their seats were. I politely pointed them in a direction I thought might help, and then noticed a gap in the fence where a security person had just walked through. So I did as anyone who wanted desperately to get a closer look into the cockpit of a Formula One car at speed would do: I walked through the gap in the fence and positioned myself in such a way that I had an amazing up-close view of the cars as they made their way through one of the left-right chicanes.

I stood there for about 30 minutes soaking up as much as I could from this vantage point, asking myself why some drivers were doing what they were doing, what made the fast ones fast, and noticed the dynamics of the cars as they passed. Even at the level Formula One drivers are at, there were things that I could notice that some could improve on. There were some drivers who had their cars at the limit through about 95 percent of the section of track I was observing. There were others who had their cars at the limit for 99 percent of the time. There was one driver who consistently had his car at the limit for 100 percent of the section, Michael Schumacher. He was doing something inside the cockpit that allowed him to have his car at the very ragged edge through the middle section of the chicane in a way that the others didn't, not even Haikkonen or Montoya. The difference at that level of driving is minute. I'm sure that had I not been able to get as close to the track as I did that day, I would never have been able to notice the difference.

So, if putting together just one section of track—driving it at the absolute limit throughout a section—is so difficult, imagine how challenging putting together an entire lap can be. And I bet you have by now, imagined putting together one perfect lap, with the car at its absolute limit every inch of the track. That's what mental imagery is all about. But if it's difficult for even Formula One drivers to have the car at the absolute limit for every inch of the track, how challenging is it for you?

SPEED SECRET

Driving the limit means driving the limit every inch of the track.

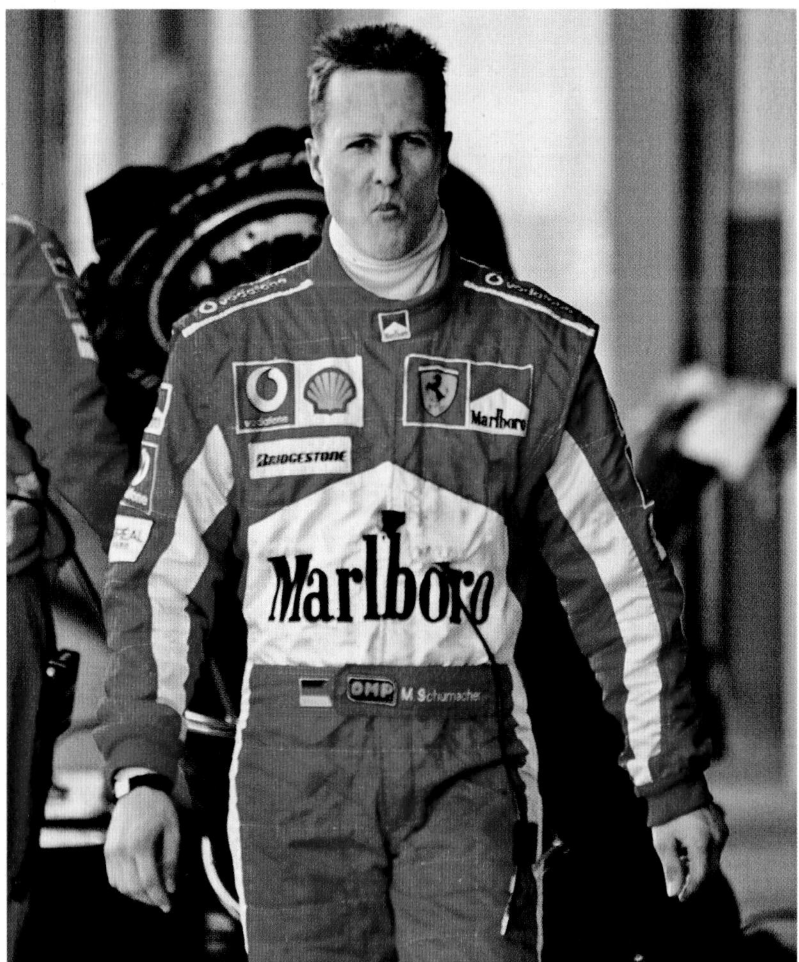

Qualifying, getting that last bit of speed out of yourself and the car for just one lap, is as much about your mindset as it is about any technique. Triggering that mindset starts before you even get in the car. Even the way you walk to the car can impact your mindset. Michael Schumacher's walk sends a message to himself and others that he means business. *Shutterstock*

I strongly recommend that when you're doing mental imagery of a perfect qualifying lap, you break it down. Break it down, almost to the point where you can see, feel, and hear yourself over every inch of the track, even if that means doing it in slow motion in your mind. If there is a piece of the track, no matter how small, where you can't imagine having the car at the limit, you won't when you get out there to qualify. And for some drivers, there are pieces of the track that they can't imagine in their minds at all; it's as if a section of the track is missing. Guess what? If that's the case for you, it's doubtful you'll be able to drive that section of track well. Do everything you can to imagine that part of the track, even if it means going out to that section and getting another view of it while other cars are on track, imagining yourself in one of them. Or walk the track at the end of a day, stopping to take a mental snap shot of specific sections, and then close your eyes and mentally rehearse it before walking on to the next section. Or if you have some in-car video, play specific sections in slow motion, and stop and replay them over and over until you've got a clear mental picture of every inch of the track.

And that's the process for putting together a lap. Get a good clear mental image of what every inch of the track looks, feels, and sounds like at the limit, build a mental trigger to launch the program, and then just let that program take over when on the track. Perhaps the most difficult part of this puzzle is that last piece: trusting the program. For many drivers, the idea of not trying and just trusting their mental programming to drive the car just doesn't seem right. They have the mental model of a race driver constantly fighting with the car, making it do what he or she wants, thinking about every single movement in the car, and consciously being in control.

Ayrton Senna was a master of trusting his programming to drive the car. One of his legendary qualifying drives was at Monaco in 1988 where he took the pole by outqualifying his McLaren teammate, four-time World Champion Alain Prost, by 1.4 seconds! Afterwards, Senna said, "And I suddenly realized I was no longer driving the car consciously. . . . I was kind of driving it by instinct, only I was in a different dimension. . . . I was just going and going, more and more and more and more. I was way over the limit but still able to find even more. Then suddenly something just kicked me. I kind of woke up and realized that I was in a different atmosphere than you normally are."

Senna got to that level more than most drivers, and if you really listen to how he described his driving it was clear that he trusted his mental programming to drive the car, even if that resulted in him being a little scared by the feeling: "It frightened me because I realized I was well beyond my conscious understanding." While he may not have truly understood what was happening, he certainly allowed himself to get there more often than most. And that may be the most important point: allowing yourself to get there, trusting that by letting go you'll actually perform better by letting your programming drive the car.

Putting a lap together, driving the car at the limit every inch of the track, is all about developing a clear mental program, triggering it, and then trusting it to do the job.

SPEED SECRET

Program every inch of track you want to drive at the limit, and that better be every single inch of track.

42 THE RACE

Before a race, think about where you are starting on the grid. Who is starting around you and what are they like to race with? Can you trust them to run wheel to wheel with you? Are they fast starters? Do they run a few fast laps, then begin to fade?

Analyze those factors and have a plan well before you head out for the start of the race.

During your first pace lap (or the first lap of a practice or qualifying session for that matter), your first priority is to get the tires and brakes up to operating temperature. Many drivers will weave back and forth across the track to heat the tires. This is great, but be careful. Often, you will end up in the "marbles" off line with cold tires. Many drivers have spun out doing this. Also, it has been known to happen where two drivers get so caught up trying to warm their tires, that they actually collide. Watch closely for what the other drivers around you are doing. Don't be surprised by someone accelerating and then braking hard.

In fact, race tires will heat up quicker from hard acceleration and braking than just weaving back and forth from side to side. A goal should be to heat the tires from the inside. If you build heat in the brakes pads, the temperature transfers through to the rotors, to the hub or uprights, to the wheel, and into the air inside the tire. This builds temperature in the carcass, not just the surface of the tire.

So weave back and forth while using the brakes with your left foot to heat them up. Accelerate hard in a straight line, getting some wheelspin, and then brake heavily. If possible, hang back a little when approaching a corner, then accelerate to take the turn quickly, even trying to work the steering wheel back and forth to scrub the front tires. At the same time, take one last good look at the track surface in case some oil or whatever was dropped on it in the previous races. If it's raining, really work the car around to feel how slippery it is. Make sure you're comfortable with what the car is going to feel like during the opening laps.

At the start, look far ahead, not just at the cars around you. If possible, watch the start of other races to see where (approximately) the starter drops the green flag. And if you are using a two-way radio, have a pit crew member watch the starter and radio you as soon as he sees the flag drop.

Sometimes you can hang back just a little from your grid position, then begin to accelerate just slightly before you think the green flag is going to drop. If you've timed it right, you will have a slight advantage on the others around you. If not, you're going to have ease off the throttle. What you don't want is for the flag to drop just as you're backing off the gas.

In fact, depending on your grid position, once you have started to accelerate, don't lift. If you do and the green flag drops, you are going to lose positions. If you try to anticipate the green and begin accelerating, stay on it (within reason, obviously). If you do this, one of two things will occur:

- The flag drops just after you begin to accelerate and you get a jump on the field.
- The starter doesn't drop the green flag and there will be a second pace lap (don't try jumping the start the second time).

Be careful going into the first turn on the first lap, as more crashes happen there than anywhere else. Having said that, it is important to get a good start. If you start too conservatively and lose contact with the lead pack of cars, you may never be able to make up for it.

SPEED SECRET

Races are not won in the first corner; however, they are often lost there.

It's usually best to run as quickly as you can for the first few laps, then settle in to a comfortable, consistent pace, all the while ready to take advantage of any opportunity to pass. Never turn down an opportunity to pass; you may never get it again.

SPEED SECRET

Most races are won in the last 10 percent of the race.

Be sure you're able to run strong at the end. Sometimes that means saving the car for the end of the race, and being a little easy on the brakes, tires, or whatever.

Never give up, no matter how far behind you are, no matter how unlikely it seems you will catch your competitor in front of you. Keep pushing until the checkered flag falls. You never know if the competition is having problems that might be terminal if they have to drive hard to fend you off. How many times have you seen the leader of a race have a mechanical problem with only a few laps to go? You will never be able to take advantage of their problems if you are not close.

SPEED SECRET

You have to be close to take advantage of luck.

The most successful racers of all time, like Jackie Stewart, Michael Schumacher, Rick Mears, Dario Franchitti, Richard Petty, and Jimmie Johnson all have one thing in common: They finish races. In fact, if you look closely they have an incredible finishing record. Never forget, "To finish first, first you have to finish."

Most of these drivers would also agree that you should attempt to win at the slowest speed possible. Some drivers are not content to just win the race. They feel they have to set lap records every lap or lap the entire field. Most of these drivers have a poor finishing record. They also have a poor winning record. All anyone remembers is who won. It doesn't matter by how much you win, just that you win.

PITTING

Experience, practice, and a little thought, not to mention a well-prepared crew, are the keys to successful pit stops. Simply put, as a driver your job is to stop the car right on the exact marks set by your crew. Stay calm while you're stopped, perform whatever functions your team requires (reset fuel counter, foot on or off brakes, and so on), and be ready to go the second the crew is finished. Be sure you know exactly what is expected of you by your team during a pit stop.

Seeing your pit while speeding down pit lane and determining exactly where to stop can be a challenge at some tracks. Know what kind of signal your crew is going to give you, and have some other form of reference point for your pit (the number of pits past pit lane entrance or from the end, in relation to the start and finish line, and so on).

One aspect of pit stops often overlooked is your "in" and "out" laps. Many drivers click into pit stop mode (mentally stopped) on the entire lap before entering the pit lane and then take forever to get back up to race speed after the stop. Instead, you want to drive flat out until the very last second before diving into pit lane, and then return to the track as quickly as possible (remembering you may be on cold tires). Watch an Indy-car race, and make note of how little time the winner spends on his in and out laps compared to other drivers and how much he gains on them during that time.

ENDURANCE RACES

Races at least three hours long and requiring a driver change are usually considered endurance races. Typically, they are 6, 12, or 24 hours long and can be either amateur or professional.

It is a good idea for any driver to compete in as many endurance races as possible, no matter what type of car it's in. In terms of seat time, you can't beat it. Often, you will drive for at least one-and-a-half-hour stints and perhaps up to three hours. It's great practice and really trains you to concentrate for a long period of time. This is going to help a lot when competing in sprint races.

Plus a driver learns to "save" the car, to not abuse it mechanically. This practice will rub off on your sprint race driving technique.

In most endurance races there will be many classes of cars competing all together. This means you will get a lot of practice passing and being passed in a relatively short period of time, perhaps as much in one race as you would in an entire season of a one-class or "spec" series.

When driving endurance races, it's important to get yourself into a rhythm early on and stick to the pace you and the team have decided on. Avoid getting caught up in a heavy battle with another car. Yes, you want to beat your competitors, but pace yourself. Sometimes, if you can't pass and pull away from a competitor, you're better off following them for a while. Often, this will result in them losing concentration and making a mistake.

Obviously, in an endurance race, pit stops are going to play a vital role. Make sure your team practices them. And practice driver changes. Often, the amount of time spent in the pits fueling and changing drivers determines the outcome of the race.

Driver changes can be difficult. The biggest problem is the varying sizes of drivers. Seating position and comfort is sometimes a compromise. But remember what I said earlier about how the seating position can affect your performance, so do everything possible to minimize the compromises.

A general rule in endurance racing is the less time spent in the pits, the better your chances of winning. I know this sounds obvious, but it's surprising how many teams seem to ignore this, instead relying on their speed on the track. There is nothing more frustrating than beating a competitor on the track, only to have them beat you overall through better pit work and strategy. Besides, it's much less expensive to improve a team's pit work speed than it is the car's speed.

THE RACER

The perfect race driver is fast. That should go without saying. What should also go without saying is that the perfect race driver is dynamite when racing wheel to wheel. Racecraft is that ability to position your car in such a way as to minimize what you lose to anyone when passing or being passed: the ability to outmaneuver other drivers. In other words, racecraft is the ability to come out on top when competing closely with other drivers, even when driving a car that is less competitive than the rest of the field. The perfect race driver has racecraft.

Experience and observation tell us that there are more drivers who are fast than there are great racers. The late, great Carroll Smith used to refer to the fact that there were drivers, and then there were racers. Drivers are people who can drive fast; a racer can win races even when he or she is driving a slower car.

Gilles Villeneuve, the Canadian Ferrari Formula One driver who died in a crash in 1982, was a racer to the greatest degree. He won races he should never have been able to win. In 1981, he won the Spanish Grand Prix in a Ferrari that should not have beaten the competition that day. It was a bullet down the straightaway, but was one of the worst handling Formula One cars ever built. When Gilles crossed the finish line that day, there was a nose-to-tail line of four other cars directly behind him (first to fifth place were separated by only 1.24 seconds), all trying to pass him—and all of them could have turned faster lap times if they had gotten by Gilles. But Gilles was a fighter, perhaps the greatest fighter in a race car ever. He would not give up, he drove that Ferrari so hard through the turns, making up for what the car was not capable of through sheer will. Gilles' racecraft is legendary.

Take the last two laps of the 1979 French Grand Prix as another example. His battle for second place with Rene Arnoux in the Renault is regarded by most as the most exciting, most brilliant couple of laps in auto-racing history. (Check it out for yourself. If you do an Internet search for Gilles Villeneuve/Rene Arnoux, you will be able to find a video of that famous battle).

SPEED SECRET

Have the mindset of a racer; be aware, attack, and work traffic to your advantage.

Racing wheel to wheel and great racecraft are mostly about your mindset. Sure, there are some techniques that a great racer will use, but when it comes down to a real battle, it's most often going to go to the driver who wants it the most. Great racers seem to just want it more than others, and therefore are able to find a way to make it work. They know when to squeeze another driver a little. They know that being right next to the other car, rather than moving closer to the inside of the turn, will intimidate the other driver, will mean that if the two cars touch, the impact will be much less and will give him a better line through the corner. They know that if they move far to the inside, they provide an opportunity for the other driver to repass them, and they really haven't taken control of the corner.

Great racers know that all you need to do is get beside the other car when trying to outbrake it. In fact, if you allow yourself to go farther, you open up the line and provide an opportunity to be passed again coming off the corner.

Great racers know when to back off just the tiniest little bit to allow themselves to get a run out of a turn to set up a pass at the other end of the straightaway. They know that if they are tucked right up tight behind another car, they are unlikely to be able to begin accelerating any earlier than the car they are trying to pass and perhaps even later than it. The key is being able to judge how far back from the other car you should be. Too far and you won't be able to make up that distance. Too close and you lose your momentum.

Great racers have an awareness of where other cars are at all times. They learn to drive with "big eyes," looking farther ahead and seeing more to the sides than other drivers. In addition to practicing seeing more when driving on the street—and that's a critical part of becoming more aware when racing—you can develop this ability through mental imagery as well. You can preplay yourself racing, always being aware of what's going on around you, seeing just the tiniest flash of movement in your mirrors and knowing how to react, making accurate predictions on when and where you're going to pass another car, setting up passes and then making them in a decisive fashion. You can, and should, program being super-aware.

Great racers seem to have luck on their side when in traffic. While racing with another driver, they will come up on another car, and it will seem as if the slower car will be working with them. The great racer will slip by just as the slower car turns in and blocks the other driver. They seem to be able to judge and time the other driver's passes so the slower car doesn't hold them up as it does other drivers. To the observer, it's as though the great racer is just luckier with traffic. But great racers know that it's due to their mindsets. They believe that other drivers will help them and not others. Yes, this is part of their belief system. They know things will go their way, and they usually do. And even when they don't, it's such a rarity that it doesn't bother them. They just continue to attack and make the next pass work.

Have you ever noticed that you can tell just by watching when a driver is in "attack mode"? A driver and car seem to have an attitude that you can see when they're on the move. If you've never noticed this, look a little closer and you'll see it. If you have noticed it, then you know why other drivers can also sense it. If you have that attitude, that mindset, that attack mode, other drivers will notice. You may be surprised at how they react. When a driver has that attitude, it's as if they are saying to everyone else, "Get out of my way. I'm coming through." And guess what happens? Without even knowing it, they tend to be a little easier to pass.

Once again, you can develop that attitude or mindset through mental imagery. Preplay races in your mind, with you having the attack-mode mindset. See, feel, and hear yourself as the "Master of Traffic."

Great racers are just that. They are racers. Again, the best are much more than just a fast driver. A great race driver can win races while driving a car that is not as fast as the competition but by being a better racer.

SPEED SECRET

How bad do you want it? Desire wins race battles.

43 THE COMPLETE RACE DRIVER

It's long been an accepted fact in motorsports that a driver needs far more than just the ability to drive fast to make it to the top. Some say that it has always been this way to some extent, but most would agree that this is truer today than ever before. To be a champion race driver in today's world of motorsport, it takes much more than just the ability to drive quickly. Today's complete race driver is a package, one made up of the ingredients shown in Illustration 43-1.

This illustration shows a virtual job description for the complete race driver, a driver that is or will be a champion, a superstar.

While this book is meant to help you develop your skills in all of these areas, not all readers will be interested in making a career as a race driver. Many drivers just want to have fun racing as a hobby. If this describes your approach to the sport, you may not think you can gain from some of the skills discussed in this chapter. My bet is, though, that with a better understanding of what it takes, you will be more successful no matter what your level of racing participation.

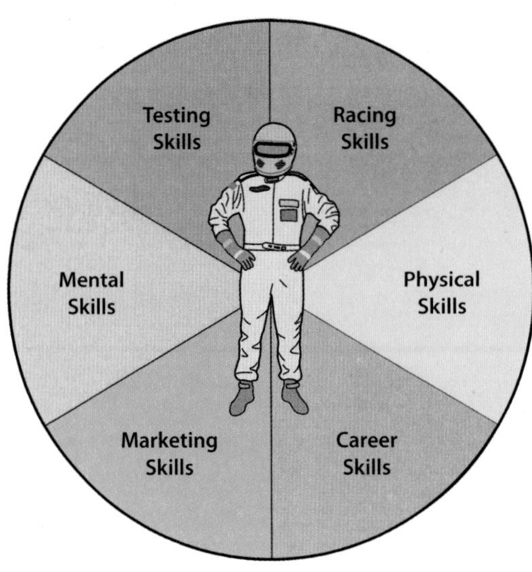

ILLUSTRATION 43-1
A champion race driver today is made of the attributes shown here. Rarely, if ever, will they be equal parts. Some ingredients are more important than others, depending on the type and level of racing, and the specific situation or team you are involved in.

The complete driver is made up of the following:

- Racing skills
- Testing skills
- Physical skills
- Mental skills
- Marketing skills
- Career skills

Take any driver currently racing in Formula One, NASCAR, or Indy cars, but one that is typically running midpack or back. Is that driver running in that position because he or she lacks the driving skills?

Look at our list of the six skills that make up the Complete Driver. Which of them is your example driver weak in? Which is the driver strong in? Now consider another driver, choosing his or her weak and strong skills. Is the driver strong in racing skills but not so in the testing and marketing skills? Or is this driver strong in marketing and career skills but not the fastest driver out there?

Consider Michael Schumacher. What were his weak and strong skills? Think about it. It's hard to find a weak skill, isn't it? He's a great racer, fast and brilliant when it comes to racing wheel to wheel. He's one of the best when it comes to testing and tuning his car. He may be the strongest ever when it comes to physical skills, and his mental skills as well. Love him or hate him, he's a marketing person's dream come true. And he certainly has proven that he knows how to be in the right place at the right time: He has career skills.

Or consider Jimmie Johnson. What are his weaknesses? Does he have one?

The point of this exercise is to demonstrate that the more complete a package you are, the more likely it is that you will be successful. Michael Schumacher and Jimmie Johnson are successful because they are so complete.

Note that when you rated the midpack drivers, you were still rating drivers that were at least in the pack. If you could rate some drivers that never even made it, it would be even easier to identify their weak points. In fact, the reason they didn't make it is just that: They had weak skill sets that let them down.

SPEED SECRET

The more complete a driver you are, the more likely you will have a successful career.

DRIVER TYPES

More than one bench-racing session has been spent arguing over who is the best race car driver of all time. You may have participated in this discussion yourself. Names like Schumacher, Senna, Andretti, Johnson, Earnhardt, Petty, Foyt, Stewart, Clark, and Fangio are most often bandied about in these sessions.

Over the past few years our Speed Secrets driver coaches have also had this discussion, but mostly for a different reason. While it's a fun topic to debate, we've looked at it from the perspective of what can we learn from these greats. Fortunately, we have a pretty interesting perspective. Many of us have raced at a high level, even competing against some of these greats. We've also coached some young drivers who have gone on to great things, such as winning the Indy 500.

Along the way, I developed the concept of the Complete Driver. But in defining what a race driver needs to focus on to become one of the great champions of all time, we also identified six types of champion race drivers. They are listed below:

- The Workhorse
- The Natural
- The Charger
- The Personality
- The Professor
- The Complete Driver

Let me define these driver types and provide you with some examples.

THE WORKHORSE

The Workhorse is a driver who has made it to the top and become a champion through a lot of hard work. These drivers have developed skills through practice and their careers through hard work and determination.

When you think of drivers such as Bobby Rahal and Jimmy Vasser, Nigel Mansell and David Coulthard, or Terry Labonte and Ricky Rudd, you think of drivers who made it through hard work. Some will say these drivers were not born with the same natural talent that others were (although that is certainly an arguable point), but that they made it through their commitment, determination, and practice.

THE NATURAL

Pure natural skill. Born with talent. That flair and skill that makes their speed seem to come so easily. That's what the Natural has.

Ayrton Senna may be the poster boy for the Natural, but others like Juan Pablo Montoya, Scott Dixon, and Kasey Kahne have shown the same innate talent.

THE CHARGER

The Charger tends to either win or crash. Attributed to big you-know-whats, bravery is often what the Charger is credited with.

Gilles Villeneuve may have been the ultimate Charger, but Paul Tracy certainly seems to have followed in his footsteps, as have Sam Hornish and Tony Stewart.

THE PERSONALITY

Could it be that some drivers have made it to the top due to marketability and personality, simply because of who they are and how they deal with people around them? Or due to their lineage, being "second-generation drivers"? While a champion must always have driving talent, there are some drivers who opened a few doors and generated some opportunities due to their personalities and the good fortune that has come with it.

Would Danny Sullivan, Jacques Villeneuve, and Michael Waltrip have had the opportunities to showcase their talents without their names and marketing abilities? Did Alex Zanardi and Damon Hill go farther in their careers due to their personalities or lineage? Without a doubt, they all have incredible driving talent, but there are certainly other attributes that have contributed to the Personality's success.

THE PROFESSOR

The Professor is the thinking driver. They often win at the "slowest possible speed," using a calculated approach. Their cars are often better setup due to their analytical approach.

Jackie Stewart's strategy was to win at the slowest possible pace to conserve his car. Alain Prost, the second most winning F1 driver ever, had the nickname "The Professor." Rick Mears may have been the best Indy-car oval racer ever due to his calculating approach to winning races. Even in the wild world of NASCAR, Alan Kulwicki won a championship by being just a little "smarter" than everyone else.

THE COMPLETE DRIVER

There have been few drivers who are the perfect combination of all of the above traits, but that is the ultimate goal of every driver, or at least it should be.

Michael Schumacher may be the most Complete Driver ever, although Mario Andretti may have been at least equal in his day. And Jimmie Johnson.

Think about Schumacher: Is he a Workhorse? Yes. He works harder at his craft than any driver. Was he born with natural talent? Certainly. Is he a Charger? Witness some of the dices he's had with Montoya, Haikonnen, and Villeneuve over the years. Is he marketable and does he have a personality? Love him or hate him, Schumacher is a somebody, and his ability to motivate people around him is legendary. A Professor? His attention to detail and his analytical approach is one of the keys to his success. The Complete Driver? Perhaps the best ever.

Although the definition of the Complete Driver has changed through the years, Mario Andretti is the only other driver in the history of the sport who we believe could rival Michael Schumacher in his completeness. He worked hard, had natural talent, was a charger, was marketable (who in the world doesn't know the name Andretti?), and was a smart driver.

Understand that I'm not suggesting that one type of driver is any better than another. Nor am I saying that because a driver is dominant in one type that he isn't also strong in another. For example, the Natural is no better than the Workhorse, just because he may (or may not) have been born with more natural talent. You can make the argument that the Workhorse, in fact, is better in that he took what he was born with and made himself as good as the Natural.

Neither is the Charger any less intelligent than the Professor, nor is the Professor any less assertive or competitive. It's just that each one has used his dominant trait to maximize his performance.

Now, if you had to pick one to be, it's obvious that being the Complete Driver is going to be advantageous. It increases your odds of making it as a professional race driver. Can you make it if you're not the Complete Driver? Well, the fact that so many champions have made it without being the Complete Driver says it can be done. However, even the champions I've used as examples for each driver type are awfully close to being the Complete Driver. It's hard to find a type or characteristic that any one of them is weak in.

When you attempt to categorize a successful driver, you will notice that some are difficult to identify as one type or another. That's because they are well-rounded, or close to the Complete Driver. The more complete a driver is, the greater his or her odds of success. Every one of the drivers I've listed here are close to being the Complete Driver. They are also all champions.

Take a look at drivers who have not won major championships. It's interesting to note that it becomes much easier to categorize them. The reason? Because they are not as complete. Coincidence? I don't think so.

All young driver hoping to make careers driving race cars need to look at their strengths and weaknesses. While the natural thought is to work on improving one's weaknesses—turning them into strengths—this is not always necessary or possible. Sure, one can make improvements, and that's important. But drivers also need to focus on taking their strengths and making them even stronger.

There is a good piece of marketing advice that goes, "Differentiate or die." Being focused on what makes you different helps define your "brand," who you are to the public, racing community, media, and sponsors, and a clearly identifiable brand makes it easier to sell yourself. It makes it easier for others to relate to and to remember you. Focusing on your strength, and emphasizing it, may make it easier to build your brand, and therefore open doors to selling your services as a driver. For example, if you have a reputation for being a thinking driver—a professor—perhaps you can emphasize this and sell yourself to teams with this reputation.

How would you define yourself? Are you very strong in one area but not so in others? Or are you a well-rounded, complete driver? Perhaps you're not sure. If so, ask people close to you and your racing experience. To what successful pro have you been compared you? Study that driver's career and see if any of the things he did to carry himself through the trial years into the successful ones would work for you. The more complete a driver you are, the more likely you are to become a great champion. What are you doing to become a more complete driver? Be honest. No matter how much natural talent you have, without being analytical, without being a charger, without having a marketable personality, and without working hard at developing all these traits, someone else with less natural ability and all the other traits will beat you. Think about that. Then ask yourself what you're doing about it.

FINAL THOUGHTS ON THE COMPLETE DRIVER

The one constant in racing is that it is constantly changing. That means that this definition of the complete race driver today may not be totally valid a year from now, let alone in five years. But if you have two of the key components today—adaptability and a burning desire to learn—you will evolve as this definition evolves.

All drivers will have their strong and weak points. You can look at this in two ways. First, if you are weak in one area, that is the area you need to work on to improve, that is, if you truly want to succeed.

On the other hand, if you have a weakness in one area, you may be able to make up for it with strengths in other areas to some extent. For example, if you are not so strong in marketing skills, you may be able to make up for it by being strong in all the other areas. If you are very strong in team building, you may be able to motivate a team member to take over some of your marketing duties.

Having said that, usually the driver who is the most complete package—the best overall compromise between all these areas—will be most successful.

SPEED SECRET

The more complete a package you are, the more successful you will be.

No, I'm not suggesting for one second that you do not need the ability to drive quickly or that these other skills can make up for a lack of speed. Racing is all about speed, and no driver is going to make it far without this ability. I wouldn't have dedicated my life to learning how to drive quickly and how to pass on this knowledge if it didn't matter. In this day and age of motorsport, however, without combining all of the other factors with speed, it is unlikely you will go far.

44 ENGINEERING FEEDBACK

No matter how fast a driver you are, if your car does not perform as well as your competitors' cars, you're at a disadvantage. In the history of auto racing, there are many drivers who could and should have been champions but were not. While it was obvious they had the speed to be a champion, there was something missing. This includes drivers such as Ronnie Peterson, Danny Ongais, and Roberto Guerrero in the past, and I have to wonder about drivers today like Dale Earnhardt Jr., Marco Andretti, and Filipe Massa.

Obviously, there are many reasons for a driver not to become a champion. Many will say the lack of success is a result of being with the wrong team at the wrong time, and that's true. But what made them the wrong team at the wrong time? Is it that these drivers, while being very fast, don't have the same ability to help develop the car as other drivers do? Is it that given the same starting point, the same car, that these drivers would not develop the car as well as the drivers that become champions? It's a fact: Some drivers just don't have the same ability to feel and communicate what the car needs to have championship-winning performance.

The most important question, then, is why? Why can some drivers feel exactly what the car is doing and then be able to describe that feel in a way that an engineer, or even the driver himself, can know what to adjust to make it better? And if a driver is lacking this ability, can it be learned? In my opinion, based on actual experience, the answer to this second question is yes. I've seen and helped drivers develop their ability to sense what the car is doing and then be able to turn that sense into words.

Breaking the problem down, there are two issues with a driver who is not good at developing a car's handling:

- The driver is not sensitive to what's happening with the car.
- The driver does not communicate what he or she feels very well.

The first issue is something that many will attribute to natural ability, but that I know can be developed by focusing on improving one's sensory input. Leaving racing for a moment, if a person loses his or her sight, over time and by focusing on improving the sense of feel, the person can read using brail. In the same way, if drivers focus on improving their sense of feel for what the car is doing, they will become better at driving. The best way I know of to do this is through the use of Sensory Input Sessions.

It's one thing to feel what the car needs, but it's another to turn that into the words to be able to get someone else to figure out what is then needed to improve it. Knowledge is important to this. The more you know about vehicle dynamics, chassis adjustments, and how chassis adjustments are made, the better you will be able to communicate. The better your technical knowledge, the more accurate your language will be, and therefore the better your communication will be. Fortunately, this can all be learned by reading a few books, such as Carroll Smith's *Prepare to Win*, *Tune to Win*, and *Engineer to Win* (these are must-reads).

One misconception that some drivers have is that they need to tell their engineer or crew what to do to fix the car. While that may be the case if you're really the engineer, if you have someone who is ultimately responsible for the actual tuning of the car, your job is simply to report what you feel. It's not to say, "Stiffen the front shock's rebound two clicks." In doing that, the engineer can only assume what you're feeling and will not learn anything while making that adjustment. And many engineers will be offended by a comment like that, as he or she will feel that you're trying to do their job. Rather, if you said, "The car understeers just after I release the brakes, the front of the car feels as though it unloads too quickly when I release the brakes. If you can control the rate the car's front end unloads, I think it will have less understeer," your engineer can determine what it needs. That's not to say you couldn't add, "It feels like if you stiffened the front rebound, it would control what I'm feeling." That would provide even more information to your engineer.

On the other end of the spectrum, there are drivers who will say, "It understeers," or worse yet, "The car sucks." If you used the second comment, I doubt you'll be much help at all. But if you use the first comment, a good engineer will then begin asking you more questions to dig deeper into what you're saying. But if not, you're going to need to ask yourself some questions. I suggest the following procedure:

QUICK DEBRIEF

- Handling: better or worse?
- If I could have the car do just one thing better, what would it be?

DETAILED DEBRIEF

- What is the car doing? Understeer, oversteer, or neutral?
- Where is the car doing it? Which turn(s)?
- Where in the turn(s)? Entry, middle, or exit?
- What am I doing when the car does this?
  - ~ Braking?
  - ~ Trail braking?
  - ~ Releasing brakes?

- ~ Coasting?
- ~ Maintenance throttle?
- ~ On power?
- ~ Slowly turning steering wheel?
- ~ Crisply turning steering wheel?
- ~ Steady steering?
- ~ Unwinding the steering wheel?

- Is it the car or me? Am I inducing the handling problem, or is it the car?

SPEED SECRET

Ask yourself, If I could have the car do just one thing better, what would it be? Then, keep digging with more questions until the answer is obvious.

By asking yourself these questions, you dig down to the core of the problem, and the solution is quite simple. The more questions you ask, the better your quality of feedback will be.

Drivers who have a reputation for being good at setting up a car really are no more naturally talented with this; they just ask themselves more questions. In doing so, they dig the answers out of themselves; they draw the feedback out of themselves. Their curiosity may be what makes them seem more naturally sensitive to what the car is telling them. Add that to a good knowledge of the technical aspects of vehicle dynamics and chassis setup and you've got what it takes to be great at setting up a car.

By focusing on taking in more sensory information through your eyes, through your body, and through your ears, you become more sensitive to what it's doing. You become more in tune with it. In other words, by doing Sensory Input Sessions. Communicating what's happening then is fairly simple. It's just replaying the turns in your mind, thinking about what happens, where it happens, what you're doing when it happens, and what could happen if you did something different.

SPEED SECRET

The better the information your senses provide to your brain, the more sensitive you'll be to what the car is doing and what it needs.

DEBRIEFING

No matter what level or type of racing you are involved in, after every session you should debrief. This may only take a minute or two, or it could last for hours. Your main objective is to determine what the objectives should be for the next session, to make further improvements.

One of the first things you should do in a debrief session is to take a track map and debrief with it. Make notes on the track map, writing down what gear you're in, and what the car is doing in the braking zone and at the entry, the middle, and exit phase of each corner. Then, rate your driving on a scale of 1 to 10 in each section of the track, with a 10 representing the car being driven at the limit and 1 being well away from the limit.

The exact number you put on "how close to the limit was I" for each section of the track is not important. Every driver will perceive the limit as something a little different, so it is not something that you could even compare from one driver to another. The goal is simply to help you become fully aware of whether you are driving every section of the track at the limit.

The interesting thing is that most drivers will have to recalibrate their ratings as they improve. Often, you will believe you are driving at a 9 or 10 on the "limit" scale for a little while. Then, with a bit more experience, and a better sense or feel for the traction limit, you will perceive that same cornering speed as only a 6 or 7. With time, what was once a 10 will only be a 7. Again, the number is not important.

It is important to go through this process prior to learning what your lap times are compared to others. Once you begin to think in terms of how you compared to the competition, the accuracy of your awareness and feedback will suffer.

As you go through this process, you will become more aware of things the car is doing. The act of writing it down leads to a fuller awareness level. Without that awareness, you will not have the information you require to make the car better, nor will you have the awareness of what you need to change to improve your driving.

Through the exercise of putting a number on how close to the limit you are driving the car in each area of the track, you will become completely aware of where there is room for improvement.

45 TEAM DYNAMICS

Is auto racing a team sport or an individual sport? I know some race drivers act as though it is entirely an individual sport, but it is definitely a team sport. Having said that, once you're in the race, barring pit stops, it is an individual sport. Sure, it took a team to get you there, but at that point it is totally up to you. Or is it? The team dynamics, the energy level within the team, the communication, and the ability of team members to work together is one of the deciding factors in how well you perform in the race.

Looking at the history of auto racing, there have been many great dynamic duos: a combination of driver and engineer and team manager that have won more than their fair share of races and championships. Colin Chapman and Jim Clark, Colin Chapman and Mario Andretti, Roger Penske and Mark Donahue, Roger Penske and Rick Mears, Ross Brawn and Michael Schumacher, Steve Challis and Greg Moore, Mo Nunn and Alex Zanardi, Mo Nunn and Juan Montoya, Ray Evernham and Jeff Gordon to name just a few of the greatest "teams." I don't think the fact that Colin Chapman, Roger Penske, and Mo Nunn are each mentioned twice is simply coincidence. These legendary team owner, managers, and engineers knew and know how to communicate with drivers. In fact, that may be the key to them becoming legendary.

COMMUNICATION

Communication may just be the most important factor in a successful driver-engineer relationship. I don't know of any engineer who can read a driver's mind, and vice versa. You must understand your preferred learning style, and the engineer yours, as this is the basis of good communications. You need to talk about how you communicate best. And more important, you need to listen. By doing that enough, it will begin to seem as though you and your engineer can read each other's mind.

The problem with some engineers is that they only hear what they want to hear. And they only want to hear what the car is doing, and that is not enough. The best ones listen to you, the driver. At the same time, you need to educate your engineer in how to listen to you and how to communicate with you. Again, I doubt whether he can read your mind. So you may have to tell him how to communicate with you.

If you want to tell him what the car is doing, because you are an auditory processor, then tell him. If you are a visual processor, then draw or write out what the car is doing. And if you are a kinesthetic processor, show him what the car is doing, even if that means driving the track in a street car or using a model to demonstrate the attitude of the car.

One sure way of destroying your ability to work with your engineer is to be unclear about what you expect of him and he of you. In other words, unclear as to the roles and responsibilities, whether that be yours and his. If you expect to get out of the car, give your feedback on the car, and then get out of the way, then tell him. If you instead want to hang around, to help with team morale, or to be available for more debriefing, let him know. More potentially great relationships have been ruined by misunderstandings about what was expected of the driver than just about anything else. And again, the opposite is true. Make sure you know what your driver expects of you.

One sure way of destroying your working relationship with your engineer is to get out of the car and say, "It sucks," and walk off! That's not a productive way of improving the car. Tell your engineer what the car is doing, not what you think of it. Also, work out with your engineer what type of information he wants. Some want you to only explain what the car is doing, leaving the work of figuring out what to change to improve it to the engineer. Other engineers like it when a driver makes suggestions about what changes could improve the car. But another way of straining your working relationship with your engineer is to tell the engineer how to do his job. If you want to do that, why have an engineer?

You may have already figured out that the key to all of this is just plain old-fashioned conversation. The more you talk and listen to your driver, the better your understanding of each other will be.

PERSONALITY TRAITS

If you think back to the chapter about personality traits, and specifically on what your behavioral profile looks like, you may begin to see where problems could occur.

For example, let's say you have a high level of dominance, are very outgoing, not so patient, and could care less about details. At the same time, being dominant is not so important to your engineer, he is more introverted and patient, and details are everything to him. Can you see where a potential clash could occur? The engineer wants the car to be dead-on perfect by the time qualifying comes around, but feels that extracting the details from you is like pulling teeth. You'd rather be out chatting with friends, competitors, or just about anyone else who will listen. But your dominating style wants to be in control of the decisions about the car setup, the team, what hotel the crew is staying in, where dinner is going to be tonight, and so on.

And that is just your behavior and your engineer's behavior. Now, mix in the rest of the race team.

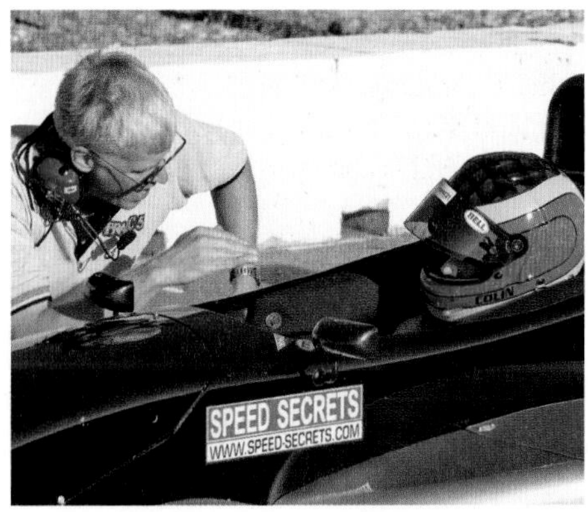

How big a problem is this? As big as you let it become. Can you change your personality traits? Yes, you can, through mental programming. And that is something you may have to do.

But within a team environment, the most important factor is not necessarily making changes to people's traits, it is being aware and understanding them. If, for example, you know your engineer's patience level, and your engineer knows yours, it makes it much easier to work together and actually complement each other.

You could have each member of your team profiled by using one of the professional firms or software packages, such as PDP or Meyers-Briggs. However, that is probably not necessary. If you simply have each member of your team do a self-evaluation, using a chart similar to the one in Chapter 27, it will provide you with what you need. Make sure that each person doing the self-evaluation fully understands what each of the traits really mean. In each category—dominance, extroversion, patience, and conformity—have each team member rate themselves, placing a mark on the scale that represents where they see themselves.

After everyone has completed the chart, sit down and talk through each person's profile. The objective is twofold. One, after some input from other people, a person may learn to adjust where they are on the scale slightly. And two, it is as important, or even more, for everyone to know what each other's profile looks like.

The key to the success of this is in the doing of the exercise. It creates an awareness and understanding for everyone involved of why individuals behave the way they do and how to manage each other in the most effective way with this information in mind.

TEAM ENERGY

I'm sure you have had the experience of one person joining a group of people, and the energy from the person bringing everyone either up or down. It's amazing the impact just one person can have on a group of people. More important, it's amazing what impact just one person can have on the performance of your driver. That one person may or may not be you.

If there is a person around your driver who could have a negative impact on your driver's energy, and therefore his performance, that person needs to be separated from the driver. Of course, this becomes very delicate when that person is a family member or close friend. If it is a crew member, it may be in the best interests of the team to remove him from the team.

In practically any close working relationship, one person will mirror the other to some extent. This certainly occurs when engineering a driver. If you show any level of frustration with your driver's performance, he will as well. If you are confident in his ability to perform, he will be too.

WORKING WITH A TEAM

Build your team. By that I mean being the team leader, motivating the entire team, and being the one they rally behind. I call this the "Schumacher effect," since Michael may be the best at this key trait and skill.

What we will discuss specifically is how to work with a teammate. At some point in your career you will most likely have one or more other drivers on the same team with you. Let's look at the issues you will need to deal with should this be the case.

The first question is should work together, sharing information with your teammate? If so, how should you go about doing that? Some drivers consider a teammate as the very first driver that must be beaten, and therefore don't give the teammate any information. Other drivers feel that they can benefit by being open with their teammate and share everything they have. Some teams insist on their drivers sharing, while others discourage it.

If the decision whether to share information with a teammate is yours and not the team's, then I believe it is something of a moral decision, and one that only you can make. You have to feel comfortable with it. If you decide to do everything you can to mess up your teammate in an attempt to gain the advantage, that's something that you need to feel good about. On the other hand, if you give everything away to your teammate and he ends up beating you, how are you going to feel?

An approach you can take is this: Be willing to share any information with your teammate that the teammate asks for, and expect and ask for the same in return. Then, make sure you ask for more from your teammate than your teammate does from you. In fact, dig as deep as you can into learning from your teammate. Learn by asking to share data, from asking questions about driving techniques, and from listening to everything said in team debriefs. Look to gain an advantage, but in a completely fair manner, and let the best (prepared) driver win.

SPEED SECRET

Turn every teammate situation into an advantage.

A quick word on team loyalty: There are opposing forces you need to balance. Being loyal, and building a reputation for being an honest and trustworthy person, will greatly increase your chances of a career in racing. If you treat people and teams like their only purpose in life is as stepping stones on your way to the top, you won't have much of a team supporting you for long. On the other hand, you should always keep in mind that there may be a better opportunity or person to help you reach your goals. Some drivers are too loyal, and their careers suffer because they failed to take advantage of someone or some team that is better. They have often become too friendly or emotionally attached to the people on the team.

As I said, these are obviously opposing forces or thoughts you need to balance. Too far in either direction and your career will not blossom as it should.

The ideal situation, and the perfect balance, is if you've built such a mutually respectful relationship with your existing team that even they will push you to leave for a better opportunity. If you've done your team-building job, they will hate to see you leave, but will wish you good luck as they encourage you to take the next step.

SPEED SECRET

Build your team relationship to the point where they will encourage you to find the best opportunity for yourself and even help you find it.

SELECTING A TEAM

Some would say that being in the position to select a team is a good problem to have, but it can be a problem.

Start by considering a team's reputation. Of course, that covers many areas, such as results, ease of working relationship, their financial dealings, and their trustworthiness.

Results are pretty easy to determine, although you should also consider how and where they got the results. Some teams will have gained a great reputation of being able to win championships in one level of the sport, but when they move to another series, they struggle. Success at one level does not guarantee it at another level. However, there are some teams that seem to be able to win at every level and in every series in which they compete. These are teams you want to run with.

When considering reputation, also look at whether the team is on its way up or on its way down. Most teams, even Ferrari, Williams, McLaren, Penske, Ganassi, Hendrick, and Roush, have their ups and downs. It's cyclical. Sometimes it's caused by changing personnel, or just finding the right combination of people. Sometimes it's due to technical reasons. Other times it's a matter of how strong the team's motivation is. The point is there are teams on their way up the cycle and others are on their way down. Guess where you want to join?

Of course, this is not an easy thing to determine. It's not something you can just ask the team owner, although doing so may give you some insight. Looking at past results, current results, and talking to other people in the sport (especially series personnel) may help you determine this important factor.

Some of the other considerations, such as the team's financial strength and ability to manage your money, and the team's ability to work with you are even more difficult to ascertain. You need to do some homework. You should talk to knowledgeable people and particularly drivers who have run with them in the past. Ask a lot of questions, of others and of the team itself. If you're the one spending the money, don't be afraid to dig deep. It's likely to be a fair amount of money, so take your time and do things right up front. If you don't, you'll regret it later.

Determine the team's motivation. Is it to win championships, help young drivers, make money, give the owner an ego boost, or what? None of these reasons are necessarily exclusive, as it's likely there will be some overlap, but there is usually one main reason or motivation for the team to be doing what it's doing. I'm not even suggesting that one reason is any better than another, but it should match or fit your objectives.

For example, if the team's existence is really a way to boost the owner's ego by getting the owner's name in the media a lot, that may not fit your objectives if your goal is to maximize the coverage you get as a result of your performance. It would likely not be a good a fit.

Is the team in this for the money? Let's hope so, at least to some point. If the team is not in it to be profitable, there's a better chance that it will struggle financially at some point during the season and your results will suffer because of it. If the team owner tells you he is not in it for the money, and it doesn't matter if he makes money, be concerned. Even if the team owner has all the money in the world, it matters. There's a reason he has that much money; he's not stupid about the way he spends it. If the race team begins costing him too much money, he will cut back. If you've been relying on his support and it goes away, you suffer.

I'm not saying that there are not teams who are willing to support your program. To some, making money is not the priority; the real motivation may be helping drivers. There are team owners who have driver development as a motivation. If you can find one, good for you. But be careful. Don't automatically think you have a team owner with this motivation when you really don't. Again, that's why you need to do your homework and dig for the real motivation behind the team. Then, see if it's a good fit for you and your program.

Once you know what the team's motivation is, your next step is to find out how motivated they are to reach your goals. If your goal is to win the championship or die, and the team is okay with just participating in the series, that's not a good match. Can you step up their motivation? Can you motivate their motivation? Can you help them become winners? While that can and has happened (look at what Michael Schumacher did first with Benetton and then with Ferrari), it's rare in the junior formula ranks. If the team isn't a winner, just realize it will be a challenge to turn them into one. Again, I'm not saying you can't do it. I'm just saying it is not easy and you need to factor that into your decision of which team you select.

The big question becomes: What is the team willing to do to help you win?

Also consider how you fit the team. How much importance does the team put on having fun? How ethical are they? Will they push the rules to win? How business-like are they? How committed are they? And how do these attributes mesh with yours? There is no right or wrong answer to any of these questions. It only matters that it's a good fit.

How well staffed is the team, and how qualified are they? I've seen teams in the junior formula ranks who have more people than some IRL teams, who don't win because of their abilities. I've also seen the opposite: one-man teams who kick butt. Others seem to have the perfect number of qualified people and yet they still don't win. These are not teams. As in any sport, sometimes the best people just don't combine well to make a great team.

What is the team's level of resources? By that I mean do they have the financial strength to quickly cover some unexpected problem, or are they going to have to cut into your budget, perhaps sacrificing something that may affect the results? If the team is on the financial ragged edge between surviving or not, do you want to take that chance? I once coached a driver whose last couple of races of the season were border-line dangerous (the team had to cut so many corners on car prep because of the lack of money). The team had not been prepared for any unexpected costs during the season. The one thing you can expect in racing is that there will be unexpected costs. Consider whether the team has the financial strength to deal with them.

Will you have a teammate with the team in question? That may or may not be a good thing, but it's something you need to consider. Teammates can be a distraction to the team, or they can help you be better. If the team does have multiple drivers, are there any written or unwritten "rules" as to who gets preferential treatment? Even if there are no written or agreed-upon driver preferences at the beginning of a season, there will always, and I mean always, end up being one driver who gets better treatment than the others. Of course, this has more to do with human nature than anything else. If one driver shows more commitment to the team, consistently gets better results, or treats the people on the team better than the other, that driver will eventually get better treatment by the team, whether the team members realize they are doing this or not. Again, it's human nature.

If you're considering joining a team where you will have a teammate, can you "take control" of the team? Can you be seen as the leader? Can you get the preferential treatment? If not, you may be better off finding another team. This is a bigger deal than it appears, so think about it. If you want to come in second, don't worry about being your team leader. If you want to win, you must be the

Teammates should be an advantage—people who push you to higher levels of performance, who you learn from, and who provide a benchmark—but often become a disadvantage if you don't learn to work with them. *Shutterstock*

team leader. Unless you're in the stage of your career where learning from a more experienced teammate may be helpful, if you're not sure you can be the leader within a team, look elsewhere.

This leads right into a sensitive issue that must be considered if the team owner or a family member drives for the team. This could be considered a conflict of interest, although it is not always the case. If the team owner drives, or has a family member driving for the team, no matter how fairly you are treated you will at some point question your treatment. It's human nature to look for something to blame when things are not going the way you want, and if there is a family member in the team, you're bound to perceive some unfairness. Our experience is that most team owners in this situation are more than fair. In fact, they often give the edge to the customer (you), but there is always perception and reality.

To avoid this situation you need to be contractually covered for every possible situation that could ever come up. But most important, you need to feel 100 percent confident with the people you're dealing with. If you don't feel you can trust the team owner 100 percent to do what is right, then no contract in the world will make it work. If the team owner or a family member is a driver in the team, be careful and protect yourself.

An obvious consideration in your team selection process has to be the equipment. If you're looking at a series where multiple chassis or engine packages are available, you had better be sure you're driving the best package. That decision is more difficult than it appears, as proven by the number of extremely experienced and knowledgeable teams who have ended up running less-than-competitive chassis and engine packages. They made a bad decision as to which would be the most competitive. You need to evaluate this decision and be totally convinced that the package you'll be driving is the best.

The equipment factor is not restricted to just the car, although that is obviously the most important. Especially if you have sponsors, though, the team's transporter and hospitality options are other factors to consider. The team's shop may be another thing to consider.

In the end, once you're convinced the team will be able to give you the "hardware"—equipment, results, personnel, etc.—your values, goals, and passion ("software") should fit the team you're considering. It needs to be a comfortable setting for you.

Once you've selected your team, the next step is to define what the common goals, objectives, and responsibilities are and ensure these are all spelled out in an agreement. It may seem obvious what the responsibilities are, but they rarely are. Who's going to arrange passes for your guests? Who's going to make sure media releases are sent out? Who's going to clean your driving suit after a weekend? Who's responsible for sending in the entry forms?

Make sure you know exactly what you're getting for each dollar you spend. Some teams include everything in their deal, while others can nickel and dime you to death if you're not careful. That's why so many team budgets that are floated around the racing community mean little to nothing. You're rarely comparing apples to apples. Make sure you know who's covering entry fees (both season-long series fees and individual event fees), crew passes and credentials, wages, tires, engine rebuilds, consumables (gears, brake pads, fluids, etc.), transport fees, crew travel, driver travel, driver's license fees, driver safety equipment, and the big one, crash damage.

BUILDING AND OPERATING YOUR OWN TEAM

Rather than rent a ride with a team, you can also put together your own team and operate it yourself. In addition to the factors that I covered earlier, here are a few more to consider.

Ask yourself what your objectives are in running your own team. Are they to make a profit? Are they simply to give yourself more control? Is it to see your own name on the team's equipment and entry? Is it to benefit you or someone else? Are there reasons you think this approach will help you win more often? If so, what are they?

Ask yourself if you want to be a team owner and manager or a driver. Which is most important? And even if driving is the priority, is it a little more important or a lot more important?

An important consideration when deciding between renting a ride with a team or starting and running your own is what to do with the car and equipment when it's time to move up to the next level. More than one driver has had his career stunted by the inability to sell the current car to afford to move up to the next series level. If you've invested in a lot of equipment (trailer, tow vehicle, pit equipment, tools, and car) for one series and you want to move up to the next, will you need to sell this equipment before doing that? And if so, how difficult will that be? Will it delay your ability to move up or at least restrict your flexibility to take advantage of an opportunity?

Having all their capital locked up in equipment that cannot be sold off in time has cost many drivers an opportunity that could have made a big impact on their career. Think about it. *Plan ahead.*

Speaking of capital investment, what equipment will you need to invest in? Will you need to buy a car or cars, trailer, tow vehicle, tools, and other equipment? Will you need to invest in a shop to work out of? If so, what else could that capital be used for? Could it be used to hire a marketing person to develop some sponsorship opportunities? What are you really investing in, a team with your name on it or your driving career?

An important piece of business advice was once given to me. It went like this, 7 out of 10 problems in business are:

1. People
2. People
3. People
4. People
5. People
6. People
7. People

In other words, the biggest challenge in building and running a successful business is the people involved. (If you're wondering what the other three are, it doesn't matter. If you select and manage your people well, everything else is easy!)

Running a race team is no different. People are the key to a winning team. So, a critical challenge you will face in building and running your race team will be to select, attract, hire, and then manage the right people. If you think shaving that last tenth of a second off your lap time is difficult, wait until you select and manage your race team personnel!

Budgeting is obviously a critical piece to successfully operate your race team. If this has never been one of your personal strengths, perhaps running a team is not meant for you. I can't give you the details in this book about how to build and maintain a budget for a team, but a good rule of thumb is to make your best estimate of the total cost for the season, then add another 50 percent. Unless you have a lot of experience budgeting for the specific series, you're going to be running, and you are good at managing a budget, the plus-50 percent rule is a minimum.

One last comment on running your own race team. Every single team owner I've ever met has at one time underestimated what it took to build and run a successful team. Obviously it can be done, by some with great difficulty, but for every one of those, there are hundreds who are not successful. And every one of the successful ones have taken more work, more money, more great people, more of everything than the owner originally planned.

SPEED SECRET

Wildly overestimate what it will take to build and run a race team, and you may actually be accurate.

If I'm sounding like I'm trying to discourage you from building and running your own race team, that's not entirely what's intended. I'm just trying to be realistic. And since a focus here is on the driver, your driving abilities and career, it only makes sense that I would focus on what's best for the driver's performance. Operating a race team will take focus away from the development of your driving skills and from your career development. That is a fact.

Sure, there are advantages to building or operating your own race team. Most racers have no problem seeing the benefits, such as the control you have, the investment in equipment that you may be able to recoup someday, and so on. It's the challenges and downside of taking this approach that is so difficult for most racers, and that's why I'm suggesting you take the time to consider the points I've made here. All of the advice delivered here comes from the same place as everything else in this book: from our own and many, many others' experience.

46 DATA ACQUISITION

Most race teams use data-acquisition equipment. This can be invaluable once you learn how to get, and interpret, all the information from it. Many people pay a lot of money for data-acquisition equipment, only to never learn how to get the most from it.

One of the real advantages of data-acquisition equipment is in using it as a driving coach. Most systems will show exactly where on the track you begin braking, your throttle position, the g-forces generated in the turns, your speed, rpm, and many engine functions. This can help you figure out where it may be possible to pick up some speed, especially if you can compare with a teammate or another driver in a similar car.

Data-acquisition systems are wonderful tools for a number of reasons. First, they often tell you something about the car and your driving that you haven't noticed. Or they are great at confirming what you already thought. A data-acquisition system can be your "driver coach" and help you determine how to go faster. And most important, they never lie. It's amazing how often you

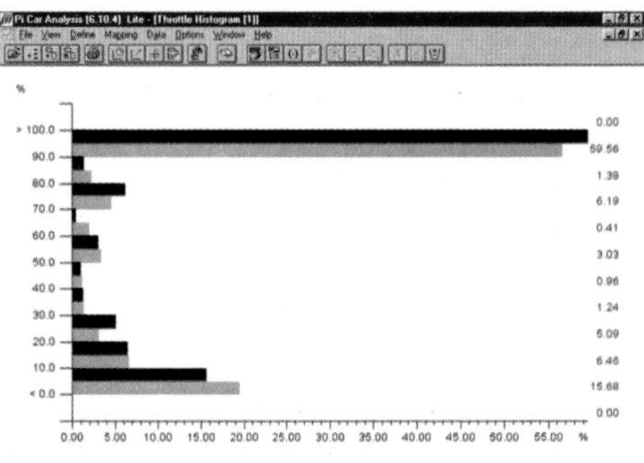

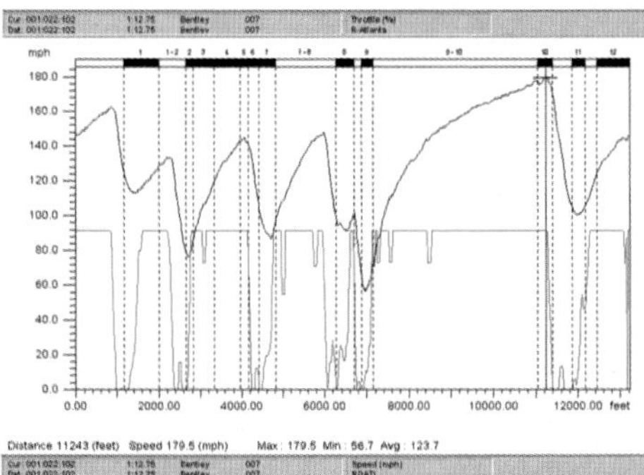

ILLUSTRATION 46-1 Understanding data-acquisition systems is a must for drivers. The upper throttle histogram graph compares the percentage of throttle openings for two separate laps. The bottom graph shows speed and throttle position over the course of a lap.

think you are taking a fast sweeper flat out, only to have the computer show that you did ease off the throttle slightly. After every practice or qualifying session or race, I sit down and go through every detail on the computer. I know it's going to help make me faster, and if I don't use it, I'll be left behind by my competitors who do.

I have no intentions here of talking about the technical side of data-acquisition systems. What I do want to discuss in this chapter is how to use them to help you as a driver. The first and most important point I want to make is that no data-acquisition system in the world, no matter how sophisticated, can ever replace your feedback. The most successful car engineers know this. They know that the driver's feedback is more important.

Am I saying your feedback is more accurate than data acquisition? No. What I'm saying is that no matter what the data says, if you feel, read, or perceive the car doing something different from what the data is saying, that is the way it is. The old saying, "Perception is reality" definitely applies in the case of race drivers.

Of course, not all engineers will agree with you on this subject. Many feel that the data is what matters most, and in many ways they are absolutely right. But in most cases, between improving a driver's ability to sense the limit and level of confidence and improving the overall performance of the car, the former will usually result in the biggest improvement. In other words, you're more likely to drive a car faster if you have confidence in it and can sense what it's telling you than you are a car that is technically faster but very edgy and difficult to read.

You might begin to think that I do not have much use for data-acquisition systems on race cars. You couldn't be farther from the truth. They are one of the most important tools a driver can use. When I'm coaching a driver, I almost demand the car has data acquisition. The most successful car engineers know that data acquisition is an extremely valuable tool to not only engineer the car, but also to help the driver.

SYNCHRONIZING YOURSELF TO THE DATA-ACQUISITION SYSTEM

The first step in using data acquisition is to "synchronize" yourself with it. Without doing this, you and the data-acquisition system will often not agree, and that can only lead to problems.

What do I mean by synchronizing? I mean training yourself to read the track and car in a way that matches what the data-acquisition system says. I also mean learning to interpret the data in a way that matches what you're reporting. When that happens, you can make great gains in the performance of both your driving and the car.

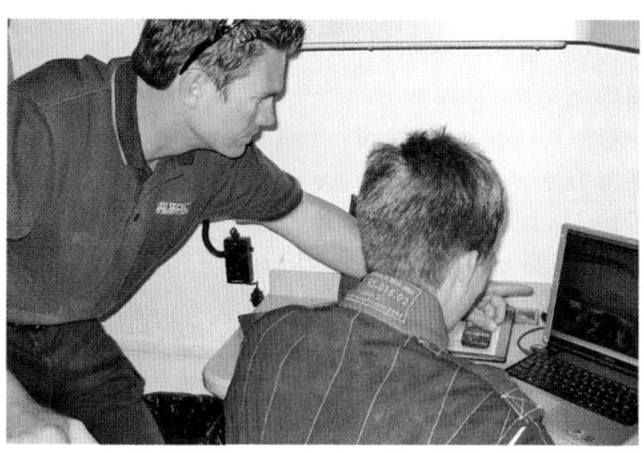

Am I saying that you and the data acquisition must always agree? No. There is nothing wrong with seeing two sides of the story: how you feel, read, and perceive it;

and how the data acquisition reports it. The information you require to make the car perform better is in the feedback coming from the two sources.

I talked about this earlier, but I'll mention it again. There are many times when you may be able to make the race car faster, but it will become less comfortable for the driver. An uncomfortable driver is a slow driver. Sometimes the data is telling you to make one change, one that would ultimately make the car quicker, and yet you want to make a change that will result in a more comfortable, confidence-inspiring car. In nine times out of ten, going with what you want will result in the biggest improvement in performance of the driver and car package.

Never lose sight of the fact that we don't race cars without drivers, and drivers don't race without cars. It is a package, one whose overall performance is not dictated by just one part of it. The car and driver package's performance is limited by the weakest half. But only one half of the package is human. It is the human element that is likely to be the most challenging component to enhance the performance of.

READING AND INTERPRETING THE DATA

While I don't intend to get into any type of discussion as to how to interpret data from a car engineering point of view, I do want to touch on a few areas that relate strictly to you as the driver. What I hope to do here is help you become aware of a few tendencies to keep an eye on, as they can tell you a lot about your driving.

Throttle-Brake-Throttle Transition

One of the most obvious driving traits identifiable with data acquisition is what you're doing with the throttle and brakes from the end of a straightaway through the exit of a corner. Here are the most common things to look for:

- Lifting off the throttle and coasting before applying the brakes
- Releasing the brakes too suddenly
- Too long a gap between full release of the brakes and application of the throttle
- Too abrupt on the application of the throttle

Look for a smooth, seamless overlap of braking and throttle, in both directions: from throttle to brake and brake to throttle. If there is any gap between the two, you're wasting time. If the application or releasing of either the brakes or the throttle is too abrupt, you will not be smooth enough to be really fast; if it is too slow and gradual, you will be too slow.

Braking Forces

Watch for inconsistent braking forces. Typically, braking forces should ramp up quickly (often to the point of there being no ramp at all; it's almost instantaneous to peak braking pressure), stay consistent at the limit, and then trail off smoothly. Some drivers initially stand on the brakes, and then begin to ease off, while others do the opposite. They take a long time to squeeze full pressure on the pedal, only getting to maximum threshold braking at the end, just prior to turning in to the corner.

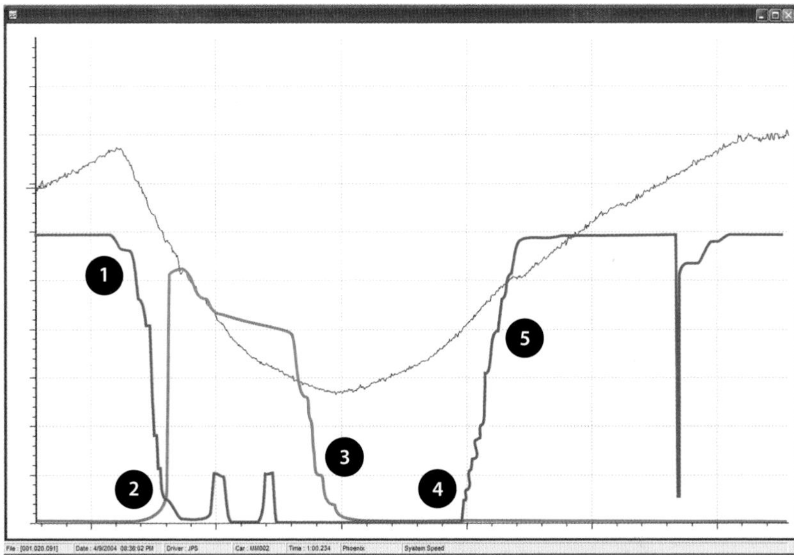

ILLUSTRATION 46-2 This data trace demonstrates a number of problems: (1) the driver eased off the throttle (the green line) too slowly and gently; (2) there's too long a gap between coming off the throttle and the beginning of braking (the red line); (3) the driver did not trail off the brakes gently enough; (4) there is too long a gap between the end of braking and the beginning of acceleration; and (5) the application of the throttle is too abrupt.

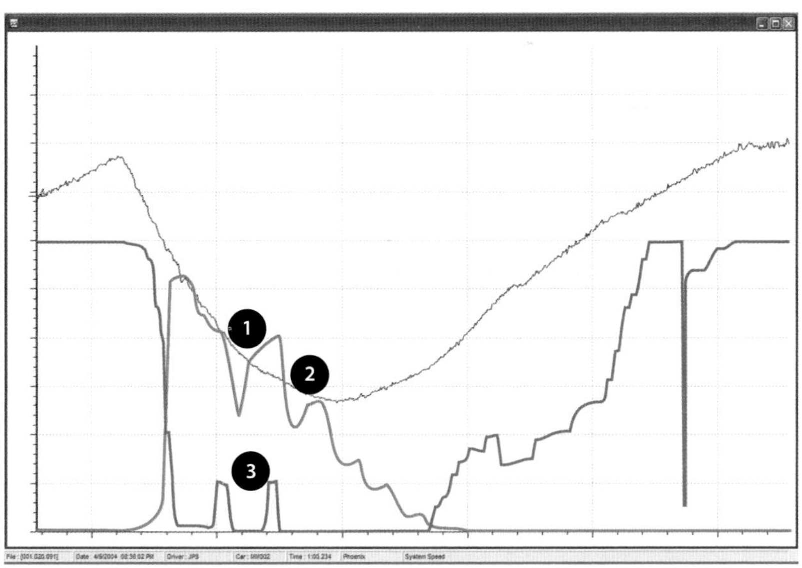

ILLUSTRATION 46-3 The braking trace (the red line) in this graph shows the driver applying inconsistent braking forces (at 1 and 2). In this case, with each downshifting blip of the throttle (3) there is a reduction in braking force.

You need to consider the type and performance of the car you're driving. If the car has a lot of aerodynamic downforce, then the initial application of the brakes should be hard and quick. From there, you should begin to trail off some pressure as the car's traction is reduced as the downforce is reduced with less and less speed.

If the car has little to no aerodynamic downforce, the braking should be relatively consistent all the way through the braking zone. The only reason you should alter the pressure through the length of the braking zone is variations in the track surface, and therefore the grip level of the track. If the braking zone is flat and smooth, so should the brake pressure trace on the data acquisition.

Steering Inputs

Often, you can detect a lack of confidence by studying the steering input trace. For example, if you're not confident with what your car will do when you turn into a corner—perhaps, fearing it will oversteer immediately, or understeer—you will turn the wheel too gradually, instead of turning it accurately and crisply. You will begin turning the steering wheel prior to the ideal turn-in point, progressively turning the wheel too slowly.

If you compare and synchronize what you report (your awareness) with the data trace (the computer's awareness), over time you will learn to recognize a lack of confidence in the car. You'll be able to look at a data trace and see that you're

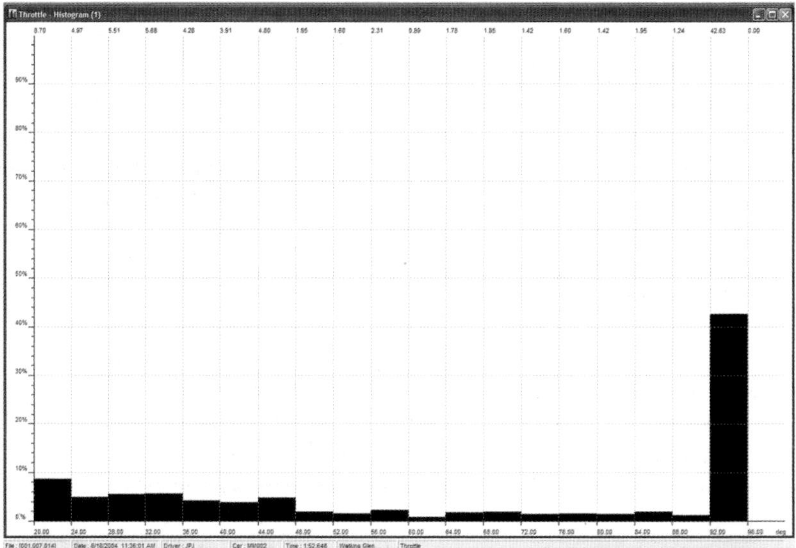

ILLUSTRATION 46-4 The throttle histogram is a useful tool to determine whether a change in car setup or driving technique results in more time spent at full throttle. But beware of reading too much into the information, as an increase in time spent at full throttle along with a reduction in time spent at part throttle could actually result in a slower lap time.

not confident with the car. Obviously, this is invaluable information, as it is not until you've identified a problem that you can find the cause and its cure, and sometimes it's not until you see the data trace that you can truly be aware of this lack of confidence in the car.

Throttle Histogram

Many drivers and engineers use the throttle histogram to determine whether a change to the car or driving technique resulted in a positive or negative change. The thinking is that if you're able to spend more time at full throttle, a greater percentage of the lap with the throttle flat to the floor, then that is an improvement.

While I agree with this in most cases where one driver's lap is compared to that driver's own previous laps, I want to warn you about going too far with it. It is possible to be at full throttle for a greater percentage of a lap and still be slower. How? By also spending much more time at no throttle, with no throttle applied whatsoever, perhaps even while braking.

When you compare the percentage of lap spent at full throttle, also compare how much of the lap is spent with no throttle applied. A 1 percent increase in the amount of lap spent at full throttle will not result in a faster lap time if the driver also spends 5 percent more time completely off the throttle.

Also, if you are comparing more than one driver using the throttle histogram, understand that driving style is a big factor. Some drivers are either on or off the throttle; while others spend less overall time at full throttle, and are still quicker, since they spend more time squeezing the throttle down in the midrange.

I once compared my throttle histogram with another driver's who was driving a similar car at Road Atlanta. He was at full throttle almost 10 percent more of the lap than I was, and yet our lap times were identical. The difference was I spent more time at part-throttle balancing the car, whereas his style was more of an on-off, point-and-shoot style of driving.

The best way to use the throttle histogram is to compare the average percent throttle. In this case, more is always better, since you are comparing the total amount of time your driver is on the throttle.

Theoretical Fastest Lap

Most quality data-acquisition systems have the capability to produce a report after a session that predicts your theoretical fastest lap. It does this by adding up the fastest times from each segment of the track throughout the session.

Although the lap time the system predicts can sometimes be a bit unrealistic, with enough laps to pick and choose from, it is a great way to evaluate how consistent you are. If the spread between your best lap time from the session and the theoretical fastest lap is less than 1 percent, that's a sign that you can consistently get the most out of the car (unless you're just plain slow, in which case you're just not using the full limits of the car). If it is much more than that, there must be some reason for the inconsistency: Either you're experimenting with your driving technique, you're not confident in what the car is doing and telling you, or you made an error on one or more laps which have skewed the data.

47 COMMUNICATIONS AND RECORDS

When you are out on the track, whether in practice, qualifying, or the race itself, it's important to know things like your lap time, position in the race or in qualifying, how far ahead or behind you are from your nearest competitors, time left in the qualifying session, and how many laps are left in the race. Usually that information is relayed to you via a radio or sometimes by a pit board.

It's important that both you and your crew member working the pit board know what each signal means. It's critical for the board man to know what information you are going to want at various times. Discuss this beforehand.

During the race, I really don't care about my lap times. All I want to know is my position, the plus or minus on the cars behind and in front of me, when to pit, and what lap I'm on. Of course, in qualifying, all I care about is my lap time and how much time left in the session. Personally, I like to have the board shown or some radio communication every lap, whether I have time to acknowledge it or not. I feel more in control knowing what's going on. I spend a lot of time before going out on the track making sure the guy with the pitboard or on the radio knows exactly what I want.

There are times where I've purposely *not* had my lap times shown or told to me during a qualifying session. It's easy to focus too much on the time. For me, that sometimes led to either trying to go faster or believing a certain time was some sort of barrier. You may want to try a qualifying session without knowing your lap times. See if it works for you.

A two-way radio is probably the best way of relaying information, as the driver can give input as well. There is often a lot of interference on the radio, however, so you can't always count on it. For that reason, many teams rely on the pit board, only using the radio as a backup for the basic information. The radio's most important use is for more detailed information such as when there is a problem with the car, when to pit, or when the green flag is dropped.

It's also a good idea to have a couple of basic hand signals that your crew will understand for problems like a tire losing air, engine problem, low on fuel, and especially, a nonfunctioning radio.

RECORDS AND NOTES

A driver should keep a record or log book with the details of each race, practice, test, or qualifying session. Use these records to learn from, looking back at them when returning to the same track again, or when you are having a problem with a specific area of your driving.

I like to write down the objectives for each session, what driving techniques or plans I need to use to achieve them, before each session. Then, after each session, I make comments on the track and conditions, what changes were made and need to be made to the car, and what the results of the session were.

How effective is a car engineer who does not keep any notes of the changes he makes to the car? Not very, right? The same thing applies to you. One of your key objectives should be to ensure you never make the same mistake twice or have to learn the same thing twice.

Therefore, you should keep extensive notes in a journal of some type. I recommend that at each track you go to, you draw your own track map. Why is that better than just using a printed track map that is supplied by either the track itself or the data-acquisition system? It's because it is important for you to draw the track as you see it, not necessarily as it really is. To back up your drawn map, you may also want to keep a copy of a printed track for extra reference and comparison.

You should make notes on the map that relate to how you drive it, such as the gears used in each turn, specific reference points ("turn in at the crack in pavement," "apex at the end of the curbing," and so on), elevation and surface changes, and good places to pass. You should make note of any particularly challenging piece of track and why. You will also want to record the date, the car, your best lap time, the fastest car's lap time, and weather conditions. This information becomes invaluable the next time you race at this track, whether you're driving the same car or not.

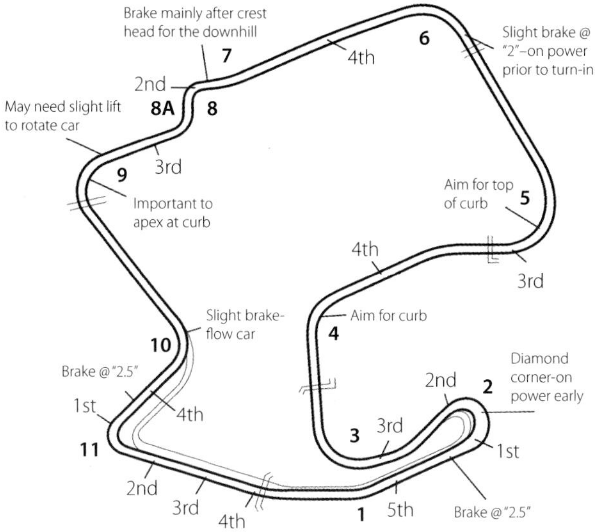

ILLUSTRATION 47-1
One of the most important tools for raising your awareness—and to improve your feedback on the car—is to make notes on a track map. These notes should include every little detail about what you see, feel, and hear—lots of references—how the car is handling, and what you're doing to the car.

You should make notes of all the things you have learned each day at the track. If that isn't a pretty extensive list, you are not doing your job. I don't care how much experience you have, you will always be learning something new, if you are aware. Obviously, writing down what you have learned will reduce the chances of you forgetting and having to learn the lesson all over again (at great expense, in most cases).

Finally, you should rate your own performance on a scale of 1 to 10 for each session on a race weekend (or the day for a test day) and make note of how you felt and what you did leading up to it. That way, over a period of time, a pattern will begin to emerge, one that spells out a routine that will lead to consistently great performances. For example, if you notice that whenever you do some type of physical warm-up prior to driving, or that a certain phrase used by you or another team member seems to lead to "9" performances, you know what must continue to be done in the future.

If you make note of your state of mind, your level of energy and intensity, what you have eaten over the past day or two, who has been around you and what they have said, how confident or nervous you felt, then it becomes much easier to develop a pre-race ritual that will lead to a great performance. Without writing it down, it is easy to miss the pattern.

48 SAFETY

Racing is dangerous, there's no doubt about it. But the danger can and should be controlled. Most drivers, myself included, have the attitude toward injury that it "may happen to others, but never to me." I think you need to have this attitude to a certain extent. If you didn't, you would probably have too much fear to drive fast.

However, that's no excuse for not taking safety seriously. Over the course of a career in racing, you are bound to have at least a crash or two. How you fare in these crashes may be a matter of how much emphasis you place on the various safety equipment and systems you use.

Being safety conscious does not mean you are a wimp. It only means you are a smart, professional-acting driver. The more emphasis you put on your safety, the longer and more successful your racing career will be.

Take a good look at how serious the drivers in Indy car and Formula One are about safety. Just because you're driving a slower car does not mean it is any safer. In fact, it's often the opposite. The safety built into Indy and Formula One cars and the safety personnel at that level are superior. So if anyone has to take safety seriously, it's a driver at the beginning of his or her career.

SAFETY EQUIPMENT

As I said, if you race for some time, the odds are you are going to be involved in some incident where you may get injured, either slightly or seriously. That is why you must pay attention to all safety equipment, yours and the car's.

SPEED SECRET

Buy the best safety equipment you can.

A bargain-priced driving suit doesn't look like such a good deal when you're lying in a hospital badly burnt. The same thing applies to helmets. Buying a cheap helmet really is false economy. I like the saying, "If you have a cheap head, buy a cheap helmet."

SPEED SECRET

If you can't afford good safety equipment, you can't afford to go racing.

After you buy the best equipment you can, take care of it. Don't drop or let your helmet sit upside down on the ground. Keep your driving suit clean; it won't be fire resistant if it's covered in grease and dirt.

Not only will having the best equipment and taking care of it help save your life, but it is also a reflection of your attitude. If you look and act professional, you may have a better chance of acquiring sponsorship or being noticed by a professional team. Plus, you owe it to your family and friends to minimize the chances of injury.

Remove all jewelry before driving. Imagine what would happen if you were in a fire with a metal chain around your neck, metal watchband on your arm, or rings on your fingers. As the metal heats up, I don't need to tell you how much worse you will be burned, not to mention the complications it causes the medical staff.

Check current regulations to ensure your equipment is up to standards. I am not going to quote the standards since they change frequently, fortunately, as they are constantly being upgraded.

Make sure you have spares of all your equipment. You spend too much on your racing to have a safety equipment failure or loss to keep you out of action. It seems rather silly to spend tens of thousands of dollars on your race program, and then not be able to race because you lost a glove or the visor on your helmet broke.

HELMET

Helmets are meant to be used once, when your head is in it. If you drop it or bang it against something, it has now been used. Helmets are designed to absorb the energy of the impact by deforming and thereby destroying its structural strength. And even though the damage may not be visible, it should be checked by the manufacturer or replaced after an impact.

Many drivers, for emotional or superstitious reasons, become attached to their helmets and don't ever want to give them up. Not a good idea. Replace your helmet every couple of years whether it has been "used" or not. They also fatigue with age, especially the inner liner.

Helmets are made of fiberglass, Kevlar, carbon-fiber, or a combination. The Kevlar or carbon-fiber ones are much lighter and a little more expensive. The extra cost is worth it. It's not only easier on your neck over the course of a long race, but it also puts less strain on your neck in a crash.

I've had many drivers tell me that because they were driving a relatively slow car, perhaps a showroom stock car, they didn't need to spend the extra on a Kevlar or carbon-fiber helmet. It doesn't matter how slow the car is; in a crash the g-forces are going to be high. The heavier the helmet, the more force on your neck. That may make the difference between being injured or not.

Take the time to ensure that your helmet fits perfectly. It should fit snugly but not too tight. You should not be able to rotate it on your head, side to side or forward-backwards, with the strap undone. There should also not be any pressure points causing discomfort.

Do not paint a helmet without knowing what you're doing. Consult the helmet manufacturer. Many a helmet has been weakened by the type of paint a driver used to make it look cool.

I believe a driver should only wear a full-face helmet, even in a closed car. They offer much more protection than an open-face helmet. However, if you choose an open-face, always wear eye protection. The bubble goggles are probably the best.

Helmets are tested and rated by the Snell or S.F.I. Foundations. The standards are upgraded every few years. The only helmets legal for use in racing are ones who have passed these tests. Check the rulebook to find out the latest standard a helmet must meet.before buying one. You'll get a great deal on an out-dated helmet, but it's not much good for anything but planting flowers in.

HEAD AND NECK SUPPORT

This section is titled Head and Neck Support for a good reason. As far as I'm concerned, there is only one system that I would trust with my life, and that's the HANS (which stands for Head and Neck Support) device. If your race series doesn't mandate the use of a HANS device, it should. And you shouldn't wait to be told to use the latest equipment that can save your life.

I know of at least half a dozen drivers, friends, who I'm not sure would be alive if they weren't wearing a HANS device. Don't be stupid; use one.

DRIVER'S SUIT

First of all, a driver's suit is not fireproof. It is fire retardant. It is designed to resist and protect you from the heat of a fire long enough to either allow you to get away from the fire or the fire to be put out. Speaking from experience, good suits do this. I'm not so sure cheap ones do (and I'm never going to find out).

Before buying a suit, make sure it fits properly. A custom-fitted suit is best. Have your measurements taken carefully, using the chart supplied by the suit manufacturer. If it doesn't fit right, send it back to have it altered. A poor fitting suit will be uncomfortable and perhaps even dangerous.

Check the rating of the suit. The S.F.I. Foundation certifies and rates driving suits, as does the F.I.A. An SFI 3.2A-1 rating theoretically gives you approximately 2 seconds of protection; a SFI 3.2A-5 gives you approximately 10 seconds; SFI 3.2A-10 about 20 seconds (double the last number of the rating spec to give you an approximation of the number of seconds before you will be burnt). Remember, this is only a guideline, no guarantees. And if a suit is not rated by S.F.I. or F.I.A., do you really want to buy it? How do you know how good it is? It may be a great price, but what price do you put on your body?

Once you have a properly fitted, good quality suit, take care of it. Make a habit of getting changed before working on your car. A driver's suit is not meant to be work coveralls. There is nothing worse than getting a fire-retardant suit covered in oil, grease, and fuel.

During practice for the Indianapolis 500 in 1993, a fuel regulator cracked while I was driving through Turn 4 at more than 200 miles per hour. The fuel sprayed forward into the cockpit and ignited. All of a sudden I was engulfed in a 2,200-degree methanol fire. Fortunately, I got the car stopped on the front straight and bailed out, while some crew members started putting the fire out. I was in this fire for close to 40 seconds, and yet the suit protected me perfectly.

My face was burnt from the heat that radiated through the visor, and from when I tried to open the visor for a second to get some air. My neck was burnt where the fire got between where I had my balaclava tucked into my underwear. And my hands were burnt pretty badly for two reasons. First, because my gloves were so soaked with sweat; my hands were steam-burnt. And second, the gloves I was wearing didn't have a Nomex layer on the palms; it was just leather.

If I hadn't been wearing such good equipment, though, I probably wouldn't be writing this now. Most of my suit was charred through to the inside layer. Even parts of my Nomex underwear were charred. But it never got all the way through to my skin.

I did learn some good lessons from that experience. I now make sure I always put on dry gloves that have a Nomex lining in them and wear a double layer balaclava that I make sure is tucked properly into my suit.

MISCELLANEOUS DRIVERS EQUIPMENT

In addition to your helmet and driver's suit, you require other equipment: driving shoes, fire-retardant gloves, balaclava, underwear, and socks. Again, the same rule applies: Buy the best, and then take care of it.

Wearing fire-retardant underwear under your suit is absolutely critical as far as I'm concerned. A two-layer suit with underwear provides better protection than a three-layer suit without underwear. I know it's tempting to not wear it on hot days, but once you're in the car and driving, you'll never know if it's hotter or not. It's also more comfortable wearing underwear between you and the suit; it helps absorb the sweat better. A two-layer, rather than single-layer, balaclava is required by most rulebooks.

Only wear gloves with a full lining of fire-retardant material between your hands and the leather palm surface. Many gloves don't have this; they only have leather on the palms. Turn them inside out to be sure. Some race-sanctioning groups require the lining, while others don't. Again, check the rulebook. Or better yet, only buy the good ones with a fire-retardant lining.

It's the same story with driving shoes. Buy real driving shoes, which have a fire retardant lining. They give a lot more protection and are much more comfortable than just using bowling shoes or runners, like some drivers do. If you're going to drive formula-type cars, you probably won't be able to work the pedals properly in the tight confines of the foot area of the cockpit with anything other than real racing shoes.

One other thing I feel should be mandatory: ear plugs. Not only will it save your hearing over the years, but it allows you to actually hear the car sounds better,

concentrate better, and means you won't fatigue as quickly. Auditory input is important feedback when driving. If your hearing is impaired, you be won't as sensitive to what the car is telling you. The best earplugs are the ones custom molded to fit your ears, although the small foam ones work quite well for most situations.

SAFETY HARNESS

The safety harnesses, or seat belts, in a race car may be the most important safety component of all. Once again, only use the best belts, and take care of them. After all, they're taking care of your life. They also help support your body in the cockpit so that you can drive most effectively.

All belts should be replaced or rewebbed at least every two years. They will seriously deteriorate, losing up to 80 percent of their effectiveness, simply by being exposed to weather and ultraviolet light. Anytime you've had a crash, replace or reweb them immediately, as they will have stretched and weakened. As well, regularly check, clean, and lubricate (if necessary) the buckle mechanism.

Belts should be tight before you start driving. Then, make sure you can tighten at least the shoulder belts while driving. Often, they seem to loosen during the course of a race. And you may be surprised at how far they actually stretch in a crash, allowing your body to impact things you would never imagine.

Anti-submarine belts not only help you from sliding forward in a crash, they also help support your body under heavy braking. So make sure they are adjusted properly, snug and comfortable.

Practice getting the belts undone in a hurry and getting out of the car quickly. This could be valuable practice. Many cars are practically impossible to get out of quickly.

Make sure the belts are mounted securely and in the right position. Sometimes the shoulder belts are mounted too far apart, which would allow them to actually slip past your shoulders in a heavy impact. Try to have them mounted so that they help hold you down, as well as prevent you from being thrown forward in a crash.

49 DRIVER AS ATHLETE

Is a race driver an athlete? This question has been knocked around for years. Who cares? All I know is it takes great physical skill and endurance to drive a race car well, not to mention the extreme mental demands.

If you want to be even the slightest bit successful in racing, you need to be in good physical condition. If you want to win, if you want to make racing your profession, then you must be in good condition.

Driving a race car requires aerobic fitness, muscle strength and flexibility, and proper nutritional habits. Without these you will be lacking in the strength and endurance to not only be successful, but also to race safely. Using the controls (steering, brakes, throttle, clutch, shifter) and dealing with the tremendous g-forces on your body demands a great deal more than most people think, especially with the extreme heat you usually have to work in.

To qualify for your racing license, and every year or two after that (depending on the level of license you have), you must have a full physical test completed by your doctor. But even though you may be healthy according to a doctor, how physically fit are you? How strong? How supple and flexible?

When your body tires during a race, it not only affects your physical abilities but also your mental abilities. When you physically tire and you begin to notice aches and pains (and even before you notice them), it distracts your mind from what it should be doing: concentrating on driving as quickly as possible.

The better conditioned your body is, the more mentally alert you will be and able to effectively deal with the stress and concentration levels. A big part of the drain on your strength is the intense and never-ending concentration you must maintain. And just a slight lapse in concentration can bring disaster. How many times have you heard the expression "brain fade" used as an excuse?

Notice how often a driver's lap times begin to progressively slow near the end of a race. The driver usually blames it on the tires "going off," the brakes fading, or the engine losing power. If the truth be known, it's usually the driver that's going off, fading, or losing power as fatigue sets in.

Drivers who claim to stay in shape simply by racing are only fooling themselves. The workout you get from racing even every weekend is not good enough. You must supplement that with a regular physical conditioning program.

PHYSICAL CONDITIONING

When you train, you become more fit. Stressing your body, in a controlled manner, through running, lifting weights, or whatever, gradually breaks down the muscle fiber. Then, with rest, the muscles heal stronger. So each time you exercise, then rest, your body becomes stronger.

Use a regular fitness training program to improve your coordination, strength, flexibility, and endurance. Sports like running, tennis, racquetball, and squash are excellent for improving your cardiovascular fitness and coordination. Added to a specially designed weight training and stretching program, these activities may mean the difference between winning and losing. Most of these will also improve your reaction skills as well.

Strength, particularly in a modern ground-effects car, is important. So weight training is a key. Keep in mind, though, you don't want to bulk up too much if driving formula-type cars, as the cockpits tend to be cramped. Concentrate on building muscle endurance as much as outright strength.

You now understand how critical being sensitive to what the car is telling you, and how important being precise in your use of the controls is. Well, try this test. Trace over a picture with a pencil, accurately and with great detail. Then do 50 push-ups. Try tracing the picture again. What happened? When the muscles in your arms tire, you lose some of the precise control. You need that precise control when driving a race car.

Your cardiovascular system takes a real workout when racing. The average person's heart rate at rest is between 50 and 80 beats per minute (BPM), less than half its maximum potential. Most athletes operate during their sport at around 60 to 70 percent of their maximum, and then often only for a few minutes at a time between rests. Studies have shown race drivers at any level often operate at close to 80 percent of their maximum BPM, for the entire length of the race.

Being aerobically fit will make the difference between winning and losing. The only way to ensure your cardiovascular system is in shape is through aerobic training: running, cycling, Stairmaster, any sport where you keep your heart rate at 60 to 70 percent of its maximum for at least 20 minutes, and preferably more.

Your reflexes can be developed. Sports such as squash, racquetball, and table tennis are great for improving your hand-eye coordination and reflexes. Computer and video games are also good for improving your mental processing and reflexes.

It's only been over the last few years that I've really begun to realize the benefits of flexibility. As part of my regular training program, I now spend quite some time stretching and working on my flexibility. Since starting this, I've had fewer muscle aches and much less cramping while driving, and I feel a lot better the day after the race.

Should you ever crash, the more flexible your body, the less chance you have of being injured. With a flexible body, your muscles will be better able to accommodate the forces from an impact.

How's your weight? If you are overweight, you owe it to not only yourself, but also to your car and team to lose weight. Why have your team work at making the car as light as possible, if you're not? But, more important, excess fat on your body works as insulation, something you don't need in the high-heat environment of a race car cockpit. Reducing your body fat content (or maintaining if you're already lean enough) should be a part of your training program.

In fact, heat is one of the race driver's worst enemies. The combination of all the fire-protective clothing, the continuous physical exertion, and the heat generated by many race cars makes for a less-than-ideal working environment. A driver's body temperature can reach more than 100 degrees.

This heat often leads to dehydration. Some drivers will lose up to 5 percent of their body weight in perspiration during a race. This can lead to weakened and cramping muscles, and less effective mental processing. In fact, studies have shown that losing just 2 percent of your body weight in sweat can reduce your work capacity by as much as 15 percent. There is only one solution for dehydration: drink fluid. Over the course of a race weekend, especially in warm weather, try to take in as much water as possible, at least 4 liters per day on race weekends.

It is well-known that an athlete's diet is extremely important to the athlete's performance. Marathon runners are famous for their carbo-loading (eating high-carbohydrate foods) prior to races. A race driver is no different. Again, if you want to win, follow a proper diet. Talk to a doctor or nutritionist. At the least, avoid foods with high-fat content on race weekends. Stick to lean meals with a good balance between carbohydrates and protein.

Finally, do you drink much? How about smoking? We all know that alcohol and cigarettes affect your health. Even if there is a one-in-a-million chance that they could slow your reactions, affect your vision, or decrease your cardiovascular level, consider whether you want to take that chance. How committed are you to being successful?

The effects of alcohol on your body and mind can last for a long time. It slows your reaction time, dulls your senses, and slows your ability to make decisions. And taking drugs to improve your performance is a major mistake. Not only will it not help, it's very dangerous.

SPEED SECRET

Given equal cars and equal skills, the fittest driver is going to win.

Often a driver with less talent and less car will win due to his fitness level. So, if you want to race, if you want to win, you owe it to yourself to be as physically fit as possible.

50 FLAGS AND OFFICIALS

Many drivers would rather not have to deal with officials, and seem to pay little attention to the flags. They're missing an opportunity. Pay strict attention to the flags shown to you by the flag marshals. They are there to assist you, to help you go as fast as possible, and to ensure your safety.

At practically every racetrack you'll ever race at, the flag marshals and officials are there as volunteers. They are there for the same reason you are: They love racing. The only difference between you and them is the area of the sport they having chosen to get involved in, for whatever reason. Often, a flag marshal is doing it because he or she can't afford to race yet, and this is better than being a spectator. In fact, working as a flag marshal can often be beneficial later in a racing career. It's great experience seeing a race from that perspective.

Without flag marshals and officials, you will not be able to race. Remember that. Don't think of flags, flag marshals, and officials as hindrances. Think of them as a way to gain an advantage.

Before you first venture onto the racetrack as a driver, it is absolutely critical that you know and understand what every flag means and how it is used. Take the time to read and understand the rulebook that you'll be racing under as the use or interpretation of a flag has been known to change or vary. Keep up to date with the latest regulations.

It's important to not only note and obey all flags, but also to "read" the flag marshals. You can really work this to your advantage. With experience, you will notice differences in the way the marshals wave a flag. If, for example, a marshal is calmly waving a yellow flag (meaning caution, slow down, there is an incident in the vicinity), it's probably not a serious incident. While your competitors are slowing up a lot, you back off a little, gaining a bit of an advantage on them. However, be prepared to slow down. And if the marshal is frantically waving a yellow flag, slow down a lot.

Having said that, remember that flag marshals risk their lives to make racing safer for you. Don't ever do anything that puts them in any greater danger than they already are. And understand that when you slow down 20 or 30 miles per hour from your racing speeds, it may seem to you like you're almost stopped. But you're probably still traveling at a very high speed with a flag marshal on or near the track assisting another driver.

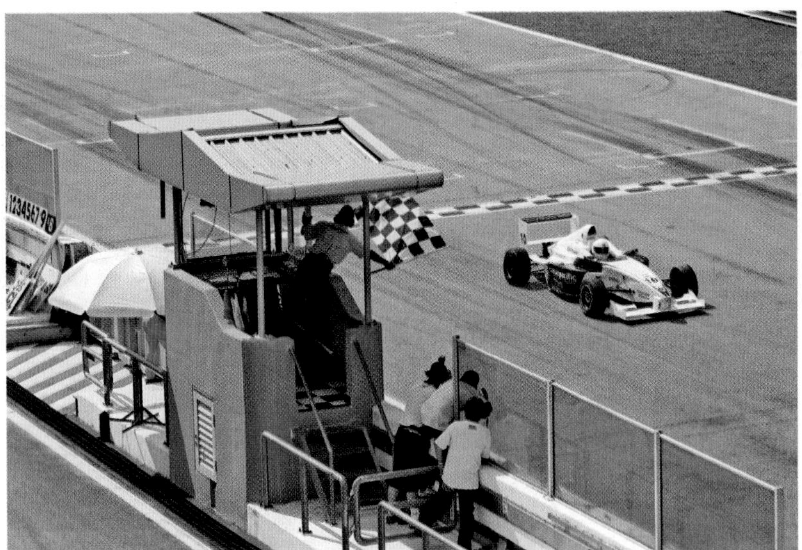

While the flag you want to see before any other is the checkered flag, every flag is a potential tool. Seeing, recognizing, and understanding a flag's meaning—even the subtleties of how it is waved—can make or break a racer. *Shutterstock*

No matter how much an official's or flag marshal's decision or action seems to be against you, try to accept it and get on with your racing. If you are sure you're being wrongly treated, take it up in the proper fashion (again, read the rulebook as to how). Don't take it out personally on them. That will only make matters worse.

The officials are only doing their job, and the better you get along with them, the more successful and enjoyable your racing will be. Often, if over a period of time you have treated the officials with respect, it may even help sway a decision in your favor.

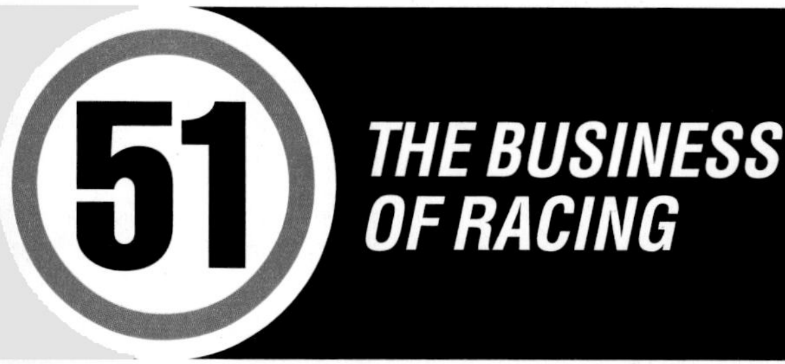

51 THE BUSINESS OF RACING

Racing today is much different from 30 or even 10 years ago. It used to be that, at the professional level of the sport, a person was chosen to drive for a team strictly on talent. Not so anymore. There are many drivers today who have the talent to win races. So when a team is looking for a driver, why not select one who is promotable and marketable—a public relations person's dream—*and* one who can bring sponsorship dollars to the team, as well as talent?

As much as this may seem unfair, it is a fact you will have to live with during your career. You can either choose to make the most of it and look at it as an extra challenge, or be miserable because you're not getting your "breaks."

You can't sit back and wait for a team to come to you, thinking that because you're a good driver you deserve it. The days of car owners coming to knock on your door are few and far between. These days, if you want a ride, you have to go after it yourself. And probably bring something to the table, as well.

I'm not saying that you will have to pay for your racing throughout your career. Even today, some drivers are selected to drive for a top team based primarily on their talent. But even they had to pay their dues. They probably had to bring personal or sponsorship money to the teams they drove for on the way to the top.

Don't ever look at this as being beneath you. If you don't believe sponsorship, professionalism, and public relations are important to your career, you'll be watching a lot of racing on television.

CAREER MOVES

As I've said before, it takes a lot more than just driving skill to be successful in racing. You must have all the right components of "the program" to be a consistent winner. Components like the right equipment (car, spares, and so on), a good crew (mechanics, engineers, team manager, even if one person handles many of these jobs on a small team), an adequate budget ("adequate" being a relative term), an appropriate testing program, and more. Then, all these components have to mesh together. It's especially important that the people work as a team. Without that, no matter how good you are, you won't win on a regular basis.

Many drivers aren't interested in trying to climb to the top of professional racing. They just want to race for fun in amateur events. There is nothing wrong with that. I know many people who have been amateur racing for years and love it for the thrill of competition, the sense of self-satisfaction, the camaraderie and friendships developed, relaxation, and so much more.

Personally, I find racing to be the most relaxing thing in the world. When I'm racing, nothing else matters. I don't care what else is happening in my life, I'm focused on racing. It allows me to forget everything else. I just relax and enjoy my driving. And, for that reason, the level of the sport you race at is not important.

No matter what level of the sport you're involved in, however, it's going to take a lot of work. As far as I'm concerned, it's worth every bit of it, every time I get behind the wheel. That said, to make it in professional racing takes much more work than amateur racing does. That's something to be aware of. If you're having a tough time managing the time and effort in amateur racing, don't think it's going to get any easier when you start racing in a professional series.

If you do aspire to reach the top of professional racing, the recommended road you take to get there can vary dramatically. It differs from driver to driver; however, there are usually some common threads. Often, it is determined somewhat by where you live (and whether you're willing to move), your personal financial position, how good you are at raising money (sponsorship, donations, whatever), your professional approach (whether or not a professional team is going to want you to drive for them), and what your ultimate goal in racing is (Formula One, Indy car, NASCAR, sports cars, sprint cars, and so on). Talk to drivers who have made it. Read biographies of the great drivers. Learn what has worked for others.

It used to be that road racers were road racers and oval track racers were oval track racers and that they never mixed. But with Indy cars running both road courses and oval tracks, many racing series have followed. More NASCAR drivers are coming from a road racing background. Today, if you don't have experience on both types of tracks, your chances of being successful in the top ranks of racing are reduced.

Look for opportunities to race on all types of tracks. Consider that when deciding which series to compete in. If a series combines both oval track and road courses, and it's your goal to move up the professional racing ladder, then choose that over a series that only races one type of track. It will pay off in the long run.

If you are lucky, you'll have to make the decision between buying your own race car or renting one from a professional race car rental business or racing school. I say lucky because many drivers cannot afford the second option and are forced to buy, manage, and maintain their own cars. There are advantages and disadvantages to both options.

First, it is best to work on your own car at least for part of your career. This helps teach you many of the basic technical aspects and makes you more mechanically sensitive to your car. In other words, you will probably learn to be a little easier on the car. The disadvantage is that you can spend so much time and concentration on the car that you spend little on your driving.

If you rent a race car from a professional race car rental business, it allows you to focus all your attention on your driving, leaving the mechanical worries to someone else, someone who is supposed to be better than you at looking after those worries. But beware. There are good car-rental businesses and bad ones.

Do your homework before choosing one. Talk to others who have used them in the past.

Assuming you find a good rental program, this does allow you to spend all your concentration on your driving. This is good, but don't forget the mechanical side. Being mechanically sensitive and able to interpret what the car is doing, and communicating that to your engineer or mechanic, may make the difference in landing a ride with a top professional team.

Choosing to compete in one of the racing school series is probably the best choice strictly from a driving perspective. They usually have instructors working as coaches during race events, which will greatly speed up your learning curve. And usually they are "spec" series, where everyone is in the same type of car. This is a good way of gauging yourself, and your progress, against others.

Again, be careful. There are good school series and bad ones. Some racing schools are only interested in your pocketbook. Again, do your homework. Talk with people who have raced in the series in the past.

Choosing between racing open-wheel (formula-type) or closed-wheel (production-based or sports racing) cars should be a decision based on where you want to go in the sport. If you are certain about only racing closed-wheel cars in your career, then stick with them. But if you're not sure where your career is going to go, then spend time racing open-wheel cars. If you only have experience in closed-wheel cars, it is more difficult to make the jump into an open-wheel car should the opportunity arise. Experience in open-wheel cars makes it easier to handle any type of car.

If you want to make a career as a professional driver, my advice is to drive every type of car you can get your hands on. Every car, from the slowest showroom stock car to the most sophisticated formula car, will teach you something different. The more you learn, the more adaptable you will be and the more successful you will be.

A good education is also important to a driver's career. Although an engineering degree will help with the technical side of racing, I believe a strong business and marketing education is perhaps more important today. With a little effort, you can learn enough of the engineering side. Today's race driver depends more on business and marketing knowledge to make a successful career in racing.

To reach the top in racing, you're going to have to sacrifice a lot. Ask yourself this: "Am I willing to give up everything to reach the top?" Are you willing to sell your street car? Your stereo? Give up your girlfriend or wife? (I'm not suggesting this is mandatory, but it has happened because of racing). If not, be truthful with yourself. Realize how much you are willing to sacrifice and realize how far that will allow you to go. There is absolutely nothing wrong with amateur racing for fun, as long as you don't fool yourself into thinking you're going to be the next world champion without sacrifice and commitment.

You have to be 100 percent totally committed if you really want to make it to the top. You will have to commit time—24 hours a day, 7 days a week—and money, usually everything you've got for a long time.

I believe anyone can be successful in racing, maybe not a superstar, but successful nevertheless if they are committed and dedicated to doing what it takes to become successful. It takes tremendous perseverance. Bobby Rahal was once quoted as saying it takes 10 percent talent and 90 percent perseverance to be successful in racing. He wasn't saying it doesn't take talent; it's just that perseverance is so important.

I can make an example of myself. Sure, there have been more successful drivers in the history of the sport, perhaps even with more talent. But I've proven you can make it to the top with hard work, perseverance, determination, sacrifice, knowledge, and maybe even some talent.

However, enjoy racing for what it is, and for whatever level you're at. You don't have to make becoming the next world champion your goal. Just do your best, and if things work, you'll make it. If not, step back to amateur racing and have fun. Keep your options open.

SPONSORSHIP

Sponsorship is what makes auto racing work. An entire book could, and has, been devoted to sponsorship (the best one I've seen is *Sponsorship and the World of Motor Racing* by Guy Edwards). So all I'm going to do is just touch on a few key points based on my experience.

First rule of sponsorship hunting: It's not what you know, it's who you know. Ninety percent of selling a sponsorship is just getting to the decision maker. Concentrate on meeting the right people. In every successful sponsor company, there is at least one key person who can see the benefits and will help push it through. You need to find that person. Talk to people you know, to see who they know.

If that was rule number one, then the following is rule one-and-a-half: It doesn't matter what you want, what counts is what the company you're approaching wants. Too many people in racing go in to a potential sponsor and tell them what they will do if the company gives them money. Then they wonder why the company said no. Put yourself in their position. Figure out what they want and how to give it to them.

Listen. Do less talking and more listening. Find out what they want; how they can use racing to benefit them. Sometimes, what they want is not necessarily what you think they need. Help them figure out what they want or need. If possible, ask them to describe to you how they would see a racing sponsorship program working for them. If they tell you, listen carefully, because they are selling themselves. If you can supply them with their vision of the sponsorship, it's almost impossible for them to say no. It was their idea.

The sponsor's name on the car—"the mobile billboard"—should be just the beginning of the program. Usually, what makes the program really work is the corporate entertainment at the track, the business partnership opportunities with co-sponsors, the employee morale programs, the public relations, media exposure, and so on. The sponsorship must be an overall marketing program, tied to the theme of a race car with the company's logo on it, that helps sell a product or service and builds corporate image.

It used to be that companies would sponsor a driver or car just for the exposure, image, or public relations value. Rarely does that cut it anymore. If it doesn't result directly in bottom-line sales, they won't go for it.

You can spend thousands of dollars on fancy-looking presentations, brochures, and packages enthusing about how great a driver you are and how this is the best marketing program the world has ever seen. But 9 times out of 10, the person making the decision is going to decide based on you as a person, and the real core of your program. Companies buy good people with good programs, not *just* a good presentation.

I'm not saying you shouldn't spend money on having a professional-looking presentation. You should. What I'm saying is if that's all you've got, you will have a tough time selling it. Spend time developing a good program that delivers value to a potential sponsor.

Often, that means using the resources of one sponsor to benefit another, and vice versa. For example, getting the local newspaper to sponsor your team strictly with advertising space equal to the dollar value you require. It has cost the newspaper little in actual dollars and in return you give logo exposure on your car along with all the other benefits you have to give a sponsor with your program. You then offer a full sponsorship to another company, *plus* the newspaper advertising, in return for the budget you require. A win-win-win program.

You will be far better off if you target specific companies you think a program such as yours can benefit from, rather than just firing off hundreds of proposals in the mail. Take the time to do research on the company, call them, and then meet personally. Using the shotgun approach is a waste of time and money.

To sell a sponsorship, as with everything else in racing, perseverance is a must. Never give up, no matter how many no's you get. But don't go blindly from one no to another. Learn from each sales attempt. Understand why they said no, and figure out how you can avoid it in the future.

In fact, selling sponsorship is a great learning experience. What you learn here will be useful in any career for the rest of your life.

You may want to use sponsor-hunting professionals, but beware. There are hundreds of so-called "professional sponsor hunters" who will waste your time, your money, and your reputation. So check them out; get references. Talk to others who have used them.

You may not have a lot of choices of who you use in the early stages of your career, so stay close to the dealings. Always remember, they are selling you and your reputation. Be sure you are comfortable with how they are doing that, how they portray your reputation, and the promises they make on your behalf.

Once you have signed a sponsor, don't just take their money and go racing. Getting the sponsor is just the beginning. You, working with the sponsor, will have to exploit the sponsorship program. If not, kiss them goodbye. It just won't work for them. If it's not improving their bottom line, they won't stay involved. And to do that, they need to have more than just their name on the side of a race car. You are going to have to work hard to give them what they want.

Communicate with your sponsors once you have them. Often, what keeps a sponsorship program going is your personal relationship with the individuals involved in making the decisions. Cultivate that, but don't be phony or try too hard. Successful businesspeople will see right through that.

Try to progress with a sponsor. It's tough to sell a million-dollar sponsor right off the bat, but over a period of time with the opportunity to show what you can do, it's possible.

In fact, it's important to educate your sponsors. You have to show them what you and racing can do—for them.

Be careful what you or your "agent" promise a sponsor, especially as far as your results go. If you promise to win every race in sight, and don't, you lose credibility and probably their support. If you promise to finish last in every race, they probably won't want to be involved. Make sure you give them realistic expectations. This also applies to the exposure and marketing results they will get from the program.

Finally, a sponsorship program must work off the track. It should be a good value to the sponsor before you and your car ever get on the track. Anything you do on the track is a bonus, especially if it's running at the front of the pack, generating that extra exposure.

My opinion on business ethics relating to sponsorship hunting is don't try to steal other driver's or team's sponsors. I believe that hurts everyone; it hurts the sport. If you go after a company already involved as a racing sponsor, usually all you accomplish is to demonstrate how unprofessional people are in motorsports. Sometimes that results in the company deciding it's not the sport they want to stay involved in. Everyone loses.

If another team's existing sponsor approaches you indicating they are dissatisfied where they are now and would be interested in hearing about what you can offer, then that's fair game. Otherwise, leave them alone. There are enough other potential sponsors out there. It's the same as driving. Concentrate on your own performance, rather than your competition's, and you will win in the end.

PROFESSIONALISM AND PERSONAL IMAGE

How you are perceived by the outside world (business community, media, and so on) and the racing community can have a great effect on your career. If you want to be a professional race driver, you must look and act professional. That means how you dress (appropriate for each occasion), your personal appearance, the way you speak, how you act in company, and so on.

Any letter or sponsorship proposal that has anything to do with you must be first-class. Often, that will be the first impression you make with a potential sponsor, team, or media person. And you know what they say about first impressions.

What you do outside of the car is just as important as what you do in the car. Remember, a big part of your job as a race driver is as a motivator and team leader. You can have all the talent in the world, but if you don't have absolutely everyone around you pulling for you—and helping you—you will not make it in this sport. There are many examples of talented drivers who have had their careers cut short by their actions outside of the cockpit.

How you "present" yourself outside of the car will play the most important role in the rides you get in the future. How you act, react, and interact with all the people around you will determine how often you win. If your actions do not motivate, if they demotivate, your mechanics, engineers, team owner, sponsors, media people, and so on, you will not get the competitive rides you need, you will lose good rides, you will lose the edge you require to win. Always remember, if you're not doing everything possible to win, some of your competition is. That is probably what will make them beat you, even if you have more natural talent. If you are unsportsmanlike outside of the car, sponsors will stay away from you, so will team owners, mechanics, the media, and everyone else you need on your side.

PUBLIC RELATIONS

Public relations is an integral part of modern-day racing. You can have all the talent in the world, but if no one knows about you, your career won't be a long one. Guaranteed, as good as you may be, there is someone else as good, or just about as good, who has a dynamite P.R. person and program letting the world know. If you want to compete on the track, you first have to compete off the track.

If you want sponsors to help support your racing program, you will have to learn all about media and public relations. Don't ever feel that promoting yourself is beneath you, or that the media should come to you because of how good you are. Those days are gone. These days a driver must not only be talented on the track, he or she must also be talented in the promotions business.

Using a professional public relations firm can be a benefit to your career if you can afford one. However, much like sponsor hunters, beware: Look for the good ones.

I strongly recommend taking a public-speaking course. If you are successful in racing and want your career to continue, you will have to make speeches at some point. Learn to make the most of the opportunity.

You will also have to do many interviews, either live on radio or TV, or with a journalist. Again, learn to make the most of it. There are courses that teach you how to be effective in getting across what *you* want to say in an interview, not just what the interviewer wants you to say.

Be yourself though, in interviews and when making speeches and public appearances. Too many drivers today have become too "polished and practiced." They sound like a "canned" press release. Let your enthusiasm for the sport and your personality shine through, and you will find the media and sponsors more interested in listening to you.

52 THE PERFECT DRIVER

How often have you seen Helio Castroneves looking unhappy? Okay, once or twice (a major personal challenge and a controversial call by an official come to mind), but it's as if it doesn't matter whether he's on the pole, leading every lap of a race and standing on the top of the podium (or climbing a fence), or he's just dropped out of a race, he sees the positive in practically everything. The key for him is that he's always improving. There is no limit.

One of the things that makes Castroneves so successful is that his attitude or mindset doesn't seem to change, no matter how successful or unsuccessful he is. If he's not on pace, he's happily focused on making things better. If he's on the pole, he's happily focused on being even better still. If he's just crashed out of a race, he's happily focused on having a better race next time. If he's just won the race, he's happily climbing a fence, and thinking about how he could have been even faster.

The perfect driver never quits trying to get still better. If you want to be a Formula One world champion, Indy 500 champion, or a NASCAR champion, that's a given. What isn't a given is the attitude that even the most low-key amateur racer needs to be a winner or to just have more fun. I can't emphasize this enough: The more you put into becoming a better driver, the more fun you will have.

Think back to the interviews you've heard with Castroneves and ask yourself how many times you've heard excuses from him. It's rare. Think about previous Penske drivers Gil de Ferran, Emerson Fittipaldi, and Rick Mears before him. How often did you hear excuses from them? Again, very rare.

Your attitude toward your racing has a big impact on your performance and the performance of those around you. Helio Castroneves has a positive impact on practically everyone around him, and that mindset likely helps his own performance. *Shutterstock*

Could it be that much of the success of the Penske teams through the years is less to do with budget and technical expertise and more to do with attitude? Could it be that every member of the team takes full responsibility for whatever happens, good or bad, and that there are no excuses? What matters to them is performing at one's best, learning how to become even better, and not blaming others for what happens.

There's no doubt that another reason Penske teams have won so often is their level of preparation. It's so legendary that the words "Penske" and "preparation" just naturally go together. Rest assured that Castroneves, de Ferran, Fittipaldi, and Mears did not begin preparing for a race when they first showed up at the track. You can be sure that they were doing mental imagery long before ever getting to the track.

Ahh, but that's all well and good for professional racers who spend all their waking hours preparing to win, you say. Let me share a few more real-life examples.

A few years ago, I began working with a club racer who had been racing for three years. Racing was and is a hobby for him, a way to break away from the stress of his business life. Over the course of about three months, with some focused coaching on specific techniques and some mental programming work on his own, this particular driver improved his best lap time at his home track by more than 4 seconds. He also improved his racecraft, learning how to race wheel to wheel with much more experienced drivers. But most important, he began to have more fun than ever before. In fact, this driver claims that he had been feeling so frustrated with his inability to improve prior to working with our coach that he was beginning to question whether to continue racing or not. Now, he can't get enough.

Another driver, a young man who had been successful in karting, began his transition to racing cars with me coaching him. In his first few car races he surprised people with his speed, his racecraft, and his overall attitude. But something happened. Either his budget dried up to the point where he could not afford a coach, or his attitude changed in a way that happens with many young, successful drivers. He thought his success was a result of his natural talent. And guess what happened? He stopped impressing people. In fact, he stopped being successful to the point where he changed his mind from thinking about becoming a professional driver to planning on going to college and simply racing for fun. I'm not saying that wasn't a good idea, since I think an education is important for race drivers, but the reason he changed his mind was not because he thought it was a good idea. The reason he changed his mind was because he wasn't willing or able to do the work it took to continue building on his early successes. The sad thing is, if he continues with that attitude, it's doubtful he will be successful regardless what he chooses to do with his life. The lesson he could have learned in this situation could have made him a winner in more places than just on the track.

A driver that I coached a few years ago had taken up racing in his mid-40s. He had been very successful in business and was now semi-retired. He had used what he thought was a coach for a season and had learned a few things, but what he really had was an instructor. You see, the fellow helping him told him what to do—he instructed him—but did not give him any long-term strategies for learning. When I provided these strategies, he went off and used them, doing mental imagery once or twice a day. His improvement was staggering, and the needle on his fun meter matched that improvement.

I must point out that the driver in this last example not only enjoyed learning, he craved it. It was like an addiction for him. His desire to learn was a critical factor to his success, as it inspired him to do whatever it took to prepare. The interesting thing is that all this preparation amounted to no more than 30 minutes a day. You could look at that and say, "I don't have that much time to put toward my racing." If so, that's okay, as long as you realize that you won't improve much without some commitment of time to prepare. It's okay as long as you don't feel bitter or frustrated that you're not improving as much or as quickly as other drivers (who probably *are* committing time for preparation).

A driver I coached a few years ago was extremely focused on his results. Great drivers focus more energy on ensuring their performance is at its maximum and trusting that will look after the results better than any other focus. They know that they can't control the results, but they can control their own performance, and ultimately it is their performance that will dictate the result. So when this driver went to a race, he had a lap time in mind, as well as a result. He would think, "If I can get into the mid-28s, that will put me in good shape to finish third or better." He had expectations going into a race weekend.

One of the first things that we worked on was changing his focus (through mental imagery) away from the results and onto his performance. Rather than doing mental imagery of him turning laps in the mid-28s, I had him do mental imagery of driving the car at the limit. So rather than having expectations, he focused on the potential, on the possibilities. Rather than focusing on the lap time or race position, he saw, felt, and heard himself performing at his best, no matter what the conditions and competition level were. He was open to whatever could happen if he performed at his best, driving the car consistently at the very limit. And not only did he enjoy his racing even more since he didn't judge himself and his abilities based on a lap time or race result, but he performed even better. He performed so well, in fact, that he began to win. Had he stayed focused on the result, it's unlikely he would have performed so well. He wouldn't have enjoyed his racing so much, and he would always have been chasing the expectations he had set for himself.

SPEED SECRET

Don't set expectations.
Focus on the possibilities and your potential.

Earlier, I suggested that the main ingredient superstars of any sport have that mere stars may not is the ability to learn quickly. In other words, it's not that they were born with any more natural talent than the rest, but it's what they've done with that natural talent. It could be that they began life with the same amount of natural talent but worked harder and smarter at developing that talent.

I still believe that. In fact, the more I coach drivers, ranging anywhere from beginners to experienced, from those who appear to be effortless and gifted to those struggling to figure it out, and from older drivers who just want to have some fun at the local club racing track to young racers who are committed to making it to Formula One, the more I stand by my initial belief. It's not how much talent you're born with that's going to make the difference; it's what you do with that talent.

To do anything with your talent, you need to begin with an open or growth-oriented mindset, one that is constantly looking for ways to improve. It's a mindset of wanting to work harder at becoming better than anyone else. It's an understanding that it's through effort that any amount of talent you currently have will turn into something special. It's knowing that, no matter how easy it's been to be successful so far, you now need to work harder than anyone else to get to the very top.

That's the attitude that every great champion or superstar has ever had. The motorsports world is littered with drivers who were successful early in their career, who developed an attitude of, "I'm great—I'm gifted—and therefore I will make it to the top with this natural talent." In my world of being able to work closely with drivers of all levels, I see far too many of these types of drivers and not enough who make the commitment to doing what it takes to become as successful as they desire. That's sad. I can't help but feel sorry for drivers who feel that they are going to make it to where they desire without doing what it takes. And it's going to take effort, it's going to take an open mind—no, a craving—for learning and improvement, and it's going to take other people to assist you.

SPEED SECRET

Be open to ever-improvement.

Before you close your mind to what I've just said and think, "Oh, that's fine for a young driver who's trying to become the next Michael Schumacher or Jimmie Johnson. I just want to have fun doing what I do at my level," think again. There are just as many drivers who race for fun who are putting in the effort to improve as there are young, up-and-coming world champions. In fact, I see it every day: drivers in amateur racing, drivers who are older, drivers who are not out to become world champions who still have the attitude that they can improve and are willing to do what it takes to do so. Why? For one reason: It's more fun!

In the end, that's why anyone should race, and why you should put in the effort it takes to be successful—to have fun.

53 THE REAL WINNER

Throughout this book I've talked about, and used as examples, a number of great race drivers, especially Michael Schumacher. Am I the president and founder of the Michael Schumacher fan club? No, but despite his struggles coming back into F1 after a three-year sabbatical, he's recognized as one of the best of all time, so who better to use as a role model or for comparison? Do I think he is something special? Yes and no.

I believe all of us—you, me, Michael Schumacher—were all born with the same amount of natural driving talent. We all have the ability to be a superstar. Yes, if you were born with the DNA makeup that resulted in you growing to be 6 feet 10 inches tall, it is doubtful you are going to make a career out of driving F1 cars. But assuming you have the basic physical design, you too can be a racing superstar. In other words, Schumacher was not born special.

If there is a difference between Michael Schumacher and you today, it is simply a result of what the two of you have done with the talent you were born with. And yes, that makes him special.

The bottom-line is this: There were a number of events in Schumacher's life that enabled him to take the basic talent he was born with and turn that into the superstar abilities he demonstrates today.

You now know the value of being integrated (whole brained) to achieving peak performance. Whereas, many children do not do enough integrating physical movement as a baby to become as integrated as possible, I suspect Schumacher did. When a small child does become integrated, he or she feels and acts more coordinated. That leads to a belief system that tells him he's coordinated, which encourages more physical movement. This belief system is reinforced by comments from outside sources (parents, friends, and so on). All of this encourages the child to do more physical movement that further enhances brain integration and so on. It becomes a self-fulfilling prophecy.

Of course, the opposite is true, as well. For example, a baby who does not do much "cross crawling" action may not become as integrated as early in life, which affects the belief system. This leads to a child who stays away from physical activities "because I'm not very good at it."

So Schumacher enters childhood at an integrated, coordinated level. I'm sure that encouraged him to participate in numerous sports, which lead to the development of his sensory input skills. The fact that his family had its own karting track certainly didn't hurt. But from what I understand about his childhood, he didn't simply spend a lot of time in a kart, he spent a lot of well-defined, strategic, deliberate practice time in a kart. In other words, he went onto the kart track with specific strategies for improving.

You see, much of the development of his natural talent was a result of the environment he grew up in.

ILLUSTRATION 53-1 Many drivers have asked whether Michael Schumacher does brain-integration exercises. I don't know if he does specific exercises, such as cross crawls, but I do know he does warm-up by practicing with a soccer ball, which is very similar and would have the same effect.

Now, I'm not suggesting that Schumacher necessarily developed these abilities at a conscious level. Actually, I'm guessing he "stumbled" into most of the techniques, just like many people do who become good at a physical task. On top of that, I suspect some techniques were taught to him, in a very specific manner.

One of the things that separates Schumacher, or any other superstar, from the rest is his ability to learn so very quickly, more quickly than most everyone else. Assuming he started his racing career with the same talent level as everyone around him, he was able to take that talent and develop it and enhance it faster. That is what gave him the edge.

Of course, that does not come from wishful thinking or without some effort. He is famous for the amount of time he spent working out physically. A similar effort went into his mental preparation. That was also the case with Ayrton Senna. When you look at all the hard work they put into developing their ability, you have to ask, "Is that natural talent or hard work that got him to where he is?"

My main point is it doesn't matter how much natural talent you believe you were born with, it is what you do with it. If you learn how to *learn* to be better faster than other people with similar talent, you will be miles ahead.

SPEED SECRET

Learn how to learn and you will never stop improving.

Of course, that is my intention as well: to continue to learn more about the art, science, traits, and techniques that lead to drivers becoming champions.

Auto racing is no different than business or life. It provides the same ups and downs, the "thrill of victory and the agony of defeat," the same lessons and emotions, good and bad, that each of us face in real life. Often though, racing provides as many of these in one season as many people face in a lifetime.

If you keep your eyes and ears open, and your mind open, you can learn many valuable lessons that will assist you in other aspects of life. Remember this in times when your racing program is not going as well as you would like. There is more to racing than what you do on the track. It's how you use what you learn on the track in your everyday life that makes you a real winner.

Through racing I have met and become friends with many of the most genuine, interesting, and exciting people in the world. I have visited places that I never would have, had I not raced. I have had the most rewarding and memorable experiences.

And finally, racing has helped make me a more complete person. It has encouraged me to be a team player; it's taught me how to work with and motivate people; to learn about business, engineering, advertising, and marketing. It has helped me to be a good money manager, to improve my public speaking, and hopefully, to be a good coach and writer.

SPEED SECRET

Have fun!

appendix A:

RESOURCES

DirtFish Rally School. www.dirtfish.com.

Driver Coach. www.apps.gedg.com.au/drivercoach.

Performance Rules! www.performance-rules.com.

PitFit Training. www.pitfit.com.

Speed Secrets Driver Development Services. www.speedsecrets.com.

Virtual GT. www.virtualgt.com.

Books

Alexander, Don. *Performance Handling.* Wisconsin: Motorbooks International, 1991.

Colvin, Geoff. *Talent Is Overrated.* New York: Penguin Books, 2008.

Csikszentmihalyi, Mihaly. *Flow.* New York: Harper & Row, 1990.

Dennison, Paul E., and Gail E. Dennison. *Brain Gym, Teacher's Edition.* California: Edu Kinesthetics, 2010.

Donahue, Mark with Paul Van Valkenburgh. *Unfair Advantage.* Massachusetts: Bentley Publishers, 2nd edition, 2000.

Dweck, Carol. *Mindset.* New York: Random House, 2006.

Edwards, Guy. *Sponsorship and the World of Motor Racing.* Surrey, UK: Hazelton Publishing, 1992.

Fey, Buddy. *Data Power: Using Race car Data Acquisition.* Tennessee: Towery Publishing, 1993.

Gallwey, Timothy. *Inner Tennis.* New York: Random House, 1974.

Gelb, Michael J., and Tony Buzan. *Lessons from the Art of Juggling,* New York, New York: Harmony Books, 1994.

Haney, Paul, and Jeff Braun. *Inside Racing Technology.* Wisconsin: Motorbooks International, 1995.

Hannaford, Carla. *Smart Moves.* Utah: Great River Books, Revised & Expanded edition, 2007.

Hannaford, Carla. *The Dominance Factor.* Virginia: Great Ocean Publishers, 1997.

Huang, Al Chungliang, and Jerry Lynch. *Thinking Body, Dancing Mind.* New York, New York: Bantam Books, 1992.

Hunter, Dr. Harlen, and Rick Stoff. *Motorsports Medicine.* Lake Hill Press, 1992.

Jackson, Susan A., and Mihaly Csikszentmihalyi. *Flow In Sports.* Illinois: Human Kinetics, 1999.

Kaplan, Robert-Michael. *The Power Behind Your Eyes.* Vermont: Healing Arts Press, 1995.

Markova, Dawna. *The Open Mind.* California: Red Wheel / Weiser, 1996.

Martin, Mark, and John Comereski. *Strength Training for Performance Driving.* Wisconsin: Motorbooks International, 1994.

Smith, Carroll. *Drive to Win.* Pennsylvania: SAE International, 1996.

Smith, Carroll. *Engineer to Win.* Wisconsin: Motorbooks International, 1985.

Smith, Carroll. *Prepare to Win.* California: Aero Publishers, 1975.

Smith, Carroll. *Tune to Win.* California: Aero Publishers, 1978.

Turner, Stuart, and John Taylor. *How to Reach the Top as a Competition Driver.* Wisconsin: Motorbooks International, 1991.

Valkenburgh, Paul Van. *Race Car Engineering and Mechanics.* California: Published by author, 1992.

Wise, Anna. *The High Performance Mind.* New York: G.P. Putnam's Sons, 1997.

appendix B:

SELF-COACHING QUESTIONS

How far ahead do I look when driving on the highway? How about when driving on city streets? On the racetrack? Can I look farther ahead?

How consistent is my corner-entry speed? Does my speed at the turn-in point vary from lap to lap by 1 mile per hour? Three miles per hour? Five or more miles per hour?

When was the last time I worked on developing, on practicing, my traction-sensing skills? When was the last time I practiced just sliding a car around, whether on a skid pad or the racetrack?

How tightly do I grip the steering wheel when driving on the street? How about when driving on the racetrack? Can I relax my grip a bit?

Where am I in the continuous learning process loop? Have I perfected the line? How about the exit phase? How is my corner entry? My midcorner speed?

What can I do to improve my line? My corner exit? My corner entry? Midcorner? Turn in later or earlier? More gently or crisper? Begin accelerating earlier, or just get on the throttle harder at the same place? Carry more speed into the corner or slow it down a mile per hour or so to get the car to turn in better? Make a smoother transition from brake to throttle? Turn the steering wheel less or begin unwinding it out of the turn earlier?

What would happen if I turned in 1 or 2 feet later? Earlier? Would I have to change my corner-entry speed to do that? Exactly where would my turn-in reference point be then?

Am I apexing too early? Too late? Is the car at the right angle when I pass the apex and pointing in the direction I want at that point?

Am I unwinding the steering from the apex on out? Am I "releasing" the car from the corner, letting it "run free" at the exit?

Which is the most important corner on the track from a lap time and speed point of view? Which is second most important? Third? And so on.

Which corner do most drivers have the most difficulty with? Which corner can I gain the biggest advantage over my competition on?

In working on the car's setup, which corner should I focus on first?

Am I using all of the tires' traction when accelerating out of the corners?

What would happen if I started accelerating sooner? If I squeezed on the throttle quicker? Am I causing the car to understeer or oversteer by accelerating too abruptly or hard? Can I squeeze on the throttle smoother?

Am I holding the car in the corner too long? Can I unwind the steering sooner?

Before I get to the apex, am I looking for and through the exit point and down the straightaway?

Can I carry 1 mile per hour more into the corner? Two miles per hour? Three miles per hour? What will happen if I carry more speed into the corner? Will I still be able to make the car turn in and "rotate" toward the apex? Will it delay when I begin accelerating?

Can I left-foot brake in my car? Do I have the sensitivity with my left foot to do it? Does my left foot have the necessary programming to do it?

Am I "snapping" my foot off the brake pedal, coming off too quickly? Can I ease off the pedal more gently? How would that feel if I did? Just how gently can I come off the brakes?

Am I easing off the brake pedal too slowly, trail braking too long? Is that causing the car to rotate too quickly or oversteer during the entry?

Am I turning the steering wheel too quickly or too slowly? Does the car respond to my initial turn of the wheel? What if I turned the wheel more quickly or slowly? Can I be smoother with the wheel? What would it feel like if I turned the wheel more smoothly? More slowly? More quickly? Do I have slow hands or quick hands?

Am I over-slowing the car on entry? Is that resulting in me getting on the throttle too hard, causing "change-in-speed oversteer"? What do I need to do to make the car turn in with more speed? Do I need to trail brake more or less? Do I need to change my line slightly and turn in earlier or later? Do I need to turn the steering wheel more crisply, or slowly and progressively?

Am I blipping the throttle enough to ensure a smooth downshift? At the right time? Am I blipping it too much, causing the car to lurch forward?

How's the car's balance during the entry phase of the corner? How about in the midcorner phase? What can I do to improve the car's balance? Ease off the brakes more gently? Be more progressive with the steering input? Squeeze on the throttle more smoothly? Make a smoother transition from braking to throttle?

ABOUT THE AUTHOR

Ross Bentley knew at the age of five that he wanted to race cars. In his late teens he began racing sprint cars and continued into Formula Ford, Formula Atlantic, Trans-Am, and eventually realized his dream of racing Indy cars. From there his career moved into sports cars and prototypes. He has used his knowledge and experience to coach drivers in road racing, oval racing, motorcycles, drifting, and even drag racing in North America and around the world.

Bentley has focused his life on learning about sports psychology, educational kinesiology, neuroscience, human learning strategies, and coaching for performance. While testing all of these techniques and strategies on himself, he won the 1998 United States Road Racing Championship, driving for the factory-backed BMW team, and the 2003 Rolex 24 Hours of Daytona. As a coach, his drivers have won at practically every level and in every form of motorsport. Bentley is a popular speaker at car club events and works with both individuals and entire groups, drivers, and instructors.

With his education and hands-on experience both as a driver and coach, Bentley is now considered the premier technical and mental coach in motorsport. He successfully applies the same performance-based approaches to the business world (coaching executives, managers, sales people, teams, and so on), and a variety of specialty driver training programs. You can access his website at www.performance-rules.com.

Bentley lives in the Seattle, Washington, area with his wife and daughter. He can be reached at ross@speedsecrets.com.

INDEX